Induction and Direct Resistance Heating

Sergio Lupi · Michele Forzan
Aleksandr Aliferov

Induction and Direct Resistance Heating

Theory and Numerical Modeling

Springer

Sergio Lupi
Michele Forzan
Department of Industrial Engineering
University of Padua
Padova
Italy

Aleksandr Aliferov
Department of Automation of Electric Technological Installations
Novosibirsk State Technical University
Novosibirsk
Russia

ISBN 978-3-319-35493-4 ISBN 978-3-319-03479-9 (eBook)
DOI 10.1007/978-3-319-03479-9

Springer Cham Heidelberg New York Dordrecht London

Softcover reprint of the hardcover 1st edition 2015

Printed on acid-free paper

Springer is part of Springer Science+Business Media (www.springer.com)

Preface

The material collected in this book reflects the historical development of the technologies of induction and direct resistance heating and represents a synthesis of the information and experience that the authors have accumulated in their activity of academic research and collaboration with industry.

The first part of the book is devoted to deepening the theoretical knowledge on the electromagnetic and thermal phenomena that determine the heating process. This theoretical approach was developed since the beginning of the twentieth century. At that time, the design of heating installations was based on analytical solutions of Maxwell's equations for simple geometries, experimental tests and the use of the "ruler" as a mean for calculations.

Although there was no discontinuity in the technological development, crucial steps were the industrial growth, in particular of the automotive industry, after World War II, the introduction of computers in universities and research institutions and the development of numerical methods for computation of electromagnetic and thermal fields.

As regards the last point, a milestone for the technologies dealt with in this book can be considered the works of Hegewaldt, Holmdal and Sundberg, Kolbe and Reiss, who in 1963, independently but almost at the same time, presented the first numerical solutions for 1D coupled electromagnetic and thermal problem in non-linear systems and a 2D numerical solution of the electromagnetic problem in a linear system.

The way for numerical calculations of heating systems was open, but it took a long time, more than 25 years, for computer simulations to become a widely used design and research tool.

Analytical and experimental methods continued to prevail in industry during the period 1960–1990, when development and use of computer simulations were mainly limited to academia and research institutions for theoretical studies. In this period several new analytical solutions have been developed in the form of infinite series of Bessel or exponential functions for different geometries, since the availability of computers allowed the calculation of a number of harmonic terms sufficient for obtaining accurate results.

Since then, more powerful computing means and increasingly sophisticated numerical procedures have been progressively available, so that analytical methods were gradually abandoned while the use of numerical programs has become customary not only in research but also in industrial design.

The content of Chap. 4 reflects this situation, since it presents a set of results which are based in part on infinite series expansion of analytical solutions and in part on the use of numerical methods.

The choice to present in the same volume results of analytical and numerical solutions stems from the belief of the authors that a profitable use of numerical methods is based on a preliminary deep knowledge of the different phenomena that affect the heating process. This knowledge is acquired in the best way from the analysis of several case studies conducted in the past.

Finally, in Chap. 5 useful information for the numerical calculation of electrotechnogical installations are given to the designer.

Contents

Abbreviations of Symbols

div	*Divergence*—scalar operator (–)
grad	*Gradient*—vector operator (–)
∇	*Nabla* operator (–)
∇^2	*Laplacian* operator (–)
x, y, z	Cartesian coordinates system (–)
r, φ, z	Azimuthal coordinates system (–)
j	Imaginary unit (–)
$\dot{X}$	Complex quantity (–)
X^*	Complex conjugate value (–)
$\bar{X}$	Vector quantity (–)
$\dot{\bar{X}}$	Complex vector quantity (–)
J_0, J_1	Bessel functions of first kind of zero and first order (–)
Y_0, Y_1	Bessel functions of second kind of zero and first order (–)
T	Absolute temperature (K)
ϑ	Celsius temperature, (°C)
t	Time (s)
τ	Time constant (s)
c	Specific heat capacity (J/(°C kg))
λ	Thermal conductivity (W/(m K))
ρ	Electrical resistivity (Ω m)
$\sigma = 1/\rho$	Conductivity (S/m)
μ	(absolute) Magnetic permeability (H/m)
$\mu_0 = 4\pi \times 10^{-7}$	Magnetic permeability of vacuum (H/m)
$\mu_r = \mu/\mu_0$	Relative (magnetic) permeability (–)
ε	(absolute) Dielectric constant (F/m)
$\varepsilon_0 = 8,86 \times 10^{-12}$	Dielectric constant of vacuum (F/m)
$\varepsilon_r = \varepsilon/\varepsilon_0$	Relative dielectric constant (–)
$\in$	Emissivity (–)
H	Magnetic field intensity (A/m)
E	Electric field intensity (V/m)

D	Electric flux density (As/m^2)
B	Magnetic flux density (T)
Φ	Magnetic flux (Wb)
A	Magnetic vector potential (V s m^{-1})
J	Current density (A/m^2)
I	Current (A)
V	Electric potential (V)
U	Electric voltage (V)
P	Active power (W)
P_u	Active power per unit axial length (–)
p	Active power per unit surface (W/m^2)
w	Volumetric power density (W/m^3)
Q	Reactive power (VAr)
q	Reactive power per unit surface (VAr/m^2)
P_a	Apparent power (VA)
$\overline{S}$	Poynting vector (W/m^2)
W_E	Energy (J)
w_E	Energy volume density (J/m^3)
$\cos\varphi$	Power factor (–)
$k = 1/\delta$	Attenuation coefficient of electromagnetic wave (1/m)
$\delta = \sqrt{\frac{2\rho}{\omega\mu}}$	Penetration depth (m)
f	Frequency (Hz)
ω	Angular velocity, angular frequency (rad/s)
R	Resistance (Ω)
r	Resistance referred to unit surface (Ω/m^2)
X	Reactance (Ω)
x	Reactance referred to unit surface (Ω/m^2)
Z	Impedance (Ω)
z	Impedance, referred to unit surface (Ω/m^2)
R_{ac}	AC resistance (Ω)
R_{dc}	DC resistance (Ω)
$k_r = R_{ac}/R_{dc}$	Skin effect coefficient of resistance (–)
$k_x = X/R_{dc}$	Skin effect coefficient of reactance (–)
v	Volume (m^3)
A	Area (m^2)
Π	Perimeter (m)
l	Length (m)
w	Width of body of rectangular cross-section (m)
$h = 2g$	Thickness of body of rectangular cross section (m)
r_i	Internal radius of a cylindrical cavity (m)
r_e	External radius of a cylindrical body (m)
$d = 2r_e$	Diameter of a cylindrical body (m)
η_e	Electrical efficiency (–)
η	Efficiency (–)

η_T	Thermal efficiency (–)
$\xi = \frac{r}{r_e}$	Dimensionless radial coordinate (–)
$m = \frac{\sqrt{2}r_e}{\delta}$	Dimensionless radius
N	Number of inductor turns (–)
n	Speed of rotation (r/s)
F	Force (N)
f_v	Specific force (N/m^3)
p_m	Magnetic pressure (N/m^2)

Chapter 1
Theoretical Background

Abstract This chapter is the introduction of the book. Induction and direct resistance heating are presented in an unified theory that, starting from the solution of Maxwell's equations, gives the basic quantities that characterizes thermal processes based on internal heating sources. The penetration depth of electromagnetic waves is a parameter that characterizes all the relevant phenomena in these heating applications. A qualitative description is provided about the several effects that affect distribution of current density and, as a consequence, of heating sources: proximity effect (that occurs between two conductors that carry electrical current), ring effect (that occur in bended conductors), slot effect (that occurs in conductor placed in the slot of a magnetic yoke), end and edge effects (that occur due to the finite length of inductor and load and their relative position). Active and reactive power in the workpiece are computed by applying Poynting's theorem for both induction and direct resistance heating.

1.1 Introduction

Direct resistance and induction heating are technologies that are successfully applied in several industrial applications. The fact that the heating sources are directly generated inside the metallic body to be heated is the main reason of their success [1]. The internal heat generation implies several positive effects, notably the possibility to achieve higher production rates, higher efficiency, higher temperature, better process control if compared with technologies that use external heating sources, like resistors furnaces or fuel burners.

Even if direct resistance has generally a higher efficiency in transforming electrical energy into heat, its application has severe limits, mostly related to the geometry of the body to be heated and the maximum current that can be supplied.

Induction heating is much more flexible than direct resistance heating and this is the main reason for the wider diffusion of this technology. Its success depends also

S. Lupi et al., *Induction and Direct Resistance Heating*,
DOI 10.1007/978-3-319-03479-9_1

on the significant improvements in power electronics supplies that occurred in recent years.

The main advantages of heating technologies with internal sources are:

- High specific heating power
- Low process time
- Precise control of the process

Moreover, since electro-technologies for heating usually improve the overall efficiency of the process, they represent a solution for a rational use of energy.

1.2 Induction Heating

Induction heating is a technology that makes use of an AC magnetic field to induce power in an electrical conductive charge [2–5].

A comprehensive description of an induction heating process is based on several physical laws: Maxwell's equations allow to determine the distributions of induced currents and power densities inside the body to be heated, while Fourier's law allows to predict the thermal field in the heated body.

Other laws and phenomena relevant to the process are: the law of electromagnetic induction, the Ohm's law, the law of Joule-Lenz and the phenomena of skin-effect, proximity-effect, ring-effect and slot-effect which are all a consequence of the law of electromagnetic induction.

1.3 Direct Resistance Heating

Direct resistance heating (or ohmic heating) is based on the flow of an electrical current through the body to be heated, which is directly connected to an electrical supply. The heating can be realized by DC or AC currents, but the industrial applications are mostly done with the alternating ones [2–7].

When a DC current flows in a straight shaped body of constant electrical resistivity and constant cross-section, the current density and, as a consequence, the power density due to Joule's effect are uniform in the cross-section. Non-uniform distribution of the current density can arise only if the body has not a straight shape (e.g. ring or bended shape) or the dimensions of its cross-section vary along the body length.

When an AC current flows in the workpiece, the current density and internal power sources distributions will be determined by the same laws already mentioned in the previous paragraph, i.e. the law of electromagnetic induction, the Ohm's law and the Joule-Lenz law.

Considerable influence on the power density distribution have also other phenomena like *skin-effect*, *proximity effect* (occurring in the heating of several workpieces placed close to one another), *ring-effect* (an effect that occurs in non-straight workpieces like bent tubes, springs, etc.) and *slot-effect* (an effect that occurs when magnetic flux concentrators or magnetic yokes, modify the current density distribution in conductors).

Moreover, during the heating of a work-piece, when temperature increases, usually the electrical resistivity increases, giving rise to non-uniform distributions of the current density and internal heating sources, both in the cases of AC or DC current supply.

Before analysing laws and phenomena which characterise induction and resistance heating processes, we will recall some basic concepts of the theory of electromagnetic fields.

1.4 Basic Electromagnetic Phenomena

In the electro-technological processes of induction and conduction heating, electromagnetic phenomena are the basis for the generation of heat and forces inside the body to be heated (in the following *workpiece*).

This paragraph describes the basic electromagnetic phenomena involved in these technologies.

1.4.1 Maxwell Equations

Maxwell's equations is a set of equations which allow to fully describe the electromagnetic phenomena and, in particular, to determine the spatial distribution of the field vectors $\overline{E}$, $\overline{H}$, $\overline{D}$, $\overline{B}$, including their time dependence, as a function of the current density $\overline{J}$ and the characteristics of the medium. These equations can be written in the form:

$$rot\,\overline{\mathrm{H}} = \overline{\mathrm{J}} + \frac{\partial \overline{\mathrm{D}}}{\partial t} \quad (Ampere's\,law) \tag{1.1}$$

$$rot\,\overline{\mathrm{E}} = -\frac{\partial \overline{\mathrm{B}}}{\partial t} = -\mu\frac{\partial \overline{\mathrm{H}}}{\partial t} \tag{1.2}$$

$$\overline{\mathrm{D}} = \varepsilon\,\overline{\mathrm{E}} = \varepsilon_0\,\varepsilon_r\,\overline{\mathrm{E}} \tag{1.3}$$

$$\overline{\mathrm{B}} = \mu\,\overline{\mathrm{H}} = \mu_0\,\mu_r\,\overline{\mathrm{H}} \tag{1.4}$$

with $\overline{\mathrm{E}}$—electric field strength; $\overline{\mathrm{D}}$—electric flux density; $\overline{\mathrm{H}}$—magnetic field intensity; $\overline{\mathrm{B}}$—magnetic induction (flux density); $\overline{\mathrm{J}}$—current density.

The quantities ε and μ, which are respectively called electrical permittivity (or dielectric constant) and magnetic permeability, are characteristics of the material;

the parameters ε_r, μ_r denote, respectively, the relative permittivity and relative magnetic permeability. These quantities have known constant values in vacuum ($\mu_0 = 4 \cdot \pi \cdot 10^{-7}, H \cdot m^{-1}$; $\varepsilon_0 = c^2/\mu_0, F \cdot m^{-1}$; $c = 299.792,458$, km/s—velocity of light). In many practical cases of induction and conduction heating they depend on the process parameters, such as the temperature or, in the case of the permeability of ferromagnetic materials, on the magnetic field intensity.

The classical interpretation of Eq. (1.1) is as follows [8]:

- the first term of the r.h.s. is the conduction current density due to an electric field $\overline{E}$ in a medium:

$$\overline{\mathrm{J}} = \sigma\,\overline{\mathrm{E}} = \frac{\overline{\mathrm{E}}}{\rho} \quad (\textit{Ohms' law}) \tag{1.5}$$

where $\sigma = 1/\rho, \Omega^{-1} \cdot m^{-1}$, denotes the electrical conductivity of the material and ρ is the electrical resistivity.

The electric field $\overline{\mathrm{E}}$ comprises a term, $\overline{\mathrm{E}}_c$, due to a voltage applied from external sources, that is the conservative term of the electric field, and a term, $\overline{\mathrm{E}}$i that depends upon the time variation of the induction field $\overline{B}$.

$$\overline{\mathrm{E}} = \overline{\mathrm{E}}_c + \overline{\mathrm{E}}_i = -\ \mathrm{grad}\ \mathrm{V} - \frac{\partial \overline{\mathrm{B}}}{\partial\, \mathrm{t}} \tag{1.6}$$

The current density due to the rotational term of the electric field corresponds to the current density $\overline{\mathrm{J}}_e$ imposed from an external source, while the other term describes the induced current. As an example, this situation corresponds in the induction heating to the inductor connected to the generator, while in the regions limited to passive conductors J_e is zero (like in the load placed inside the inductor).

- the second term $\partial\overline{\mathrm{D}}/\partial t$ is the displacement current density which represents the dominant mechanism of electric conduction in dielectric materials.

For most applications of induction and conduction heating of metals, where steady-state AC fields vary sinusoidally in time and the frequency is less than 10 MHz, the conduction current density $\overline{J}$ is much greater than the displacement current density $\partial\overline{\mathrm{D}}/\partial t$, so that the last term on the r.h.s. of Eq. (1.1) can be neglected.

With these hypotheses, using the complex exponential form ($e^{j\omega t}$), Maxwell's equations can be re-written as follows:

$$\mathit{rot}\ \dot{\overline{\mathrm{H}}} = \sigma\ \dot{\overline{\mathrm{E}}} \tag{1.7a}$$

$$\mathit{rot}\ \dot{\overline{\mathrm{E}}} = -j\,\omega\,\mu\ \dot{\overline{\mathrm{H}}} \tag{1.7b}$$

with $j = \sqrt{-1}$; ω—angular frequency, rad/s.

Applying the rotational operator to each member of Eqs. (1.7a) and (1.7b) and taking into account Eq. (1.5), we can write

$$rot\ rot\ \dot{\bar{\mathrm{E}}} + j\,\omega\,\mu\,\sigma\ \dot{\bar{\mathrm{E}}} = 0 \tag{1.8}$$

$$rot\ rot\ \dot{\bar{\mathrm{H}}} + j\,\omega\,\mu\,\sigma\ \dot{\bar{\mathrm{H}}} = 0 \tag{1.9}$$

Using the rules of vector algebra and taking into account that the electrical field intensity $\dot{\bar{\mathrm{E}}}$ and the magnetic field intensity $\dot{\bar{\mathrm{H}}}$ satisfy zero divergence conditions, it is possible to obtain the differential equations for the current and field distributions in electrical conductors (passive conducting regions):

$$\nabla^2 \dot{\bar{\mathrm{E}}} + \dot{k}^2\ \dot{\bar{\mathrm{E}}} = 0 \tag{1.10}$$

$$\nabla^2 \dot{\bar{\mathrm{H}}} + \dot{k}^2\ \dot{\bar{\mathrm{H}}} = 0 \tag{1.11}$$

where $\dot{k}^2 = -j\,\omega\mu\,\sigma = \frac{2}{\delta^2}\,(1 - j)$—eddy current constant, m^{-2}; δ—penetration depth of the electromagnetic field, m; ∇^2—is the Laplacian, which has different forms according the coordinates.

1.4.2 Theorem and Poynting's Vector

The Poynting's vector $\dot{\bar{\mathrm{S}}}$ is defined as:

$$\dot{\bar{\mathrm{S}}} = \dot{\bar{\mathrm{E}}} \times \bar{\mathrm{H}}^{*}. \tag{1.12}$$

The Poynting's vector describes the change of energy density in space. It can also interpreted as describing the flow of energy at any point in space since it specifies the power density (in VAm^{-2} or Wm^{-2}) and the direction of flow.

The active power dP flowing through an elemental surface dA with normal $\bar{\mathrm{n}}$ is

$$d\mathrm{P} = \mathrm{Re}\left\{\dot{\bar{\mathrm{S}}}\right\} \cdot \bar{\mathrm{n}}\,dA = \mathrm{Re}\left\{\dot{\bar{\mathrm{S}}}\right\} \cdot \cos\vartheta\ dA$$

Thus considering the rate of the energy flow out from a volume v enclosed by a closed surface A (Fig. 1.1), relating it to the electric and magnetic fields and denoting with $\dot{\mathrm{P}}_{\mathrm{S}}$ the complex power, the rate of complex power flow is:

$$-\dot{\mathrm{P}}_{\mathrm{S}} = \oint_A \dot{\bar{\mathrm{S}}} \cdot d\bar{A} = \int_v div\ \dot{\bar{\mathrm{S}}}\,dv \tag{1.13}$$

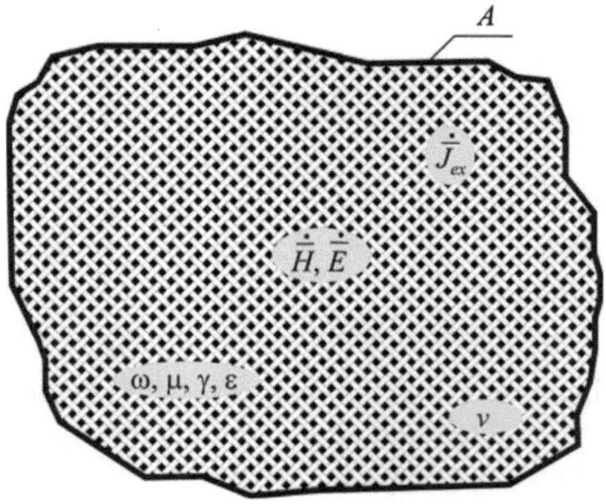

Fig. 1.1 Volume v enclosed by a surface A

Expressing the term *div* $\dot{\bar{S}}$ by Eq. (1.12), on the basis of vector identities and Maxwell's equations it is possible to show that Eq. (1.13) can be re-written as follows:

$$\oint_A \dot{\bar{S}} \cdot d\bar{A} = \int_v div\, \dot{\bar{S}}\, dv$$
$$= -\frac{\partial}{dt}\int_v \left(\frac{1}{2}\,\mu H^2\right) dv - \frac{\partial}{\partial t}\int_v \left(\frac{1}{2}\,\varepsilon E^2\right) dv - \int_v \dot{J} \cdot \bar{E}^* \, dv \qquad (1.14)$$
$$= -\frac{\partial}{\partial t} W_m - \frac{\partial}{\partial t} W_e - \int_v \dot{J} \cdot \bar{E}^* dv$$

where is

$W_m = \int_v \left(\frac{1}{2}\,\mu\, H^2\right) dv$ the energy in the magnetic field;

$W_e = \int_v \left(\frac{1}{2}\,\varepsilon\, E^2\right) dv$ the energy in the electric field;

$\int_v \dot{J} \cdot \bar{E}^* \, dv$ term which represents *either* the power dissipated as ohmic losses *or* the power generated by a source inside v.

If it represents the power dissipated by ohmic losses, the term $\dot{J} \cdot \bar{E}^*$, which represents ohmic losses per unit volume—can be written as

$$w = \dot{J} \cdot \bar{E}^* = \sigma\, \dot{E} \cdot \bar{E}^* = \sigma\, E^2 = \rho\, J^2, \text{Wm}^{-3}. \qquad (1.15)$$

Poynting's theorem allows to calculate the complex impedance of various electrical devices.

If, for example, the volume considered is a conductor carrying the current $\dot{I}$, then the total complex power in the volume is defined as:

$$\dot{P}_S = I^2\ \dot{Z}, \tag{1.16}$$

and the impedance of the conductor is equal to

$$\dot{Z} = -\frac{\oint \dot{\bar{S}}\ d\bar{A}}{I^2}. \tag{1.17}$$

Time-Averaged Poynting's Vector for Time-Periodic Sinusoidal Electromagnetic Fields

For time-periodic sinusoidal electromagnetic fields, the average power flow per unit time can be found by treating electric and magnetic fields as complex vectors as follows:

$$\begin{aligned} \dot{P}_S &= \dot{\bar{E}} \times \dot{\bar{H}} \\ &= \Re e\left\{\dot{\bar{E}}\right\} \times \Re e\left\{\dot{\bar{H}}\right\} \\ &= \Re e\left\{\dot{\bar{E}}\ e^{j\omega t}\right\} \times \Re e\left\{\dot{\bar{H}}\ e^{j\omega t}\right\} \\ &= \frac{1}{2}\Re e\left\{\dot{\bar{E}} \times \dot{\bar{H}}^*\right\} + \frac{1}{2}\Re e\left\{\dot{\bar{E}} \times \dot{\bar{H}}\ e^{j2\omega t}\right\} \end{aligned} \tag{1.18}$$

The average over time is given by

$$\left\langle \dot{\bar{P}}_S \right\rangle = \frac{1}{T}\int_0^T P_S(t)\ dt = \frac{1}{T}\int_0^T \left[\frac{1}{2}\Re e\left(\dot{\bar{E}} \times \dot{\bar{H}}^*\right) + \frac{1}{2}\Re e\left(\dot{\bar{E}} \times \dot{\bar{H}}\ e^{j2\omega t}\right)\right] dt$$

Since the second term is a sinusoidal curve

$$\Re e\left\{e^{j2\omega t}\right\} = \cos 2\omega t,$$

its average value is zero, thus giving

$$\left\langle \dot{\bar{P}}_S \right\rangle = \frac{1}{2}\Re e\left\{\dot{\bar{E}} \times \dot{\bar{H}}^*\right\} = \frac{1}{2}\Re e\left\{\left(\dot{\bar{E}}\ e^{j\omega t}\right) \times \left(\dot{\bar{H}}^*\ e^{-j\omega t}\right)\right\} \tag{1.19}$$

Example 1 [8] Round wire of radius r_e and resistance R_1 per unit length carrying direct current I_z (Fig. 1.2)

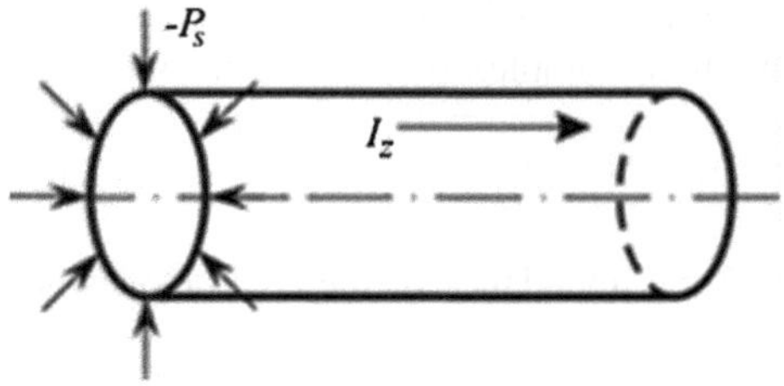

Fig. 1.2 Direction of Poynting's vector P_S in a conductor carrying DC current I_z

- The electric field in the wire, from Ohm's law is: $E_z = I_z R_1$
- The magnetic field at the surface is: $H_\varphi = I_z/(2\pi r_e)$
- The Poynting's vector (1.12) is everywhere radial, directed toward the axis and has the value:

$$S = -E_z H_\varphi = -\frac{R_1 I_z^2}{2\pi r_e}$$

1.4.3 Phenomena Affecting the Current Density Distribution in Conductors

- ***Skin-effect in cylindrical conductors***

The skin-effect is the effect which produces an uneven distribution of the alternating electrical current in the cross-section of a conductor or the magnetic flux in the cross section of a magnetic core. The current density in the conductor or the magnetic flux in the core decreases from the surface towards the internal part of the body.

In a conductor in which DC current flows, the current density J is constant in the cross-section and is given by the ratio:

$$J = \frac{I}{A},$$

with I—current in the conductor, A; A—conductor's cross-section, m^2.

A different situation occurs when in the conductor an AC current flows, as illustrated in Fig. 1.3a, where the current I flows in an infinitely long metallic cylinder.

The direction of lines of magnetic flux density $\dot{\overline{B}}_I$ produced by the current I, is determined by the rule of advance of a right-hand screw.

The AC magnetic flux density $\dot{\overline{B}}_I$ induces in the conductor the eddy current i, which in turn produces a reaction magnetic field, characterised by the vector $\dot{\overline{B}}_i$.

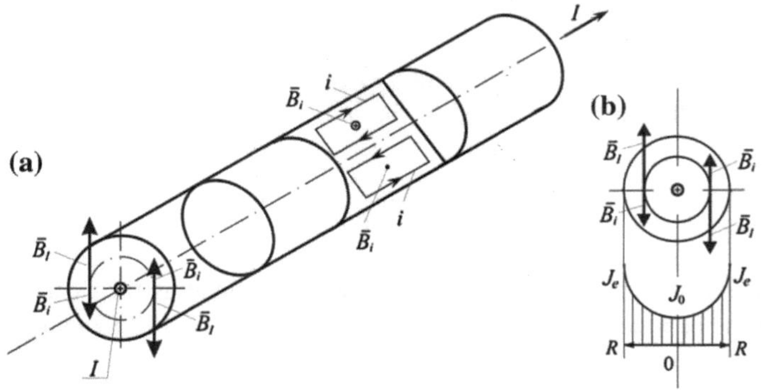

Fig. 1.3 Skin-effect in a cylindrical conductor. **a** path of induced current i, **b** distribtion of current density J in the conductor cross-section [2]

According to Lenz's law, the magnetic field $\dot{\overline{B}}_{\mathrm{i}}$ counteracts to the process that has created it. This means that the direction of the magnetic induction vector, produced by the current i, will be opposite to that of the magnetic induction created by current I. The direction of the path of the induced current i, encircling the lines of magnetic induction, also obeys the right-hand screw law.

Therefore, the induced eddy currents in the conductor i will increase the total value of the current near the surface and weaken it in the central part of the conductor, producing in this way an uneven distribution of current density in the cross-section. This phenomenon is illustrated qualitatively in Fig. 1.3b.

- ***Penetration depth of the electromagnetic field***

In case of a pronounced skin-effect (or in the conducting semi-infinite body) the reduction of the current density from the surface towards the interior of the conductor follows the exponential law:

$$J = J_e e^{-\frac{y}{\delta}}, \tag{1.20}$$

where J—current density at a distance y from the conductor's surface, A/m^2; J_e—current density at the conductor's surface, A/m^2; δ—penetration depth of the electromagnetic wave, m; δ is a quantity with dimension of a length, which characterizes the degree of attenuation of electromagnetic field in the conductor.

From Eqs. (1.10) and (1.11) the penetration depth in a homogeneous and linear conducting medium is

$$\delta = \sqrt{\frac{2}{\omega\,\mu\,\sigma}} = \sqrt{\frac{2\,\rho}{\omega\,\mu}} = \sqrt{\frac{2\,\rho}{\omega\,\mu_0\,\mu_r}},\ \mathrm{m}, \tag{1.21}$$

where $\omega = 2\pi f$—pulsation of the electromagnetic field, rad/s; μ_r—relative magnetic permeability.

Introducing in Eq. (1.21) the values $\omega = 2\pi f$ and $\mu_0 = 4\pi \cdot 10^{-7}$, H/m, the penetration depth can be expressed in the more convenient form:

$$\delta = 503 \sqrt{\frac{\rho}{\mu_r f}}, \text{ m.} \tag{1.22}$$

From Eq. (1.20), the penetration depth can be defined as the distance from the surface of the conductor at which the current density is reduced to a value $e = 2.718$ times lower than the current density at the surface.

According to Eq. (1.22), the value of the penetration depth varies with the square root of electrical resistivity and inversely with the square root of relative magnetic permeability and frequency. Therefore, it depends upon the electromagnetic properties of the conductor's material and the applied frequency and, at the same frequency, it has different values in different materials [4, 5, 9].

Since the volumetric power density depends on the square of J, according with Eq. (1.20), its distribution from the surface towards the interior of the conductor can be expressed by the equation:

$$w = w_e\, e^{-\frac{2y}{\delta}} \tag{1.23}$$

with $w_e = \rho\, J_e^2$.

The curves of Fig. 1.4 show the commonly assumed classical distributions of current and power density versus the distance from surface (in current penetration depths) due to the skin effect, obtained from Eqs. (1.20) and (1.23).

The analysis of the curves shows that at a depth greater than 3 times δ, the values of all the quantities of interest (current density, volume specific power and also magnetic field intensity) are practically negligible.

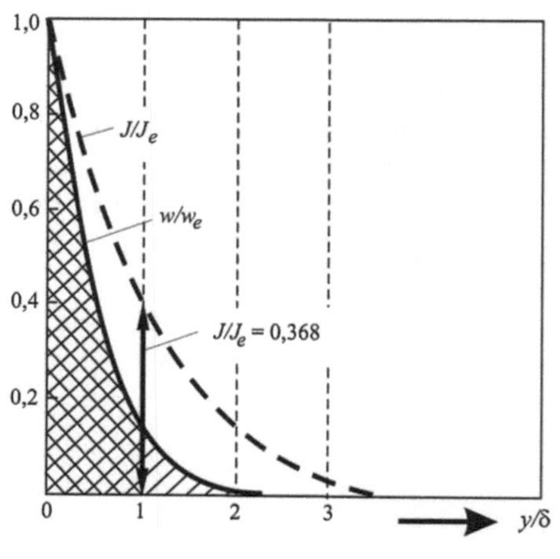

Fig. 1.4 Current density (J) and power density (w) distributions as a function of the distance from the surface of the workpiece, according to Eqs. (1.20) and (1.23)

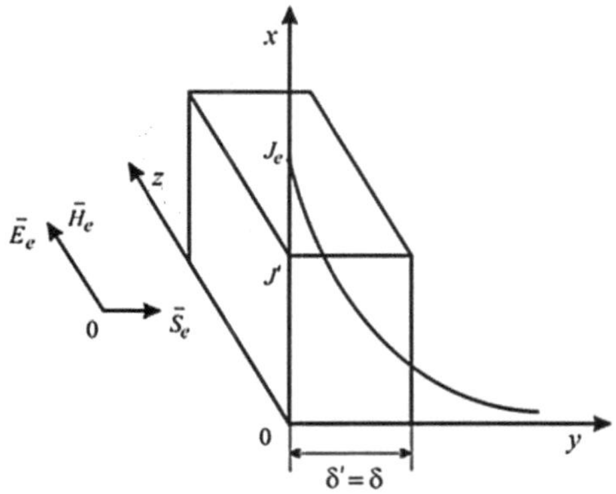

Fig. 1.5 Equivalent surface active layer

It can be easily demonstrated that the ratio of the power released in the surface layer of thickness equal to the penetration depth δ to the total active power released in the conducting semi-infinite body, is equal to [2]:

$$\frac{P_{s\delta}}{P_{tot}} = 0.865,$$

i.e. in the surface layer of thickness equal to the penetration depth is allocated 86.5 % of the total active power dissipated in the semi-infinite conducting body.[1]

Considering that nearly all the active power is concentrated in the surface layer of thickness δ, for practical calculations can be sometimes useful to make reference to the concept of an *equivalent surface active layer.* The equivalent surface layer is obtained by replacing the actual exponential current density distribution in the heated body by a fictitious distribution, supposing that in the equivalent surface active layer of finite thickness δ' a current density J' is uniformly distributed, while it is equal to zero in the rest of the cross-section (see Fig. 1.5).

The thickness of the layer δ' and the effective value of the current density J', are determined by the condition that the total active power released in the conductor with exponential distribution of the current density must be equal to the one dissipated in the equivalent active layer.

Choosing arbitrarily the value of δ' equal to the penetration depth of the electromagnetic wave δ, it's easy to demonstrate that the current density at the surface of the body with exponential distribution and the current density in the equivalent active layer are related by the expression[2] $J_e = \sqrt{2}J'$ [2, 7].

---

[1] This ratio is valid only for the exponential distribution of the current density J inside the body.

[2] The active layer thickness δ' and the current density in the active layer J' are introduced by an artificial engineering operation.

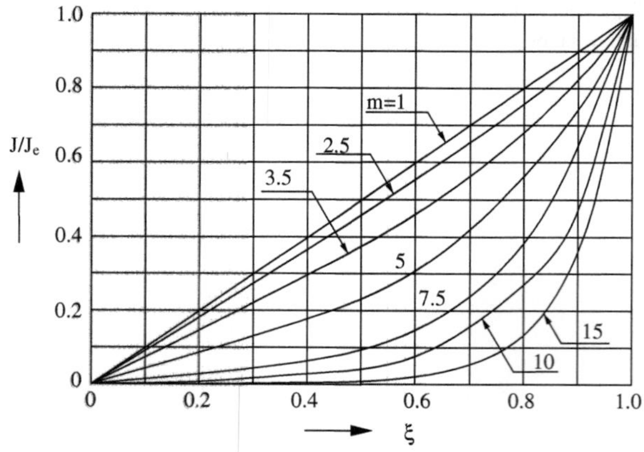

Fig. 1.6 Radial distributions of induced current density in cylindrical workpieces ($\xi = r/r_e$; $m = \sqrt{2}\ (r_e/\delta)$; r_e—external radius of cylinder) [6]

- ***Non-exponential distribution of heat sources***

As stated before, the exponential distribution described by Eq. (1.20) occurs in a homogeneous and linear body in case of pronounced skin effect (e.g. at relative high frequency).

At lower frequencies, also in case of homogeneous and linear materials, the distribution differs from the exponential one depending on the frequency, the material characteristics and the geometry of the body to be heated.

A typical example occurs in the longitudinal flux induction heating of non-magnetic cylindrical work-pieces, where—with constant resistivity—the radial distributions of the induced currents are those illustrated in Fig. 1.6 as a function of $\xi = r/r_e$. In this case the exponential distribution is reached in practice only for values of $m = \frac{\sqrt{2} r_e}{\delta} \geq 12$.

In many applications of rapid induction heating or direct resistance heating, during the thermal transient—due to the skin effect—the temperature in different points of the work-piece cross-section may have different values and the local resistivity, which is temperature dependent, varies from point to point.

This reflects on the distribution of the specific power per unit of volume, given by Eq. (1.23), and the distribution of heat sources in the cross section, which may differ significantly from the exponential distribution even in case of pronounced skin effect.

This phenomenon is more pronounced in ferromagnetic work-pieces, where the current density distribution is influenced not only by the temperature dependent resistivity, but also by the local values of magnetic permeability, which depend not only on the temperature but also on the local magnetic field intensity.

Figure 1.7 shows, as an example, how the radial distribution of specific power modifies during direct resistance heating of a magnetic steel rod. The curves—which relate to different rod's surface temperatures—show that at the beginning of heating the power distribution is nearly exponential. At subsequent instant of heating transient—due to variations of resistivity and permeability with temperature and local magnetic field—the power distribution remains highly non-uniform below Curie point;

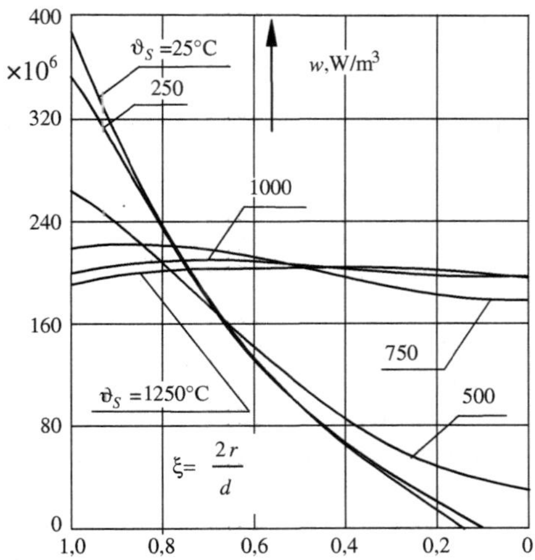

Fig. 1.7 Power density distributions along radius in DRH of a cylindrical rod, at different instants when the surface temperature is ϑ_s. ($r_e = d/2 = 30$ mm; $\xi = r/r_e$; $\ell = 6$ m; U = 100 V; steel C45) [6]

but finally undergoes a dramatic change above that temperature, where it becomes more "flat" and is influenced mainly by the radiation losses from the rod surface.

A particular phenomenon, known as "*magnetic wave*", occurs in induction surface hardening where the power density distribution along the radius/thickness may show a peculiar shape, significantly different from the commonly assumed exponential distribution. Here, as usual, the power density has its maximum value at the surface, and decreases toward the core. But then, at a certain distance from the surface, the power density suddenly starts to increase again, reaching a maximum value before it finally declines (Fig. 1.8) [6, 10].

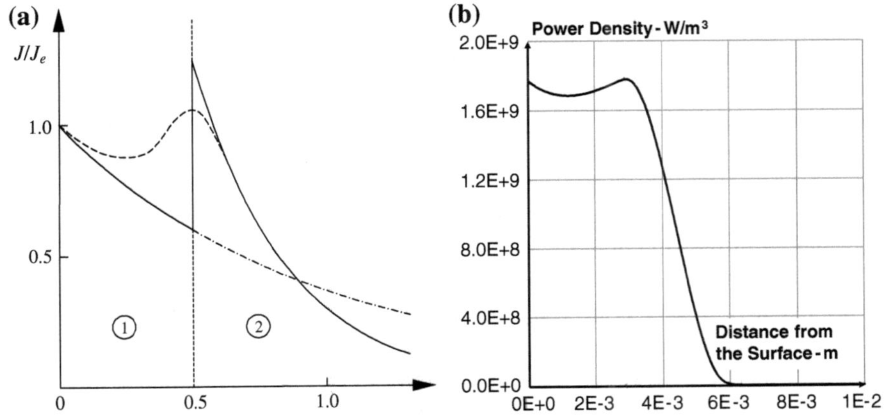

Fig. 1.8 **a** Schematic representation of the magnetic wave phenomenon; **b** radial distribution of power density in induction heating at 4 kHz of a carbon steel shaft, 100 mm diameter, with $H_0 = 240$ kA/m (rms value)

The main cause of this behavior relates to the phenomenon that, in induction surface hardening, carbon steel retains its magnetic properties in the core (region 2 in Fig. 1.8a), while a surface layer (region 1) which is at higher temperature becomes non-magnetic being heated above A_{c3}, the critical austenitization temperature.

In this situation, the current density distribution can be roughly schematized by two exponentials corresponding to two different values of δ caused by the material characteristics of the magnetic and non-magnetic regions, as it is illustrated in Fig. 1.8a.

In particular, in some cases, the value of current and power density, in correspondence to the surface of separation of the two layers, can result to be higher than the one at the surface.

However, given that the values of ρ and μ gradually change from layer to layer, the distributions result more complex, as indicated qualitatively by the dashed curve in the same figure and the phenomenon can manifest itself with the maximum values in an internal layer of the workpiece and not at its surface.

Figure 1.8b shows an example of the radial distribution of power density during the transition of Curie point, in the induction heating of a steel billet.

- ***Proximity effect***

The proximity effect is a surface effect consisting in the re-distribution of the current density in certain areas of the conductor as a result of the overall interaction of the electromagnetic fields produced by all current carrying conductors, comprising the conductor under consideration.

In a single conductor carrying AC current, the current is distributed unevenly over the cross-section, symmetrically about the axis of symmetry for conductors of circular cross section or the planes of symmetry for conductors of rectangular cross section (Fig. 1.9).

If alternating currents flow in two conductors, the current density distribution is affected by the distance between them and the actual direction of currents.

If the currents in the conductors flow in opposite directions (with a phase difference of 180°), the highest current density will be concentrated on the sides facing each other (Fig. 1.10a). If the currents have the same direction (i.e. with phase difference of 0°), the maximum current density will concentrate on the far side of the conductors (Fig. 1.10b).

To explain this phenomenon, we can use the same approach used for the explanation of the skin-effect.

The redistribution of the current density in one conductor is due to the influence of the magnetic field produced by the current in the other conductor.

Consider for example the case of oppositely directed currents (Fig. 1.10a). The magnetic field produced by the current flowing in the conductor on the right, penetrates the left conductor in the direction characterized by the vector of the

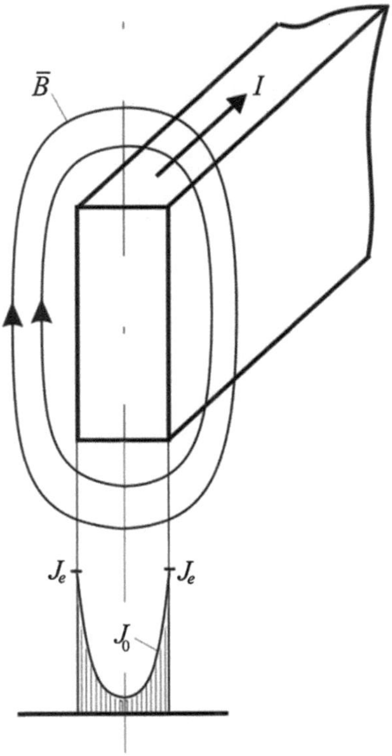

Fig. 1.9 Skin effect in a single conductor [2]

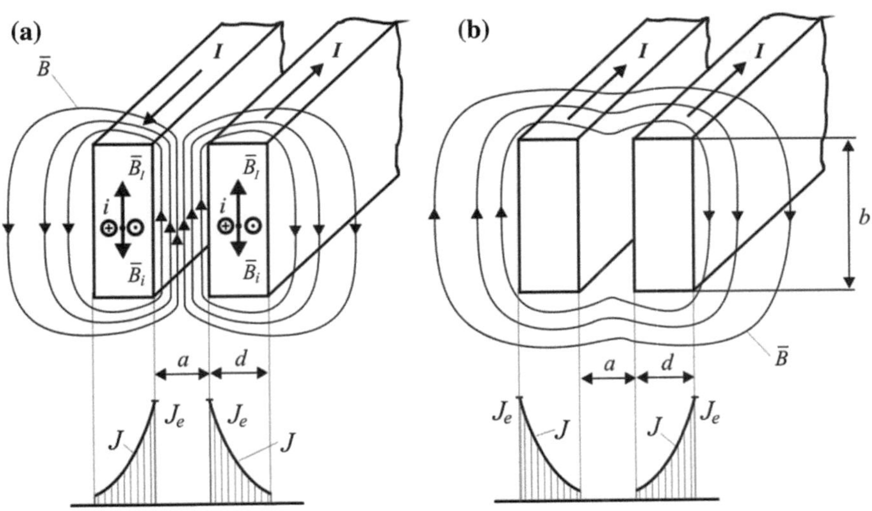

Fig. 1.10 Proximity effect [2]

magnetic induction $\dot{\bar{B}}_I$. This magnetic field, being external to the left conductor, by the law of electromagnetic induction will produce in it electromotive forces (*e.m.f. s*), which will induce in the conductor eddy currents *i*, whose direction is determined by Lenz's rule.

According to the Lenz's rule the induced eddy current has a direction to create a magnetic field $\dot{\bar{B}}_i$ that must oppose the external magnetic field $\dot{\bar{B}}_I$ that has produced it.

Thus, in the left conductor will be induced the eddy current *i,* which increases the current *I* at the surface facing the second conductor, and decreases it at the surface far from the second conductor.

Arguing similarly, we can show that in the second conductor (on the right) the direction of the eddy current *i* will coincide with the direction (or the phase) of the current *I* at the surface facing the left conductor, and will be in the opposite direction at the surface far from the left conductor.

Similarly, we can explain the redistribution of the current density resulting when the currents in the conductors are in the same direction (Fig. 1.10b).

When considering a current-carrying conductor at the surface of an electrically conductive body, in which initially no current flows (like in the induction surface hardening of steel products), the conducting body is penetrated by an external magnetic field created by the current *I* and characterized by the vector of the magnetic induction $\dot{\bar{B}}_I$ (Fig. 1.11).

The alternating external magnetic field creates in the body an electrical *e.m.f.*, which produces the eddy current i_2, whose direction is determined by the Lenz's rule. According to this rule the intrinsic magnetic field $\dot{\bar{B}}_I$ of the eddy current must counteract the external magnetic field $\dot{\bar{B}}_I$.

For the rule of the right-hand screw, the direction of the eddy current *i,* induced in the conducting body, is opposite to the direction of the current *I* in the current-carrying conductor. Thus, the phase of the current flowing in the active conductors

Fig. 1.11 Proximity effect [2]

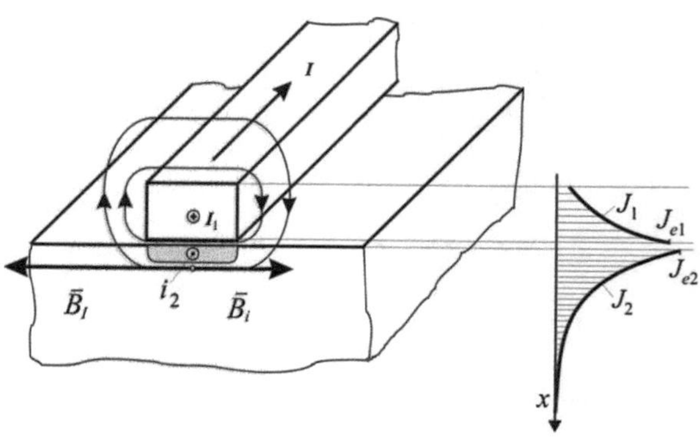

(such as the conductor of the inductor) and the phase of the eddy current induced in the conducting body (the heated workpiece) will differ by 180°.

In both conductors, i.e. the inductor and in the heated body, an uneven distribution of the current density will result. This effect corresponds to the proximity effect in two parallel conductors, with the maximum of the current density located at the surfaces of the inductor and the heated body.

The proximity effect depends on the ratio *a/d* (see Fig. 1.10a) and is more pronounced as the distance between the conductors decreases (i.e., for smaller ratio *a/d*) and the greater is the thickness of the conductor in comparison with the penetration depth.

- ***Ring effect***

In bent or ring shaped conductors, carrying DC or AC current, the highest current density is concentrated on the inner surface of the ring as illustrated in Fig. 1.12a, b. This phenomenon is called ring- or coil-effect.

The presence of such uneven distribution can be explained by the asymmetry of the magnetic field. The current is concentrated near the inner surface of the ring, flowing in the section of the conductor having the smallest reactance (and

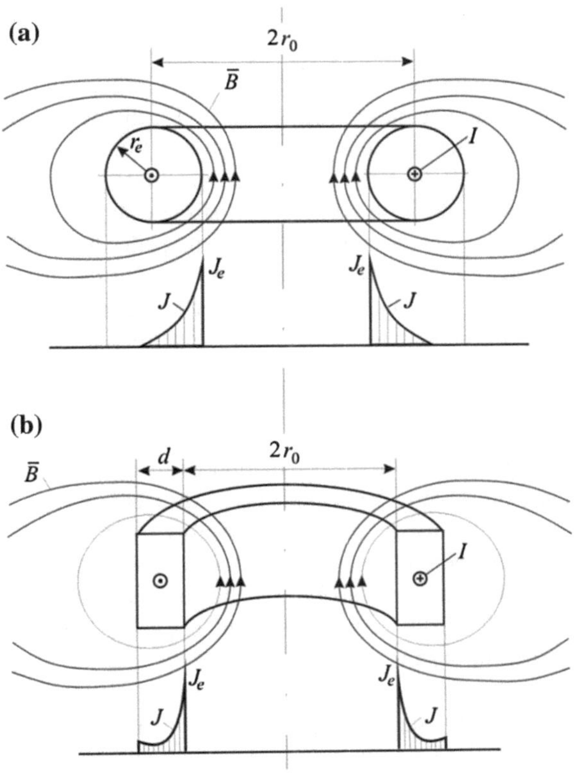

Fig. 1.12 **a** Ring effect in a conductor of circular cross-section **b** Ring effect in a conductor of rectangular cross section [2]

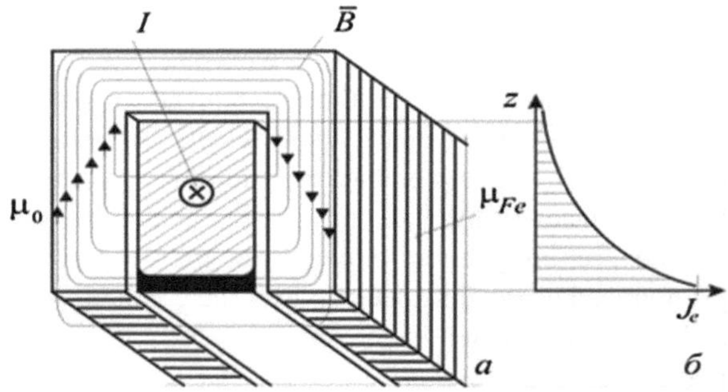

Fig. 1.13 Slot effect [11]

resistance). The ring effect is stronger the greater is the ratio of the curvature radius r_0 and the radial thickness of the conductor d (i.e., the larger the ratio d/r_0—for a conductor of rectangular section, and the larger the ratio $r_1 r_0$—for a conductor of circular cross section).

- ***Slot effect***

The presence of a magnetic core has a significant effect on the current density distribution in a conductor carrying AC current.

If the conductor is placed in an open slot of a magnetic circuit (Fig. 1.13), we observe the phenomenon of an unilateral skin effect—the slot effect.

The highest current density will be concentrated in the part of the conductor located near the open side of the slot. For a qualitative description of the slot effect, the conductor cross section can be ideally subdivided into two parts. One part is closer to the magnetic material, so it is 'linked' with a magnetic flux density that is higher than the one close to the open side of the slot.

Therefore, in the inner zone, stronger counter-electromotive forces are induced and, as a consequence, the current paths have higher reactance values and are crossed by currents of lower intensity. On the contrary, the current density in the regions of the conductor with lower reactance, i.e. in the area at the open side of the slot, becomes more intense than in the depth of the slot.

- ***Load and coil end- and edge-effects***

As previously underlined, the distributions of current and power densities are fundamental in the determination of the transient and final temperature pattern in the workpiece to be heated. These distributions are affected not only by skin-effect and heat transfer conditions but also by the electromagnetic field distortion that occur in the end regions of inductors and at the edges of workpieces. These field distortions are responsible for the so-called *end-* and *edge-effects.*

These effects have a different impact in the direct resistance heating and the induction heating applications.

In direct resistance heating of bars, edge-effects occur only in limited areas near the contact system, where the current density lines have a complex two dimensional space pattern also in axis-symmetric geometries.

In induction heating end- and edge-effects can act differently depending on the application, the type of load, magnetic or non-magnetic, cylindrical or with rectangular cross-section (slabs), and the relative lengths of inductor and load.

Referring the reader for a more detailed discussion to the large bibliography on this subject [10, 12], we will report here only two examples in order to attract the reader's attention on the importance of these phenomena.

Longitudinal end- and edge-effect

These effects appear typically when heating cylinders with solenoid coils or circular plates with pancake coils.

As shown by the sketch of Fig. 1.14a, depending on the frequency, the inductor coil overhang Δa, the work-piece and coil geometrical dimensions and the material characteristics, the end effect can produce under heating or overheating of the work-piece extremities.

Figure 1.14b shows, as an example, the influence of the coil overhang on the distribution of the surface power density along the length of a non-magnetic cylinder in case of a pronounced skin-effect [10].

It also shows that an appropriate choice of the above parameters allows to obtain a reasonable uniform power density distribution, like in the curve with overhang Δa_4.

More complex is the end-effect phenomenon in magnetic materials, since in this case two opposing factors act on the distortion of the electromagnetic field.

On one side the demagnetizing action of the induced currents tends to expel the magnetic field; on the other side the magnetizing effect of the material tends to concentrate the magnetic field lines within the work-piece.

As a result of these factors, unlike with non-magnetic materials, in ferromagnetic cylinders the workpiece ends may be either overheated or under-heated even in magnetic fields produced by coils with large overhangs.

Edge effect in rectangular slabs

In the induction heating of rectangular slabs occurs the overlap of the transversal and longitudinal edge-effects, which are determined by the finite size of the workpiece.

With reference to Fig. 1.15a, considering a slab located in an uniform magnetic field, oriented along the z-axis, and assuming that the size (ℓ) and (w) are much greater than (h), both edge-effects are determined only by the parameter (h/δ) and have a two-dimensional behaviour. The only exception occurs at the three-edge corners, where the field is fully three-dimensional [12].

The same figure shows the distributions along the x and z axes of the ratio (p/p_c) of the power density induced on points of the upper surface of the slab and its value in the central point of the centre line of the longer side.

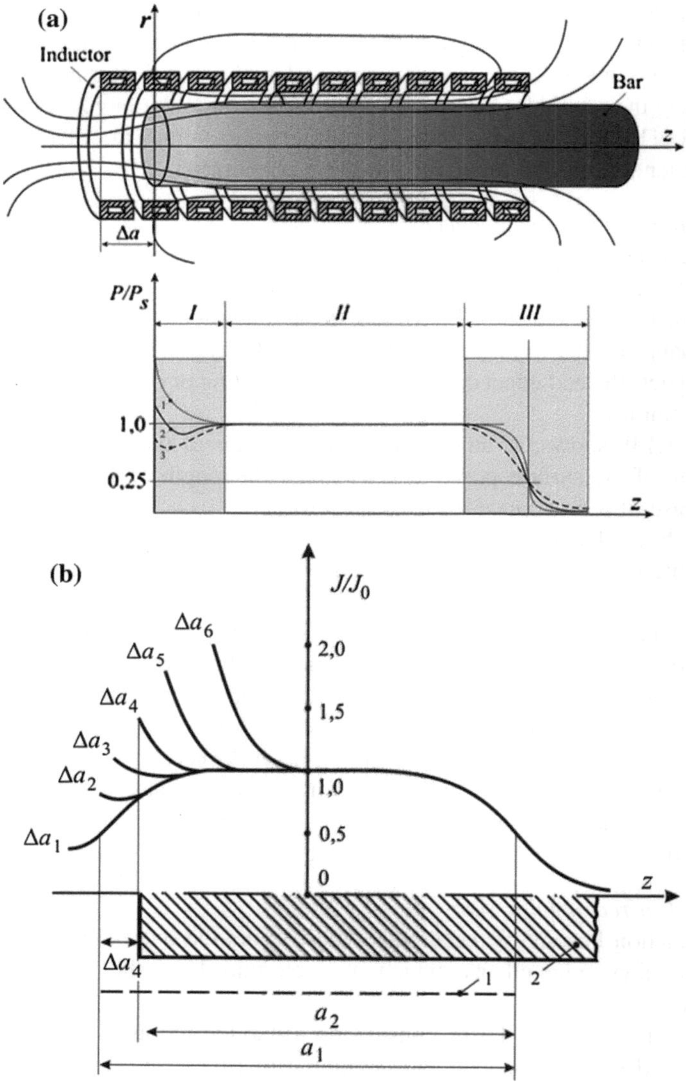

Fig. 1.14 Longitudinal end-effect in induction heating of cylinders: **a** sketch of power density distribution along the length; **b** influence of different coil overhangs on current density distribution [10, 13]

In Fig. 1.15a the distributions are plotted for $h/\delta = 4$ (solid lines) and $h/\delta = 1$ (dashed lines).

These distributions are typically characterised by three zones, i.e. a central "*regular*" zone, where the values of the induced power density are constant and correspond to the ones in an infinite plate excited with the same surface magnetic

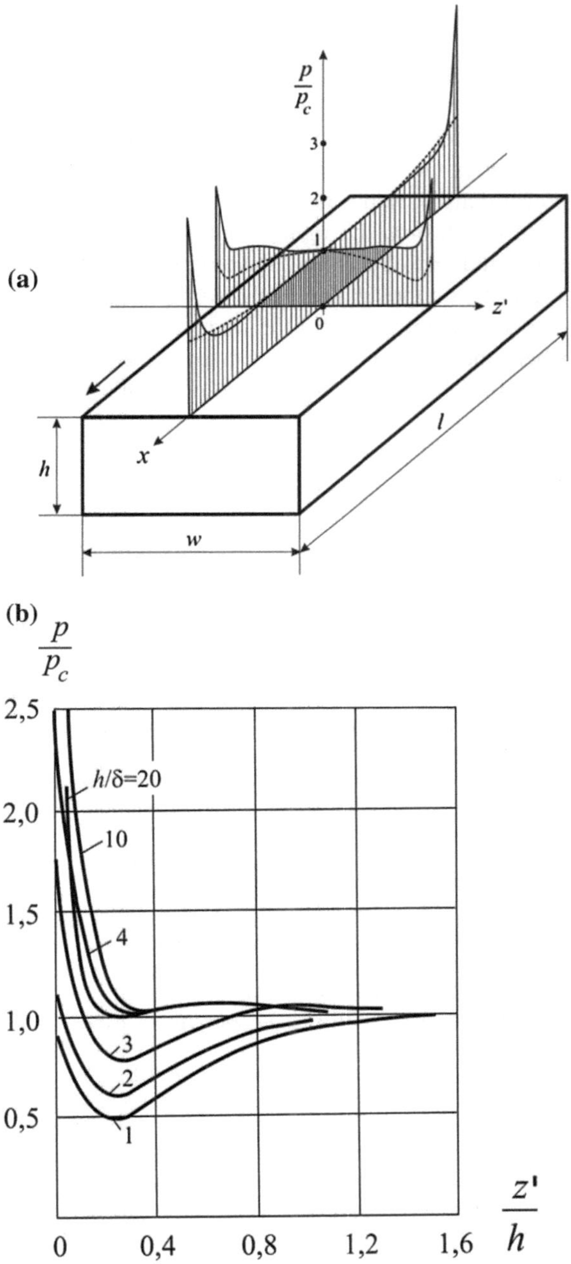

Fig. 1.15 Edge effects in induction heating of slabs heated in longitudinal uniform magnetic field: **a** relative power density distribution along slab length and width (*solid lines* h/δ = 4) (*dashed lines* h/δ = 1); **b** relative power density distribution along slab width for different values of h/δ (z′—distance from slab edges) [12]

field intensity, and the areas of "*transverse edge effect*" and of "*longitudinal edge effect*".

The variations of values of (p/p_c) as a function of the relative distance (z'/h) from the edges of the slab are given in Fig. 1.15b for different values of *(h/δ)*. They are characterised near the edges by a region with induced power values lower than the ones in the "regular" zone (under-heated area) and an external area with higher power density values (possible overheating).

In conclusion, the figures underline on one hand the strong influence that the edge- and end-effects may have on the power density distributions and, as a consequence, on the heating of the workpiece; on the other hand, that—in some cases—a convenient choice of the frequency allows to obtain an induced power pattern where the overheated areas can compensate to a certain extent the under-heated ones, by taking advantage of the material's thermal conductivity.

References

1. Mühlbauer, A. (2008). *History of induction heating & melting* (202 pp.). Vulkan-Verlag GmbH. ISBN 978-3-8027-2946-1.
2. Aliferov, A., & Lupi, S. (2004). *Direct resistance heating of metals* (223 pp.). Novosibirsk: NGTU. ISBN 5-7782-0475-2 (in Russian).
3. Aliferov, A., & Lupi, S. (2010). *Induction and direct resistance heating of metals* (411 pp.). Novosibirsk: NGTU. ISBN 978-5-7782-1622-8 (in Russian).
4. EDF. (1996). *Induction, Conduction électrique dans l'industrie* (780 pp.). Paris (France): CFE DOPEE85.
5. Rudnev, V., & Totten, G. E. (Eds.). (2014). *ASM Handbook. Volume 4C—induction heating and heat treatment* (820 pp.). Materials Park, OH 44073-0002: ASM International. ISBN-10 1-62708-012-0.
6. Lupi, S. (2005). *Elettrotermia* (453 pp.). Padua (Italy): Libreria Progetto (in Italian).
7. Nacke, B., Baake, E., Lupi, S., Forzan, M. et al. (2012). *Theoretical background an aspects of electrotechnologies—Physical principles and realization* (356 pp.). Basic I, St. Petersburg: Publishing house of ETU. ISBN 978-5-7629-1237-2.
8. Ramo, S., Whinnery, J. R., & Van Duzer, T. (1965). *Fields and waves in communication electronics* (754 pp.). N.Y: Wiley.
9. Kuvaldin, A. B. (1976). *Low-temperature induction heating of steel* (111 pp.). Moscow: Energia (in Russian).
10. Rudnev, V., Loveless, D., Cook, R., & Black, M. (2003). *Handbook of induction heating* (777 pp.). N.Y.: Marcel Dekker Inc. ISBN 0-8247-0848-2.
11. Fomin, N. I., & Satylobskij, L. M. (1979). *Electrical furnaces and induction heating installations* (247 pp.). Moscow: Metallurghia (in Russian).
12. Nemkov, V., & Demidovich, V. B. (1988). *Theory and calculation of induction heating installations* (280 pp.). Leningrad: Energoatomiszdat. ISBN 5-283-04409-2 (in Russian).
13. Slukhotskii, E., Nemkov, V. et al. (1981). *Induction heating installations* (328 pp.). Leningrad: Energoisdat (in Russian).

Chapter 2
Electromagnetic Field in Workpieces with Flat Surfaces

Abstract This chapter deals with the induction or direct resistance heating of flat bodies. Analytical solutions are derived for some typical configurations in order to calculate the main process parameters, like active and reactive power in the workpiece and power sources distributions [1]. The chapter is subdivided into four paragraphs and in each paragraph the solution is provided both for induction and direct resistance heating. In the first paragraph, solutions are presented for heating bodies of infinite dimensions having linear material properties. In the second chapter the same geometrical configuration is studied for ferromagnetic bodies. In the third paragraph, the solution of the electromagnetic problem is firstly obtained for the induction heating of a two-layers slab, than a workpiece of rectangular cross-section with finite dimensions is taken into account in a longitudinal flux system. The direct resistance heating of a finite dimension cross-section is then discussed. The last paragraph details the induction and resistance heating of finite cross-section loads, evaluating the current distribution near to different shaped edges.

2.1 Semi-infinite Body of Homogeneous Material

2.1.1 Induction Heating

To describe how an electromagnetic wave propagates inside an electrically conductive body, we consider a semi-infinite metal body as in Fig. 2.1 infinitely extended in the directions *x*, *y*, *z*, with its surface along the coordinates *x* and *z* and depth along y.

We use the notations: $\overline{H}_e, \overline{E}_e$—vectors of magnetic and electric fields intensities in air at the external surface of the heated workpiece, and $\overline{H}, \overline{E}$—values of the same quantities inside the body.

Nowadays, the solution of a partial differential equation system (PDE) can be computed by using numerical methods, without the need to make almost any assumption to simplify the complexity of the problem. Nevertheless, analytical solutions of the differential Eq. (2.1a) are meaningful to properly understand the

S. Lupi et al., *Induction and Direct Resistance Heating*,
DOI 10.1007/978-3-319-03479-9_2

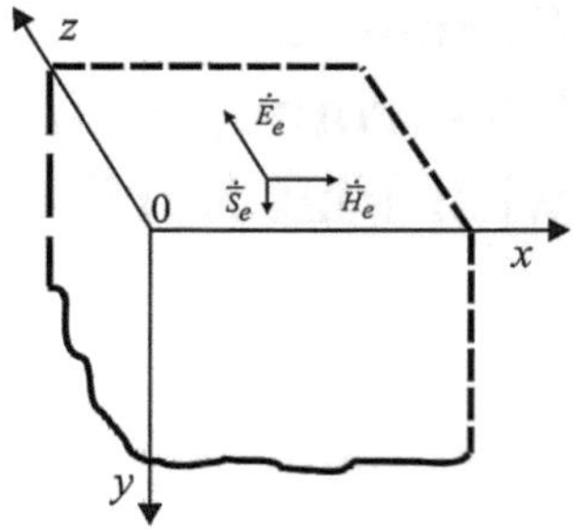

Fig. 2.1 Schematic of a semi-infinite homogeneous body

electromagnetic phenomena that occur in direct and induction heating [1]. Electric and magnetic fields are described as time dependant vector fields. This means that, for instance, the electric field is defined on a certain point P_0 (x_0, y_0, z_0) by the values of three spatial quantities (E_x, E_y, E_z) and each spatial quantity is also time dependant. An analytical solution that satisfies Eq. (2.1a) exists only for very simple geometries, for material with linear and isotropic properties and for time independent (DC) or with sinusoidal variation fields.

The equations for the study of electromagnetic processes are the basic equations of the electromagnetic field, i.e. the Maxwell equations

$$\left.\begin{aligned} rot\,\overline{H} &= \overline{J} + \tfrac{\partial \overline{D}}{\partial t} \\ rot\,\overline{E} &= -\tfrac{\partial \overline{B}}{\partial t} \\ div\,\overline{D} &= \rho_c \\ div\,\overline{B} &= 0 \end{aligned}\right\} \tag{2.1a}$$

For the solution of system (2.1a), we make the following assumptions:

$$\overline{J} \gg \frac{\partial \overline{D}}{\partial t} \quad div\,\overline{J} = 0$$

- the medium is homogeneous, i.e. its properties, electrical resistivity and magnetic permeability, are constant ρ = const., μ_r = const.
- the inductor current is sinusoidal, so that in linear material field quantities $\overline{H}, \overline{E}, \overline{B}$ are also sinusoidal:

$$\overline{H} = \overline{H}_m \cdot \sin \omega t, \overline{B} = \overline{B}_m \cdot \sin \omega t, \overline{E} = \overline{E}_m \cdot \sin(\omega t + \varphi)$$

- the sinusoidal variation of vectors $\overline{H}, \overline{E}, \overline{B}$, allows us to write them in complex form

$$\dot{\overline{H}} = \overline{H}_m \cdot e^{j\omega t}, \dot{\overline{B}} = \overline{B}_m \cdot e^{j\omega t}, \dot{\overline{E}} = \overline{E}_m \cdot e^{j(\omega t+\varphi)}.$$

Equation (2.1a) can be written as follows, introducing the complex notation for quantities which are sinusoidal function of time:

$$\left.\begin{array}{l} rot\,\dot{\overline{H}} = \frac{\dot{\overline{E}}}{\rho} \\ rot\,\dot{\overline{E}} = -\mu\frac{\partial\dot{\overline{H}}}{\partial t}\; rot\,\dot{\overline{E}} = -j\,\omega\mu\,\dot{\overline{H}} \\ div\,\dot{\overline{B}} = 0 \\ div\,\dot{\overline{E}} = 0 \end{array}\right\} \tag{2.1b}$$

Substituting the value of the electric field from the first equation into the second, we obtain

$$\rho\cdot rot(rot\,\dot{\overline{H}}) = -\mu\frac{\partial\dot{\overline{\mathrm{H}}}}{\partial t} = -j\omega\mu\,\dot{\overline{\mathrm{H}}}.$$

Consider the left side of this equation. Since is

$$\begin{array}{l} rot\;(rot\,\dot{\overline{H}}) = grad\;(div\,\dot{\overline{H}}) - \nabla^2\dot{\overline{H}} \\ div\,\dot{\overline{H}} = 0, \end{array}$$

the system (2.1b) can be transformed into the equation

$$\nabla^2\dot{\overline{H}} = \frac{\mu}{\rho}\frac{\partial\dot{\overline{H}}}{\partial t} = \mathrm{j}\,\omega\frac{\mu}{\rho}\dot{\overline{H}}, \tag{2.2}$$

with $\nabla^2\dot{\overline{H}}$—the Laplace's operator.

Taking into account that is

$$\frac{\partial\dot{\overline{H}}}{\partial t} = \frac{\partial}{\partial t}(\overline{H}_m\cdot e^{j\omega t}) = j\omega\cdot\overline{H}_m\cdot e^{j\omega t} = j\omega\cdot\dot{\overline{H}},$$

Equation (2.2) is transformed as follows:

$$\nabla^2\dot{\overline{H}} = j\frac{\omega\mu}{\rho}\dot{\overline{H}}. \tag{2.3a}$$

In Cartesian coordinates the Laplace's operator has the form

$$\nabla^2\dot{\overline{H}} = \frac{\partial^2\dot{\overline{H}}}{\partial x^2} + \frac{\partial^2\dot{\overline{H}}}{\partial y^2} + \frac{\partial^2\dot{\overline{H}}}{\partial z^2}.$$

Since the plane wave of the electromagnetic field, incident on the body, is coincident with the plane xoz, and the intensity of magnetic field $\dot{\overline{H}}$ is constant in any plane of coordinate y, then $\partial^2\dot{\overline{H}}\big/\partial x^2 = \partial^2\dot{\overline{H}}\big/\partial z^2 = 0$ and the Laplace's operator is equal to

$$\nabla^2\dot{\overline{H}} = \frac{\partial^2\dot{\overline{H}}}{\partial y^2}. \tag{2.3b}$$

As a result of the substitution of (2.3b), the Eq. (2.2) is transformed into

$$\frac{\partial^2\dot{\overline{H}}}{\partial y^2} = j\frac{\omega\mu}{\rho}\cdot\dot{\overline{H}}$$

Introducing the quantity $k^2 = j\omega\sigma\mu$, we obtain:

$$\frac{\partial^2\dot{\overline{H}}}{\partial y^2} = k^2\dot{\overline{H}}. \tag{2.3c}$$

The general solution of this linear differential equation of second order is

$$\dot{\overline{H}} = \dot{\overline{C}}_1 \cdot e^{-ky} + \dot{\overline{C}}_2 \cdot e^{ky}, \tag{2.4a}$$

and the spatial modulus of the magnetic field intensity vector is

$$\dot{H} = \dot{C}_1 \cdot e^{-ky} + \dot{C}_2 \cdot e^{ky}, \tag{2.4b}$$

with $\dot{\overline{C}}_1, \dot{\overline{C}}_2, \dot{C}_1, \dot{C}_2$—integration constants, $\dot{\overline{C}}_1, \dot{\overline{C}}_2$—complex vectors, and $\dot{C}_1, \dot{C}_2$—moduli of the complex vectors.

We are able now to write the solution of the electric field intensity in a general form. From the first equation of the system (2.1b) is

$$\dot{\overline{E}} = \rho \cdot rot\,\dot{\overline{H}}$$

In a cartesian coordinate system is

$$rot\,\dot{\overline{H}} = \overline{u}_x \cdot rot_x\,\dot{\overline{H}} + \overline{u}_y \cdot rot_y\,\dot{\overline{H}} + \overline{u}_z \cdot rot_z\,\dot{\overline{H}}$$

with $\overline{u}_x$, $\overline{u}_y$, $\overline{u}_z$ unit vectors along the coordinates x, y, z;

$$\left.\begin{aligned} rot_x\,\dot{\overline{H}} &= \frac{\partial\dot{H}_z}{\partial y} - \frac{\partial\dot{H}_y}{\partial z} \\ rot_y\,\dot{\overline{H}} &= \frac{\partial\dot{H}_x}{\partial z} - \frac{\partial\dot{H}_z}{\partial x} \\ rot_z\,\dot{\overline{H}} &= \frac{\partial\dot{H}_y}{\partial x} - \frac{\partial\dot{H}_x}{\partial y} \end{aligned}\right\} - \textit{components of vector rot } \dot{\overline{H}} \textit{ in cartesian coordinates}$$

Since in the case considered here the vector of the magnetic field intensity is directed along the axis ox (see Fig. 2.1), then its components along the axes oz and oy are equal to zero, i.e. $\dot{H}_z = \dot{H}_y = 0$.

In view of the fact that for a fixed value of y the derivative $\partial\dot{H}_x/\partial z = 0$, it follows that

$$\frac{\partial\dot{H}_z}{\partial y} = \frac{\partial\dot{H}_y}{\partial z} = \frac{\partial\dot{H}_x}{\partial z} = \frac{\partial\dot{H}_z}{\partial x} = \frac{\partial\dot{H}_y}{\partial x} = 0.$$

Therefore

$$rot\,\dot{\overline{H}} = \overline{u}_z \cdot rot_z\,\dot{\overline{H}} = -\overline{u}_z \cdot \frac{\partial\dot{H}_x}{\partial y}$$

From Ampere's law the electric field intensity can be calculated from the relationship

$$\dot{\overline{E}} = -\overline{u}_z \cdot \rho \cdot \frac{\partial\dot{H}}{\partial y}. \tag{2.5a}$$

Since the vector $\dot{\overline{H}}$ is directed along x, here and in the following we will omit the index x. Then, by substituting Eq. (2.4a) in the obtained expression, we have

$$\dot{\overline{E}} = -\overline{u}_z \cdot \rho \cdot \frac{\partial}{\partial y}\left(\dot{C}_1 \cdot e^{-ky} + \dot{C}_2 \cdot e^{ky}\right),$$

and

$$\dot{\overline{E}} = \overline{u}_z \cdot k \cdot \rho \cdot \left(\dot{C}_1 \cdot e^{-ky} - \dot{C}_2 \cdot e^{ky}\right). \tag{2.5b}$$

We can determine now the integration constants $\dot{C}_1$ and $\dot{C}_2$. When the coordinate y increases, the first term of Eqs. (2.4a) and (2.5b) decreases, while the second ones are increasing. These formulas physically say that the incident wave is attenuated when it penetrates into the conductive body, while the reflected one exhibits an opposite behavior. The first term $\dot{C}_1 \cdot e^{-ky}$ describes the incident wave, the second term $\dot{C}_2 \cdot e^{ky}$—the reflected one.

In the case of propagation of an electromagnetic wave inside a semi-infinite plane there is not a reflective surface and the reflected wave is absent; therefore the integration constant $\dot{C}_2$ must be equal to 0.

Equations (2.4a) and (2.5b) become therefore:

$$\dot{\overline{H}} = \dot{\overline{C}}_1 \cdot e^{-ky}, \tag{2.6}$$

$$\dot{\overline{E}} = \overline{u}_z \cdot k \cdot \rho \cdot \dot{C}_1 \cdot e^{-ky}. \tag{2.7}$$

The integration constant $\dot{\overline{C}}_1$ can be determined from the boundary conditions. On the surface, at $y = 0$, the amplitude of the magnetic field, $\dot{\overline{H}} = \dot{\overline{H}}_e$ is known. In the hypothesis that $\dot{\overline{H}}_e$ is a vector tangential to the metal surface the magnetic field intensity does not vary crossing the surface between the conductive and dielectric media. Therefore, $\dot{\overline{C}}_1 = \dot{\overline{H}}_e$ and $\dot{C}_1 = \dot{H}_e$.

Thus

$$\dot{\overline{H}} = \dot{\overline{H}}_e \cdot e^{-ky} \tag{2.8}$$

$$\dot{\overline{E}} = \overline{u}_z \cdot k \cdot \rho \cdot \dot{H}_e \cdot e^{-ky}. \tag{2.9}$$

In these equations, the damping coefficient of the electromagnetic wave is equal to

$$k = \sqrt{\frac{j\omega\mu}{\rho}} = \sqrt{2j}\sqrt{\frac{\omega\mu}{2\rho}} = \frac{1+j}{\delta} = \frac{\sqrt{2}}{\delta} e^{j\frac{\pi}{4}}. \tag{2.10}$$

Substituting Eq. (2.10) in (2.8) and (2.9), we obtain the equations for calculating the instantaneous values of the vectors of the magnetic and electric field intensities:

$$\begin{aligned}\dot{\overline{H}} = \dot{\overline{H}}_e \cdot e^{-ky} &= \overline{u}_x \cdot H_{em} \cdot e^{j\omega t} \cdot e^{-(1+j)\frac{y}{\delta}} \\ &= \overline{u}_x \cdot H_{em} \cdot e^{j\omega t} \cdot e^{-\frac{y}{\delta}} \cdot e^{-j\frac{y}{\delta}} = \overline{u}_x \cdot H_{em} \cdot e^{-\frac{y}{\delta}} \cdot e^{j\left(\omega t - \frac{y}{\delta}\right)}\end{aligned}$$

i.e.

$$\begin{aligned}\dot{\overline{H}} &= \overline{u}_x \cdot H_{em} \cdot e^{-\frac{y}{\delta}} \cdot e^{j\left(\omega t - \frac{y}{\delta}\right)}, \\ \dot{\overline{E}} &= \overline{u}_z \cdot \frac{1+j}{\delta} \cdot \rho \cdot \dot{\overline{H}}_e \cdot e^{-(1+j)\frac{y}{\delta}} \\ &= \overline{u}_z \cdot \frac{\sqrt{2}}{\delta} \cdot e^{j\frac{\pi}{4}} \cdot \rho \cdot H_{em} \cdot e^{j\omega t} \cdot e^{-\frac{y}{\delta}} \cdot e^{-j\frac{y}{\delta}} \\ &= \overline{u}_z \cdot \frac{\sqrt{2}}{\delta} \cdot \rho \cdot H_{em} \cdot e^{-\frac{y}{\delta}} \cdot e^{j\left(\omega t - \frac{y}{\delta} + \frac{\pi}{4}\right)}\end{aligned} \tag{2.11}$$

and

$$\dot{\overline{E}} = \overline{u}_z \cdot \frac{\sqrt{2}}{\delta} \cdot \rho \cdot H_{em} \cdot e^{-\frac{y}{\delta}} \cdot e^{j\left(\omega t - \frac{y}{\delta} + \frac{\pi}{4}\right)}. \tag{2.12}$$

In trigonometric form Eqs. (2.11) and (2.12) become:

$$\dot{\overline{H}} = \overline{u}_x \cdot H_{em} \cdot e^{-\frac{y}{\delta}} \cdot \sin\left(\omega t - \frac{y}{\delta}\right) \tag{2.13}$$

$$\dot{\overline{E}} = \overline{u}_z \cdot \frac{\sqrt{2}}{\delta} \cdot \rho \cdot H_{em} \cdot e^{-\frac{y}{\delta}} \cdot \sin\left(\omega t - \frac{y}{\delta} + \frac{\pi}{4}\right) \tag{2.14}$$

The amplitudes of the electric and magnetic field intensities are

$$\left.\begin{array}{l} H_m = H_{em} \cdot e^{-\frac{y}{\delta}} \\ E_m = H_{em} \cdot \frac{\sqrt{2}}{\delta} \cdot \rho \cdot e^{-\frac{y}{\delta}} \end{array}\right\}. \tag{2.15}$$

Figure 2.2 shows the attenuation of the amplitude of the modules of vectors $\dot{H}$ and $\dot{E}$ in a semi-infinite metal body (curves 1 and 2) and the graphs of variations in time of the modules of the vectors $\dot{H}$ and $\dot{E}$ (curves 3, 4)

The amplitudes of the magnetic and electric field intensities decay exponentially as the electromagnetic wave penetrates in the conducting body, with a decay rate proportional to the damping factor $k = (1 + j)/\delta$, which is an index of the phenomenon of skin effect in conductors.

In real cases, where the resistivity ρ is dependent on temperature and the relative permeability is a function of magnetic field intensity and temperature, the above dependencies are only approximate. The instantaneous values of the current density

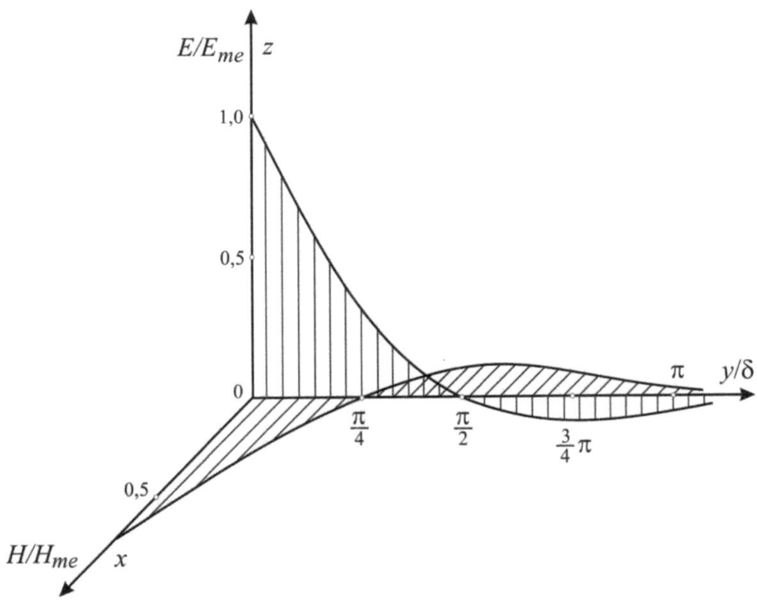

Fig. 2.2 Damping in a semi-infinite metal body of modules of H and E

vector $\dot{\bar{J}} = \dot{\bar{E}}\big/\rho$ in the thickness of the semi-infinite metal body, can be evaluated from Eq. (2.12); we can write:

$$\dot{\bar{J}} = \bar{u}_z \cdot \frac{\sqrt{2}}{\delta} \cdot H_{em} \cdot e^{-\frac{y}{\delta}} \cdot e^{j\left(\omega t - \frac{y}{\delta} + \frac{\pi}{4}\right)}. \tag{2.16a}$$

and the modulus of the instantaneous values of the current density can be written in exponential form as

$$\dot{J} = \frac{\sqrt{2}}{\delta} \cdot H_{em} \cdot e^{-\frac{y}{\delta}} \cdot e^{j\left(\omega t - \frac{y}{\delta} + \frac{\pi}{4}\right)}, \tag{2.16b}$$

or in trigonometric form as

$$\dot{J} = \frac{\sqrt{2}}{\delta} \cdot H_{em} \cdot e^{-\frac{y}{\delta}} \cdot \sin\left(\omega t - \frac{y}{\delta} + \frac{\pi}{4}\right). \tag{2.16c}$$

The amplitude J_m and the r.m.s. value J of the modulus of the current density vector inside the semi-infinite metallic body therefore are:

$$J_m = \frac{\sqrt{2}}{\delta} \cdot H_{em} \cdot e^{-\frac{y}{\delta}}, \tag{2.16d}$$

$$J_m = \frac{1}{\delta} \cdot H_{em} \cdot e^{-\frac{y}{\delta}}. \tag{2.16e}$$

By introducing the notation

$$J_{em} = \frac{\sqrt{2}}{\delta} \cdot H_{em}, \tag{2.16f}$$

for the amplitude of the current density at the surface of the body, Eq. (2.16d, e) can be written in a more convenient form

$$J_m = J_{em} \cdot e^{-\frac{y}{\delta}}, \tag{2.17a}$$

$$J_m = \frac{1}{\delta} \cdot H_{em} \cdot e^{-\frac{y}{\delta}}. \tag{2.17b}$$

$$J = \frac{J_{me}}{\sqrt{2}} \cdot e^{-\frac{y}{\delta}}. \tag{2.17c}$$

Figure 2.3 shows the distribution of instantaneous values of the modulus of the current density vector in the semi-infinite metal body (curve 1), while the dotted line (curve 2) shows the variation of its peak values.

The maximum amplitude of the modulus of the current density vector occurs at the surface (y = 0).

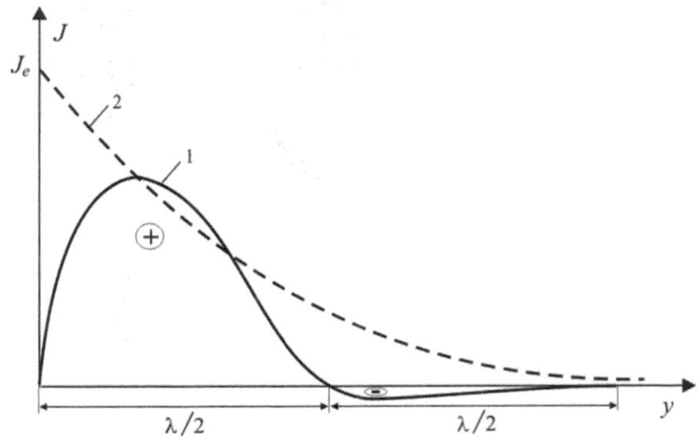

Fig. 2.3 Distribution of instantaneous values of modulus of vector of the current density (*curve 1*) and of its amplitude (*curve 2*)

As it penetrates into the body, the amplitude decreases exponentially. In a metallic medium the electromagnetic field decays almost completely within a single wavelength, which is linked to the penetration depth by $\lambda = 2\pi \cdot \delta$.

Moreover, since not only the amplitude, but also the phase of the current changes along the thickness from point to point, it may occur that in some regions of the body the current flows in forward direction, while in some other regions it flows even in the opposite direction.

The total electrical resistance of one square meter of the metallic semi-infinite body, or the so called *wave impedance*, is equal to

$$\dot{Z}_{e1} = R_{e1} + jX_{e1} = \frac{\dot{E}}{\dot{H}} = \frac{\dot{H}_e \cdot \rho \cdot \frac{1+j}{\delta} \cdot e^{-\frac{1+j}{\delta}y}}{\dot{H}_e \cdot e^{-\frac{1+j}{\delta}y}} = \frac{\rho}{\delta}(1+j)$$

and the corresponding resistance and reactance are therefore given by

$$R_{e1} = \frac{\rho}{\delta} \quad X_{e1} = \frac{\rho}{\delta}$$

Then the power factor of a semi-infinite metallic body is:

$$\cos\varphi = \frac{R_{e1}}{\sqrt{R_{e1}^2 + X_{e1}^2}} = \frac{1}{\sqrt{2}} = 0.707 \tag{2.18}$$

It is noteworthy that the power factor does not depend on the metal characteristics and frequency and is equal to 0.707.

2.1.1.1 Electromagnetic Energy in the Body

The electromagnetic wave carries out energy, defined by the vector of the energy flow—the Poynting vector

$$\overline{S} = \left[\overline{E} \times \overline{H}\right] \tag{2.19a}$$

If the modules of electric and magnetic field vectors vary in time sinusoidally, the average value of the energy flux vector in a period is

$$\dot{\overline{S}} = \frac{1}{2}\left[\dot{\overline{E}} \times \dot{\overline{H}}^*\right], \tag{2.19b}$$

with $\dot{\overline{H}}^*$—conjugate of the complex vector $\dot{\overline{H}}$.

Substituting in Eq. (2.19b) the expressions (2.11) and (2.12) and taking into account that the product of a complex variable by its complex conjugate is equal to the square of the modulus of the complex number, we obtain:

$$\begin{aligned}\dot{\overline{S}} &= \frac{1}{2}\left\{[\overline{u}_z \times \overline{u}_x] \cdot \dot{H}_e \cdot \frac{\sqrt{2} \cdot e^{j\frac{\pi}{4}}}{\delta} \cdot \rho \cdot e^{-\frac{1+j}{\delta}y} \cdot \dot{H}_e^* \cdot e^{-\frac{1-j}{\delta}y}\right\} \\ &= H_{em}^2 \cdot \frac{\sqrt{2}}{2\delta} \cdot \rho \cdot e^{-\frac{2y}{\delta}} \cdot e^{j\frac{\pi}{4}}[\overline{u}_z \times \overline{u}_x]\end{aligned}$$

with $\dot{H}_e = H_{em} \cdot e^{j\omega t}$, $\dot{H}_e^* = H_{em} \cdot e^{-j\omega t}$.

Since the vector product $[\overline{u}_z \times \overline{u}_x] = \overline{u}_y$, then

$$\dot{\overline{S}} = \overline{u}_y \cdot H_{em}^2 \cdot \frac{\sqrt{2}}{2\delta} \cdot \rho \cdot e^{j\frac{\pi}{4}} \cdot e^{-\frac{2y}{\delta}} \tag{2.20a}$$

or

$$\dot{\overline{S}} = \overline{u}_y \cdot H_{em}^2 \cdot \frac{1+j}{2\delta} \cdot \rho \cdot e^{-\frac{2y}{\delta}}. \tag{2.20b}$$

This vector gives the real and imaginary parts of the flow of energy of the alternating magnetic field into the volume per unit time. The real part of this vector gives the active power which is transformed into heat in the metal.

Thus, the active and reactive power per unit surface at a depth y is

$$p = q = H_{em}^2 \cdot \frac{\rho}{2\delta} \cdot e^{-\frac{2y}{\delta}}. \tag{2.21}$$

Accordingly, at the surface (y = 0), it is

$$p_e = \frac{\rho}{2\delta} \cdot H_{em}^2 \,\mathrm{W/m^2}.$$

The specific active power at a depth equal to the penetration depth δ, is

$$p_\delta = H_{em}^2 \cdot \frac{\rho}{2 \cdot \delta} \cdot e^{-2} = 0.135 \cdot \frac{\rho}{2 \cdot \delta} \cdot H_{em}^2$$

Hence, the heat generated in the external surface layer of the metal of thickness δ is:

$$p_{e\delta} = p_e - p_\delta = 0.865 \cdot p_e.$$

This result is identical to the one obtained in Sect. 1.4.3, for a cylindrical geometry.

Thus, in a conducting medium 86.5 % of the total active energy carried by the electromagnetic wave is transformed into heat inside a surface thickness equal to the penetration depth δ.

The derived expressions are very convenient in the engineering practice, because in many cases of induction heating, the thickness h of the heated workpiece is much larger of the penetration depth δ.

Taking into account that is $\mu = \mu_0 \cdot \mu_r$, $\mu_0 = 4\pi \times 10^{-7}$ and $\delta = 503 \cdot \sqrt{\rho/\mu_r \cdot f}$, when $\delta \gg h$, we can calculate the distribution of active power in the heated body, with the formula:

$$p = 10^{-3} \cdot H_{em}^2 \cdot e^{-\frac{2y}{\delta}} \cdot \sqrt{\rho \mu_r f} \tag{2.22}$$

We apply now Eq. (2.22) to the case of a planar inductor, located in front of the surface of a semi-infinite metallic body, by expressing the magnetic field intensity $\dot{H}_e$ at the surface of the body in terms of the current flowing in the inductor.

Figure 2.4 shows the relative positions of the inductor and the heated body.

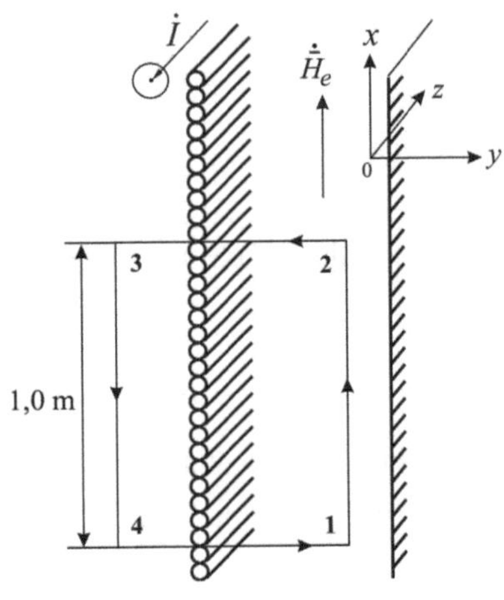

Fig. 2.4 Mutual arrangement of flat infinite inductor and semi-infinite conducting body

Along the surface of the body in the directions ox and oz the inductor has infinite dimensions. It has N_1 turns per unit length (along ox).

This ideal coil produces an alternating magnetic field, $\dot{\bar{H}}_e$ directed along the ox in the air-gap between the inductor and the body.

In this ideal geometry, the computation of the line integral along a closed path "1–2–3–4–1" is very easy because, due to the infinite length of the coil, the magnetic field along path "3–4" is zero. The only non-null term of the line integral is the one due to the field $\dot{\bar{H}}_e$, along the path "1–2" inside the air-gap.

It results:

$$\int_{\ell} \dot{\bar{H}} \cdot d\bar{\ell} = \dot{H}_{12} \cdot \ell_{12} + \dot{H}_{23} \cdot \ell_{23} + \dot{H}_{34} \cdot \ell_{34} + \dot{H}_{41} \cdot \ell_{41} = \dot{H}_e$$

with $\dot{H}_{12}$, $\dot{H}_{23}$, $\dot{H}_{34}$, $\dot{H}_{41}$—modules of the magnetic field vector on the paths 1–2, 2–3, 3–4, 4–1, and $\ell_{12}, \ell_{23}, \ell_{34}, \ell_{41}$—lengths of these paths, respectively.

By Ampere's law, this integral is equal to the current linked with the contour of integration, i.e.

$$\oint_{\ell} \dot{\bar{H}} \cdot d\bar{l} = \dot{I} \cdot N_1,$$

and finally

$$H_{em} = I_m \cdot N_1 = \sqrt{2} \cdot I \cdot N_1, \tag{2.23}$$

where I_m, I—amplitude and r.m.s value of the inductor current; N_1—number of turns per unit length of the inductor, H_{em}—maximum value of the magnetic field intensity at the surface of the body to be heated.

Substituting Eq. (2.23) into (2.20a) and (2.21) we obtain the formula for calculating the vector describing the flux of the electromagnetic energy in the semi-infinite metal body

$$\dot{\bar{S}} = 2 \cdot (N_1 \cdot I)^2 \cdot \frac{1+j}{2\delta} \cdot \rho \cdot e^{-\frac{2y}{\delta}} \tag{2.24}$$

The active and reactive components p and q of the complex vector describing vector the flux of electromagnetic energy therefore are:

$$p = q = 2 \cdot (N_1 \cdot I)^2 \cdot \frac{e^{-\frac{2y}{\delta}}}{2\delta} \cdot \rho. \tag{2.25}$$

Taking into account the expression of the penetration depth, we can rewrite Eqs. (2.24) and (2.25) as follows

$$\dot{\bar{S}} = 2 \times 10^{-3} \cdot (N_1 \cdot I)^2 \cdot \sqrt{\rho\mu_r f} \cdot (1 + j) \cdot e^{-\frac{2y}{\delta}} \tag{2.26}$$

$$p = 2 \times 10^{-3} \cdot (N_1 \cdot I)^2 \cdot \sqrt{\rho\mu_r f} \cdot e^{-\frac{2y}{\delta}}, \tag{2.27a}$$

$$q = 2 \times 10^{-3} \cdot (N_1 \cdot I)^2 \cdot \sqrt{\rho\mu_r f} \cdot e^{-\frac{2y}{\delta}}. \tag{2.27b}$$

At the surface of the semi-infinite metallic body (i.e. at y = 0) the active and reactive components of the flux of electromagnetic energy are

$$p_e = 2 \times 10^{-3} \cdot (N_1 \cdot I)^2 \cdot \sqrt{\rho\mu_r f}, \tag{2.28a}$$

$$q_e = 2 \times 10^{-3} \cdot (N_1 \cdot I)^2 \cdot \sqrt{\rho\mu_r f}. \tag{2.28b}$$

Since in the metallic semi-infinite body the electromagnetic wave decays completely, Eq. (2.28a, b) represent respectively the specific active power dissipated in the body, and the specific reactive power in the volume of the body per square meter of its surface, i.e.

$$P_a = \sqrt{p^2 + q^2} = \sqrt{2} \cdot p.$$

The power factor is $\cos\varphi = \frac{p}{P_a} = \frac{1}{\sqrt{2}} = 0.707$, i.e. the transfer of energy in a metallic semi-infinite body occurs with a power factor which does not depend on the characteristics of the metal and the frequency.

These results allow to compute the integral parameters of the equivalent series circuit of the inductor-load system. The description of methods of calculation of the equivalent circuit parameters is dealt with in Chap. 4.

The temperature field in the heated body depends on the distribution of internal power sources, w, given by Joule's law:

$$w = \rho \cdot J^2,$$

where

$$J = \frac{J_m}{\sqrt{2}} = \frac{\sqrt{2}}{\sqrt{2}} \cdot \frac{H_{em}}{\delta} \cdot e^{-\frac{y}{\delta}} = \sqrt{2} \cdot (N_1 \cdot I) \cdot \frac{e^{-\frac{y}{\delta}}}{\delta}.$$

Therefore

$$w = 7.9 \times 10^{-6} \cdot (N_1 \cdot I)^2 \cdot \mu_r \cdot f \cdot e^{-\frac{2y}{\delta}}. \tag{2.29}$$

2.1.2 Resistance Heating

In this paragraph we will consider the distribution of the electric and magnetic fields in a semi-infinite slab due to the flow of a sinusoidal current.

With reference to the schematic of Fig. 2.5, we make the following assumptions:

1. the heated workpiece has finite width w, finite length ℓ and infinite thickness. This assumption is valid in metal workpieces when the thickness of the workpiece is much greater than the penetration depth δ.
2. edge and end effects are neglected, i.e. it's assumed that current density and magnetic field intensity are evenly distributed across the width and the length of the workpiece.
3. a sinusoidal current flows in the workpiece.
4. electrical resistivity ρ and relative magnetic permeability μ_r of the body are constant.

We make reference to a Cartesian coordinate system such that the plane xoz is coincident with the surface of the body, and the axis oy is directed from the surface towards the inside.

At the surface of workpiece the complex vectors of the magnetic field intensity and the current density are $\dot{\bar{H}}_e$ and $\dot{\bar{J}}_e$.

We neglect the edge effects that may occur at the end surfaces of coordinates $z = 0$ and $z = l$ and at the side surfaces of coordinates $x = 0$ and $x = w$.

Then the variation of the magnetic field vector $\dot{\bar{H}}$and the current density vector $\dot{\bar{J}}$ will occur only along the coordinate y.

With these assumptions, the vector $\dot{\bar{H}}$ has only the component $\dot{H}_x$, and the vector $\dot{\bar{J}}$ only the component $\dot{J}_z$. On the surface of a thin layer of thickness dy at distance y from the plane xoz, the magnetic field intensity is $\dot{H}_x$ and the current density $\dot{J}_z$; at the surface of the layer of coordinate $y + dy$, the magnetic field intensity is $\dot{H}_x + d\dot{H}_x$ and the current density $\dot{J}_z + d\dot{J}_z$.

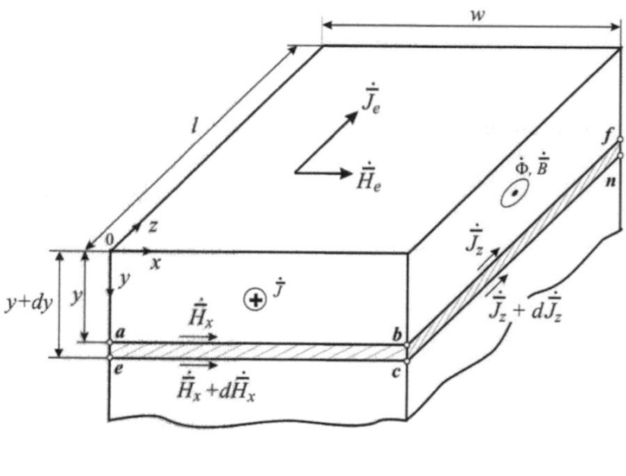

Fig. 2.5 Flow of alternating current in semi-infinite slab

Applying the law of the total current in integral form $\oint \dot{H} d\ell = \dot{I}$ along the path a–b–c–e we obtain:

$$\dot{H}_x w - \left(\dot{H}_x + d\dot{H}_x\right) w = \dot{J}_z(wdy), \tag{2.30}$$

with $\dot{H}_x$—magnitude of the component of the magnetic field intensity, directed along the x coordinate and variable along the coordinate y; $\dot{J}_z$—component of the current density directed along the coordinate z and variable along the coordinate y.

From Eq. (2.30) we obtain

$$\frac{d\dot{H}_x}{dy} = -\dot{J}_z. \tag{2.31}$$

The law of the electromagnetic induction (*Faraday's law*) can be written in integral form:

$$\oint \dot{E}\, dl = -d\dot{\Phi}/dt.$$

Electric and current density fields are related by the Ohm's law for specific quantities $\dot{E} = \rho \dot{J}$, and since the induction field has only the component along the coordinate x, the flux of magnetic induction can be calculated as $d\dot{\Phi} = B\ell dy$.

By substitution, we obtain:

$$\rho \dot{J}_z \ell - \rho\left(\dot{J}_z + d\dot{J}_z\right)\ell = -\frac{d\dot{B}_x}{dt} \ell dy.$$

Considering that $\dot{B}_x = \mu \dot{H}_x$ and $\frac{d\dot{H}_x}{dt} = j\omega \dot{H}_x$, it results

$$\rho \frac{d\dot{J}_z}{dy} = j\omega\mu \dot{H}_x. \tag{2.32}$$

Substituting $\dot{H}_x$ from Eq. (2.32) into (2.31), we have

$$\frac{d^2 \dot{J}_z}{dy^2} = j\frac{\omega\mu}{\rho} \dot{J}_z. \tag{2.33a}$$

With the notation $k^2 = j\omega\mu/\rho$, Eq. (2.33a) takes the form of a linear differential equation of second order

$$\frac{d^2 \dot{J}_z}{dy^2} = k^2 \dot{J}_z, \tag{2.33b}$$

which has the following general solution

$$\dot{J}_z = \dot{A}e^{-ky} + \dot{C}e^{ky}.$$

The integration constants can be determined from the boundary conditions:

(a) for y → ∞ the current density $\dot{J}_z$ cannot have an infinite value, so it must be $\dot{C} = 0$;
(b) for y = 0 the current density is $\dot{J}_z = \dot{J}_e$, therefore $\dot{A} = \dot{J}_e$.

The solution takes the form

$$\dot{J}_z = \dot{J}_e e^{-ky}. \tag{2.34}$$

with: $\dot{J}_e = J_{me}e^{j\omega t}$.

Since

$$k = \sqrt{j\frac{\omega\mu}{\rho}} = \sqrt{2j}\sqrt{\frac{\omega\mu}{2\rho}} = \frac{1+j}{\delta},$$

Equation (2.34) can be written as

$$\dot{J}_z = J_{me}e^{-\frac{y}{\delta}}e^{j\left(\omega t - \frac{y}{\delta}\right)}, \tag{2.35a}$$

or in trigonometric form

$$\dot{J}_z = J_{me}e^{-\frac{y}{\delta}}\sin\left(\omega t - \frac{y}{\delta}\right). \tag{2.35b}$$

In Eq. (2.35a, b) J_{me} is the amplitude of the current density at the surface of the body.

From Eq. (2.35b) we can draw the following conclusions:

- in the thickness of a semi-infinite slab, the current density decreases from the surface with exponential law;
- penetrating in the semi-infinite body, the phase angle of the current density starts to lag with respect to the phase of current density at the surface by an amount y/δ;
- the penetration depth δ is the distance from the surface of a semi-infinite body at which the current density amplitude is 2.718 times lower that the surface value.

The value of the total current flowing in the semi-infinite slab can be determined as follows:

$$\begin{aligned}\dot{I} &= \int_0^\infty \dot{J}_z \cdot w \cdot dy = \int_0^\infty J_{me} e^{j\omega t} e^{-\frac{y}{\delta}(1+j)} \cdot w \cdot dy \\ &= \dot{J}_e \frac{w\delta}{1+j}\left[e^{-\frac{y}{\delta}(1+j)}\right]_0^\infty = \dot{J}_e \frac{w\delta}{1+j} = \dot{J}_e w\delta \frac{1-j}{2} \\ &= \frac{\dot{J}_e}{\sqrt{2}} w\delta e^{-j\frac{\pi}{4}} = \frac{\dot{J}_e}{\sqrt{2}} w\delta e^{j\left(\omega t-\frac{\pi}{4}\right)} \\ &= \frac{\dot{J}_e}{\sqrt{2}} w\delta \sin\left(\omega t - \frac{\pi}{4}\right)\end{aligned} \tag{2.36}$$

Since $\dot{I} = I_m \sin(\omega t)$, then is $I_m = \frac{J_{me}}{\sqrt{2}} w\delta$.

Euler's identity, that links trigonometric and exponential functions has been applied in Eq. (2.36):

$$\frac{(1-j)}{\sqrt{2}} = \cos\frac{\pi}{4} - j\sin\frac{\pi}{4} = e^{-j\frac{\pi}{4}}.$$

Therefore, the total current lags the current density at the surface of an angle of 45°.

We can now evaluate the specific power delivered to the unit surface of the semi-infinite slab as follows:

$$\begin{aligned}p &= \int_0^\infty \rho\left|\dot{J}_z\right|^2 dy = \rho J_e^2 \int_0^\infty e^{-\frac{2y}{\delta}} dy = -\frac{\rho\delta}{2} J_e^2 \left(e^{-\frac{2y}{\delta}}\right)\Big|_0^\infty \\ &= \frac{\rho\delta}{2} J_e^2\end{aligned} \tag{2.37}$$

where:

$J_z = J_e e^{-\frac{y}{\delta}}$ module of the current density, dependent on the coordinate y and numerically equal to the r.m.s. value of the current density at the particular value of y;

J_e r.m.s. value of the current density at the surface of the body

- *Resistance R_e, internal reactance X_{ie} and impedance $\dot{Z}_e = R_e + jX_{ie}$ of the semi-infinite slab.*

The equivalent series impedance $\dot{Z}_e$ can be calculated from the voltage applied to the body $\dot{V} = \rho\dot{J}_e\ell$, and the total current $\dot{I} = \dot{J}_e w\delta\frac{1-j}{2}$ (Eq. 2.36):

$$\dot{Z}_e = \frac{\dot{V}}{\dot{I}} = \frac{2\rho\dot{J}_e\ell}{\dot{J}_e w\delta(1-j)} = \frac{\rho\ell}{w\delta}(1+j) = \frac{\rho\ell}{w\delta} + j\frac{\rho\ell}{w\delta}. \tag{2.38a}$$

The real part of the impedance is the equivalent series resistance, $R_e = X_{ie} = \frac{\rho\ell}{w\delta}$, and the module of the impedance is:

$$|\dot{Z}_e| = \sqrt{2}\frac{\rho\ell}{w\delta}. \tag{2.38b}$$

The power factor of the semi-infinite slab therefore then results

$$\cos\varphi = \frac{R_e}{\sqrt{R_e^2 + X_{ie}^2}} = \frac{1}{\sqrt{2}} = 0.707. \tag{2.39}$$

Finally, it should be noticed that the power factor of the semi-infinite slab has the same value of the power factor of the semi-infinite metallic body heated by induction (see Eq. 2.18).

2.2 Semi-infinite Body of Ferromagnetic Material

In the previous paragraph with reference to non-magnetic materials it has been shown that, in a semi-infinite metallic body (for induction heating) or a semi-infinite slab (for resistance-heating), the distributions of the electromagnetic field quantities (E, H, J) and the values of integral parameters (Z, X, R, cos φ) are the same.

As it will be seen in this paragraph, also in semi-infinite ferromagnetic bodies heated by induction or by direct resistance the relationships for the above quantities are the same. Therefore, we do not introduce any distinction between induction and resistance heating.

In the induction heating of magnetic steel workpieces from the ambient temperature to the final temperature of the process, in most cases in the range 700–1,250 °C, the electrical resistivity increases from 2 to 7 times, and the relative magnetic permeability μ_r decreases by 5–1,000 times. Moreover, μ_r is a function also of the magnetic field intensity H.

The current density distribution in the cross-section of a ferromagnetic workpiece and its resistance and internal reactance could be computed applying the formulas of Sect. 2.1 with $\rho = \text{const.}$ and $\mu_r = \text{const.}$, assuming for the resistivity its average value in the heating interval and for the permeability the value corresponding to the surface magnetic field intensity. But the results obtained in this way should be considered only a rough approximation of the reality.

A better representation of the phenomena occurring in the process of induction and resistance heating of ferromagnetic workpieces to a final temperature above the magnetic transition (the Curie point), can be obtained subdividing the process into three steps, corresponding to the so-called "*cold-*", "*intermediate-*" and "*hot-stage*". Each stage is characterized by a different distribution of the heat sources in the cross-section of the workpiece [7].

In the "*cold*" phase, the calculation can be made using the electromagnetic model of the semi-infinite conductor, with constant resistivity equal to its value at 650 °C. As regards the relative permeability, it depends by the local magnetic field intensity that, in turn, is a function of the coordinate y; i.e. is $\mu_r = f(H)$ and $H = g(y)$.

The "*intermediate*" phase starts when a surface layer of thickness h_C is above Curie temperature ($\vartheta_{curie} \approx 750\,°\mathrm{C}$), while the inner part of the cross-section is at a lower temperature, which decreases with the distance from the surface. As a consequence, ρ and μ_r vary from point to point and, in particular, μ_r changes nearly abruptly at the internal boundary of the surface layer.

The "intermediate" phase can be represented schematically with a two-layers electromagnetic model, in which the surface layer is characterized by $\rho = \text{const.} \approx 1.0 \times 10^{-6}$ Ωm and $\mu_r = 1$; in the second layer the material properties are a function of the coordinate y, i.e. of the distance from the surface of the body. It can be assumed $\rho = \text{const.} = 6.0–6.5 \times 10^{-7}$ Ωm, and $\mu_r = f(y)$.

In the "*hot*" phase the boundary at Curie temperature, separating the external non-magnetic layer and the internal magnetic core, is so far from the external surface that it can be supposed that the electromagnetic wave decays completely in the first layer.

With this hypothesis, in the hot stage, it is possible to consider the heated body as homogeneous, with material properties above the Curie point. In this stage it is applicable the computational model of the single-layer semi-infinite body with properties of the hot material ($\rho = \text{const.} \approx 1.0 \times 10^{-6}$ Ωm; $\mu_r = 1$).

It can be assumed that such stage is reached when the thickness h_C of the first layer satisfies the condition $h_C > 3\ \delta_H$, with δ_H penetration depth of the electromagnetic wave in the "hot" layer.

In many heat treatment processes of ferromagnetic workpieces (e.g. annealing after mechanical deformation, tempering after quenching, heating for hot working, etc.) it is often required to limit the temperature differential inside the heated body to values not exceeding 5 % of the final heating temperature. In these cases, the sequential solution of the electromagnetic and thermal problems with analytical methods may not allow to reach the required accuracy using the subdivision into the three stages previously described. Therefore, to improve accuracy it is necessary to choose lower temperature subdivision intervals, and consistently obtain a solution for each of them.

Nowadays modern numerical simulation tools allow to subdivide the thermal transient into very small time steps, minimizing the error due to the use of averaged values of the relevant material properties.

We consider now an analytical solution for the distributions of electric and magnetic fields and current density in the "cold" stage of heating a ferromagnetic semi-infinite body.

In this solution, the variation of the relative permeability μ_r with the local magnetic field intensity H is taken into account by means of an approximate representation of the real magnetization curve.

Such approximation is based on the following assumptions [5]:

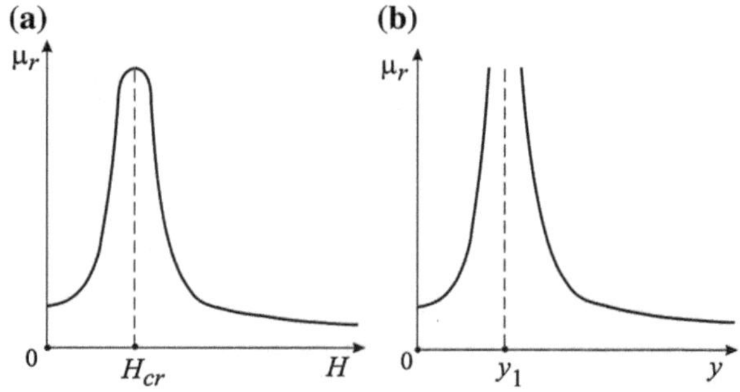

Fig. 2.6 Distribution of relative permeability: **a** real—as a function of magnetic field intensity; **b** model—as a function of the distance y from the surface

1. The real curve of first magnetization $B = f(H)$ (Fig. 2.6a) is approximated by a parabolic piecewise function (Fig. 2.6b)

$$B = K_1 \cdot H^{1/n},$$

where n and K_1 are chosen according to the steel grade and the magnetic field intensity saturation value (see Ref. [12]).

In particular, for $H_{me} > H_{cr}$ it is $4 < n < 20$; for $H_{me} < H_{cr}$, $n \leq 1$.

The condition $H_{me} > H_{cr}$ corresponds to strong magnetic fields, $H_{me} \leq H_{cr}$ to weak magnetic fields.

Due to the fact that in the practice of resistance and induction heating usually is $H_{me} >> H_{cr}$, in these applications the parabolic dependence with $4 < n < 20$ should be used.

2. The characteristics $B(t)$ and $H(t)$ as a function of time are replaced by their first harmonics B_{m1} and H_{m1}, and the permeability is calculated from their amplitudes

$$\mu_r = \frac{B_{m1}}{\mu_o H_{m1}}$$

In this way the dependence of permeability on time is eliminated, and the permeability becomes a function only of the amplitude of the first harmonic of the magnetic field intensity.

3. The introduction of the function $\mu_r = f(H_{m1})$ in the system of Maxwell's equations does not allow to obtain a solution in general form. Therefore, instead of this function, it is introduced the function $\mu_r(y)$ which satisfies the following conditions:

- for $H_{me} > H_{cr}$ the function $\mu_r = f(y)$ first increases, reaches a maximum and then decreases;
- for $H_{me} < H_{cr}$, $\mu_r = f(y)$ decreases monotonically.

The required behavior is fulfilled by the function

$$\mu_r = f(y) = \frac{\mu_e}{\left(1 - \frac{y}{y_1}\right)^2}, \tag{2.40}$$

with μ_e—relative permeability at the body surface, calculated with the first harmonics of H_{me} and B_{me}; y—distance from the surface; y_1—distance from the surface to the point corresponding to the maximum of the relative permeability.

With the above assumptions, the system of Maxwell's equations for the semi-infinite body is transformed into a second order differential equation of the form:

$$\frac{d^2\dot{H}}{dy^2} = j\frac{\omega\mu_0\mu_e}{\rho}\frac{\dot{H}}{(1 - y/y_1)^2}.$$

Note in this formula and the following ones only the first harmonic describing the magnetic field intensity is considered.

The solution for the vector $\dot{H}$ can be written as:

$$\dot{H} = \dot{H}_e\left(1 - \frac{y}{y_1}\right)^{\dot{\alpha}_1}, \tag{2.41}$$

with:

$$\dot{\alpha}_1 = \alpha_1' + j\alpha_1'' = \frac{1}{2}\left(1 + \sqrt{\frac{\sqrt{1 + 64k_e^4 y_1^4} + 1}{2}}\right) + j\frac{1}{2}\sqrt{\frac{\sqrt{1 + 64k_e^4 y_1^4} - 1}{2}}$$

The module of the complex quantity $\dot{\alpha}_1$ is

$$|\dot{\alpha}_1| = \sqrt{(\alpha_1')^2 + (\alpha_1'')^2},$$

where:

$\alpha_1' = \frac{2n}{n-1}$ coefficient of the real component of the magnetic field intensity;

$\alpha_1'' = \alpha_1'\sqrt{\frac{n+1}{2n}}$ coefficient of the imaginary component of the magnetic field intensity; $k_e y_1 = \frac{\sqrt[4]{8n(3n+1)^2(n+1)}}{2(n-1)}$;

$k_e = \frac{1}{\delta_e}$ attenuation constant of the electromagnetic field, evaluated with constant magnetic permeability $\mu_r = \mu_e$, not dependent on y.

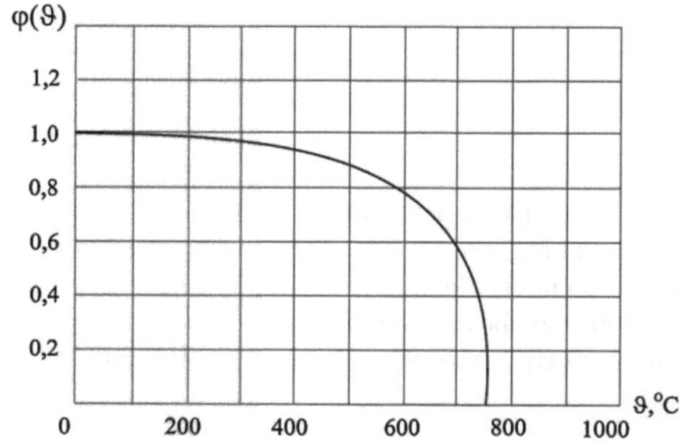

Fig. 2.7 Diagram of φ(t) as a function of temperature

For the calculation of the penetration depth in the ferromagnetic body, we will consider the variation of permeability with temperature ϑ and magnetic field H used in [3].

The penetration depth is calculated with the permeability μ_e at the surface of the workpiece, i.e.:

$$\delta_e = 503\sqrt{\frac{\rho(\vartheta)}{\mu_e f}},$$

where μ_e is given by the relationship

$$\mu_e = 1 + (\mu_{e20} - 1)\varphi(\vartheta),$$

with

μ_{e20} value of relative permeability at the surface, at 20 °C;
$\varphi(t)$ function given in Fig. 2.7, which takes into account the transition of the material from ferromagnetic to non-magnetic state at Curie point.

From Eq. (2.41) the electric field intensity can be determined with the relationship:

$$\dot{E} = -\rho\frac{d\dot{H}}{dy} = \rho\frac{\dot{\alpha}_1}{y_1}\dot{H}_e\left(1 - \frac{y}{y_1}\right)^{\alpha_1 - 1}$$
$$= \rho k_e \dot{H}_e \frac{\dot{\alpha}_1}{k_e y_1}\left(1 - \frac{k_e y}{k_e y_1}\right)^{\alpha_1 - 1}$$

Since for $y = 0$ it is

$$\dot{E} = \dot{E}_e = \rho k_e \dot{H}_e \frac{\dot{\alpha}_1}{k_e y_1}, \tag{2.42}$$

the electric field vector can be expressed as:

$$\dot{E} = \dot{E}_e \left(1 - \frac{k_e y}{k_e y_1}\right)^{\alpha_1 - 1},$$

and the current density can be determined by the equation

$$\dot{J} = \frac{\dot{E}}{\rho} = \dot{J}_{e(\mu\mathrm{var})} \left(1 - \frac{k_e y}{k_e y_1}\right)^{\alpha_1 - 1}, \tag{2.43}$$

where:

$\dot{J}_{e(\mu\mathrm{var})} = \frac{\dot{E}_e}{\rho} = \dot{H}_e \frac{k_e \dot{\alpha}_1}{k_e y_1}$ complex value of the current density at the surface of the workpiece, taking into account the dependence $\mu_r = f(y)$.

Then the module of the complex value $\dot{J}_{e(\mu\mathrm{var})}$ becomes

$$\left|\dot{J}_{e(\mu\mathrm{var})}\right| = \left|\dot{H}_e\right| k_e \frac{|\dot{\alpha}_1|}{k_e y_1}.$$

Finally, with $\mu_r = \mathrm{var} = \mu_e (1 - y/y_1)^2$, the module of the current density at the surface is

$$\left|\dot{J}_{e(\mu\mathrm{var})}\right| = \left|\dot{H}_e\right| \frac{1}{\delta_e} \frac{|\dot{\alpha}_1|}{k_e y_1}. \tag{2.44a}$$

In case of resistance heating, taking into account that for a semi-infinite slab of width w, according to the Ampere's law, the magnetic field intensity at the surface of the workpiece is

$$\dot{H}_e = \dot{I}/w,$$

the module of the current density at the surface can be expressed as:

$$\left|\dot{J}_{e(\mu\mathrm{var})}\right| = \frac{|\dot{I}|}{w\delta_e} \frac{|\dot{\alpha}_1|}{k_e y_1}, \tag{2.44b}$$

with: $\dot{I}$—value of the current flowing in the workpiece.

In the induction heating of a semi-infinite workpiece, at the surface of which is placed a flat inductor with N_1 turns per unit length, the magnetic field intensity at the workpiece surface is $\dot{H}_e = N_1 . \dot{I}_1$, with $\dot{I}_1$—current flowing in the inductor.

The module of the current density at the surface of the workpiece therefore is:

$$\left|\dot{J}_{e(\mu\text{var})}\right| = \frac{|\dot{I}|N_1}{\delta_e} \cdot \frac{|\dot{\alpha}_1|}{k_e y_1}. \tag{2.44c}$$

The comparison of the values of current density at the surface of the workpiece, evaluated with magnetic permeability constant or variable in the cross-section, shows the following.

In the case of $\mu_r = \mu_e =$ const. (Eq. (2.36) for resistance heating and (2.23) for induction heating) the modules of the current density at the surface are:

- for resistance heating

$$\left|\dot{J}_{e(\mu\cos t)}\right| = \sqrt{2}\frac{|\dot{I}|}{w\delta_e}, \tag{2.45a}$$

- for induction heating

$$\left|\dot{J}_{e(\mu\cos t)}\right| = \sqrt{2}\frac{|\dot{I}_1|N_1}{\delta_e}. \tag{2.45b}$$

For the same magnetic field intensity at the surface of the workpiece, the ratio of the modules of the current densities is the same for resistance and induction heating, for ferromagnetic and non-magnetic materials, and is equal to

$$\frac{\left|\dot{J}_{e(\mu\text{var})}\right|}{\left|\dot{J}_{e(\mu\cos t)}\right|} = \frac{|\dot{\alpha}_1|}{\sqrt{2}k_e y_1}. \tag{2.46}$$

with:

$$|\dot{\alpha}_1| = \sqrt{(\alpha_1')^2 + (\alpha_1'')^2};\quad \alpha_1' = \frac{2n}{n-1};\quad \alpha_1'' = \alpha_1'\sqrt{\frac{n+1}{2n}}$$

$$k_e y_1 = \frac{\sqrt[4]{8n(3n+1)^2(n+1)}}{2(n-1)};\quad k_e = \frac{1}{\delta_e}.$$

Table 2.1 gives relevant parameters and the ratio $\left|\dot{J}_{e(\mu\text{var})}\right| / \left|\dot{J}_{e(\mu\cos t)}\right|$ for different values of n.

Table 2.1 Influence of parameter "n" on characteristics of electromagnetic field in a ferromagnetic workpiece

n	$\frac{n-1}{n}$	α_1'	α_1''	$\lvert\dot{\alpha}_1\rvert$	$k_e y_1$	$\frac{\lvert\dot{\alpha}_1\rvert}{k_e y_1}$	$\frac{\lvert\dot{J}_{e(\mu\text{var})}\rvert}{\lvert\dot{J}_{e(\mu\cos t)}\rvert}$
∞	−1	2	1.414	2.449	1.457	1.681	1.189
9.40	−0.894	2.238	1.665	2.789	1.701	1.639	1.159
9.26	−0.892	2.242	1.669	2.791	1.705	1.639	1.159

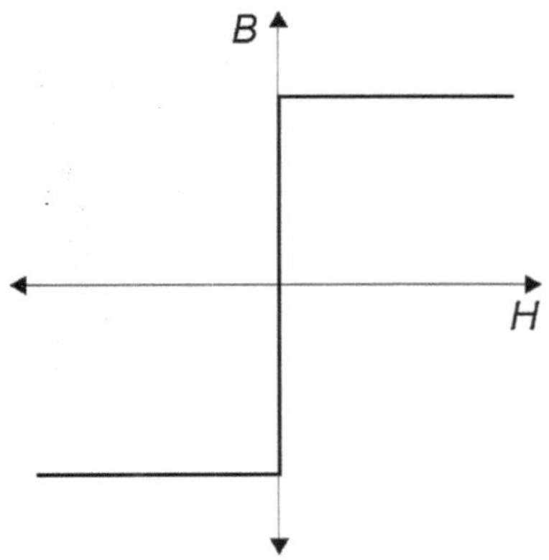

Fig. 2.8 Magnetization curve of ideal ferromagnetic material

In particular, the values of n are:
$n = 9.26$ for Cr–Mo steels [12],
$n = 9.4$ for construction steels [6],
$n = \infty$ for ferromagnetic material, with ideal magnetisation curve of Fig. 2.8.

The values given in Table 2.1 for the ratio

$$\left|\dot{J}_{e(\mu\text{var})}\right| / \left|\dot{J}_{e(\mu\cos t)}\right|$$

indicate that, for a broad class of structural and Cr-Mo steels, the use of the calculation formulas with permeability approximated by a parabolic function of the coordinate y, gives values of the current density at the surface of the workpiece higher of about 16–19 % in comparison with those calculated with $\mu_r = const.$

The comparison of the ratios $\left|\dot{J}_{e(\mu\text{var})}\right| / \left|\dot{J}_{e(\mu\cos t)}\right|$ (given in Table 2.1) calculated for real steels and material with idealised curve of magnetisation (with $n = \infty$), shows that they differ by no more than 2.5–3 %. Therefore, with sufficient accuracy for engineering calculations, it's possible to calculate the current density distribution in a semi-infinite ferromagnetic workpiece, using the following parameters: $n = \infty$; $k_e y_1 = 1.457$; $\left|\alpha_1'\right| = 2$.

These results have been confirmed by numerical calculations [8].

With this assumption, Eq. (2.43) can be re-written as follows:

- for direct resistance heating

$$\dot{J} = 1.68 \frac{I}{w\delta_e}\left(1 - \frac{y}{1.457\delta_e}\right) e^{j\omega t} \tag{2.48}$$

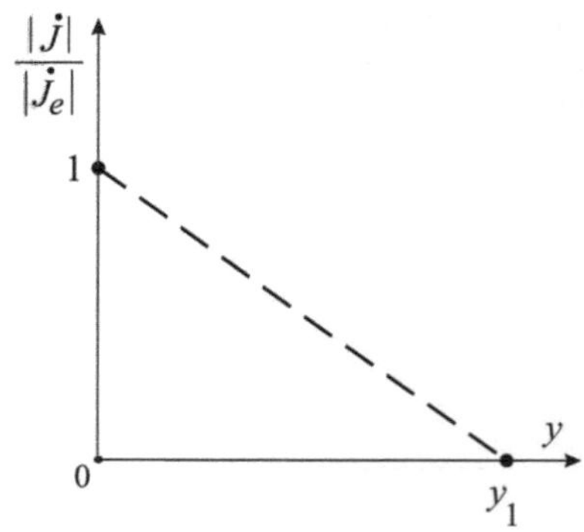

Fig. 2.9 Module of the current density as a function of the coordinate y in a semi-infinite ferromagnetic workpiece with $\mu_r = f(y)$

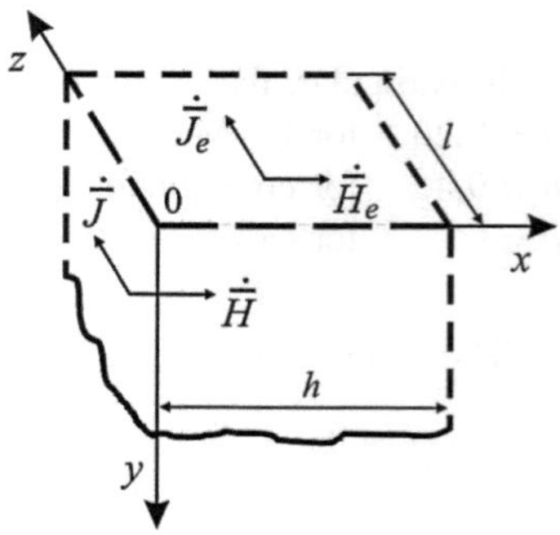

Fig. 2.10 Schematic of the semi-infinite body

- for induction heating:

$$\dot{J} = 1.68\frac{I_1 N_1}{\delta_e}\left(1 - \frac{y}{1.457\delta_e}\right)e^{j\omega t}. \tag{2.49}$$

Figure 2.9 shows the variation of the module of the current density $|\dot{J}| = f(y)$, taking into account the parabolic approximation $\mu_r = f(y)$.

We determine now the impedance $\dot{Z}_{ef} = R_{ef} + jX_{if}$ of the semi-infinite ferromagnetic workpiece in the case of induction and resistance heating respectively.

For solving this problem we consider a semi-infinite ferromagnetic workpiece with length l and width w (as in Fig. 2.10). The relative permeability varies in the thickness by the parabolic law (2.40). Edge effects on the faces of the slab at $x = 0$ and $x = w$ are neglected.

At the surface of the body the magnetic field intensity is $\dot{H}_e = H_{me}e^{j\omega t}$.

Then the voltage evaluated along a line of current of length l, lying on the surface of the body, is:

$$\dot{U}_e = \ell \dot{E}_e.$$

The value of the current, flowing through an element of width *w*, is equal to

$$\dot{I} = w\dot{H}_e.$$

The impedance of a ferromagnetic slab of length *l* and width *w* (in case of resistance-heating), and of a ferromagnetic semi-infinite workpiece of length *l* and width *w* (in the case of induction heating) will be equal to

$$\dot{Z}_{ef} = R_{ef} + jX_{if} = \dot{U}_e/\dot{I}.$$

Therefore, using previous equations and taking into account Eq. (2.42), we obtain

$$\dot{Z}_{ef} = \frac{\ell \dot{H}_e}{w\dot{H}_e}\rho k_e \frac{\dot{\alpha}_1}{k_e Y_1} = \rho \frac{\ell}{w\delta_e} \frac{2(n-1)}{\sqrt[4]{8n(3n+1)^2(n+1)}} \left(\alpha_1' + j\alpha_1''\right),$$

from which results

$$R_{ef} = \rho \frac{\ell}{w\delta_e} \frac{4n}{\sqrt[4]{8n(3n+1)^2(n+1)}}; \tag{2.50a}$$

$$X_{if} = \rho \frac{\ell}{w\delta_e} \frac{4n\sqrt{\frac{n+1}{2n}}}{\sqrt[4]{8n(3n+1)^2(n+1)}} = \rho \frac{\ell}{w\delta_e} \sqrt[4]{\frac{8n(n+1)}{(3n+1)^2}}; \tag{2.50b}$$

$$\begin{aligned} \dot{Z}_{ef} &= \sqrt{R_{af}^2 + X_{if}^2} e^{j\varphi} \\ &= \rho \frac{\ell}{w\delta_e} \frac{4n\sqrt{1+\frac{n+1}{2n}}}{\sqrt[4]{8n(3n+1)^2(n+1)}} e^{j\varphi} \\ &= \rho \frac{\ell}{w\delta_e} \sqrt{2}\sqrt[4]{\frac{2n}{(n+1)}} e^{j\varphi}; \end{aligned} \tag{2.50c}$$

$$\left|\dot{Z}_{ef}\right| = \sqrt{R_{af}^2 + X_{if}^2} = \rho \cdot \frac{\ell}{w\delta_e} \cdot \sqrt{2} \cdot \sqrt[4]{\frac{2n}{n+1}}, \tag{2.50d}$$

with $\varphi = arctg\frac{X_i}{R_a}$.

According to Eq. (2.38a), is

$R_e = X_{ie} = \frac{\rho\ell}{w\delta_e}$ internal inductance and resistance, calculated by the model of a semi-infinite body with constant relative permeability $\mu_r = \text{const.}$,

$\left|\dot{Z}_e\right| = \frac{\rho\ell}{w\delta_e}\sqrt{2}$ modulus of the impedance of an element of a semi-infinite body of length l and width w.

Then the relative values of the impedances are:

$$\frac{\left|\dot{Z}_{ef}\right|}{\left|\dot{Z}_e\right|} = \sqrt[4]{\frac{2n}{n+1}};$$
$$\frac{R_{ef}}{R_e} = \frac{4n}{\sqrt[4]{8n(3n+1)^2(n+1)}}; \quad \frac{X_{if}}{X_{ie}} = \sqrt[4]{\frac{8n(n+1)}{(3n+1)^2}} \tag{2.51}$$

The power factor of the semi-infinite ferromagnetic slab then results:

$$\cos\varphi = \frac{R_{ef}}{\left|\dot{Z}_{ef}\right|} = \frac{2n}{\sqrt[4]{4n^2(3n+1)^2}}. \tag{2.52}$$

Table 2.2 gives the ratios $\left|\dot{Z}_{ef}\right|/\left|\dot{Z}_e\right|$, R_{ef}/R_e and X_{if}/X_{ie} and the power factor $\cos\varphi$ for the values of n specified in Table 2.1, for ferromagnetic workpieces with different magnetic properties.

The previous results can be summarized as follows:

1. The model of semi-infinite ferromagnetic body for calculation of J, E, H, R, X, Z in solid workpieces gives identical values for induction and resistance-heating.
2. The use of the parabolic approximation

$$\mu_r = \frac{\mu_{re}}{\left(1 - \frac{k_e y}{k_e y_1}\right)^2}$$

allows to obtain an accurate distribution of the current density $J = f(y)$, and to evaluate its value J_{me} at the surface of the workpiece with discrepancies of about 5–10 % from experimental values.
3. With this approximation of μ_r the current density value at the surface is 1.19 times higher than that calculated with $\mu_r = \text{const}$.
4. The current density $J = f(y)$ obtained with the parabolic approximation decreases from the surface to the core with a linear law. This distribution $J = f(y)$ is more realistic and quite different from the exponential decay corresponding to $\mu_r = \text{const}$.

Table 2.2 Parameter 'n', and dimensionless resistance, reactance, impedence and power factor for different ferromagnetic materials

Material	n	$\frac{R_{ef}}{R_e}$	$\frac{X_{if}}{X_{ie}}$	$\frac{\left\|\dot{Z}_{ef}\right\|}{\left\|\dot{Z}_e\right\|}$	$\cos\varphi$
Ferromagnetic with ideal characteristic $B = f(H)$	∞	1.37	0.971	1.19	0.816
Construction steel	9.4	1.315	0.979	1.159	0.803
Cr–Mo steel	9.26	1.315	0.979	1.159	0.802

5. For high magnetic field intensities, when $H_{me} > H_{cr}$, the value of ratios $|\dot{Z}_{ef}|/|\dot{Z}_e|$, R_{ef}/R_e and X_{if}/X_{ie} are nearly the same for various ferromagnetic materials. Moreover, the results obtained with $n = \infty$ can be used for all ferromagnetic materials with an accuracy within 3–4 % on the values of integral parameters.
6. The power factor in semi-infinite bodies of real ferromagnetic materials differs by no more than 2 % from that in a workpiece with ideal magnetic properties.

2.3 Infinite Metal Slab of Homogeneous Material

2.3.1 Induction Heating by a Flat Inductor on One Side of the Workpiece

Figure 2.11 shows the layout of such system with the inductor in front of one surface of the workpiece. The inductor has infinite dimensions along the axes *ox* and *oz*, and N_1 turns per unit length along *ox*. The inductor generates a magnetic alternating field, which is characterized in the gap between inductor and workpiece by the magnetic field intensity $\dot{\bar{H}}_e$.

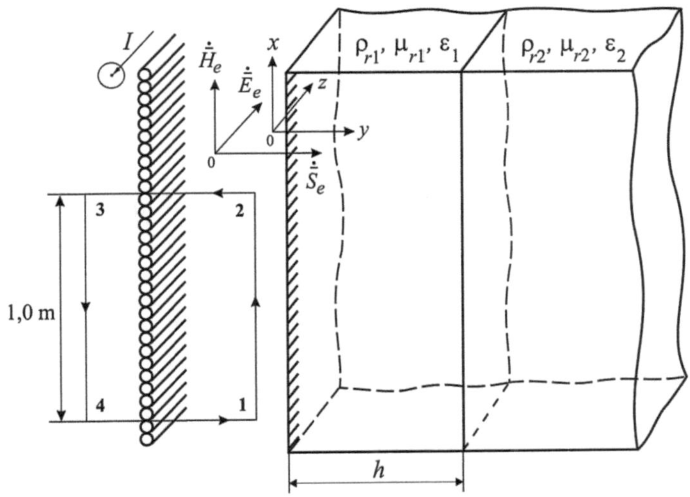

Fig. 2.11 Relative positions of flat inductor and metallic slab

In this configuration a plane electromagnetic wave is incident on one surface of the slab, while there is a semi-infinite dielectric medium on the other side. The dielectric medium can be air or a liquid non-conducting medium (e.g. in chemical reactors).

When the thickness h of the workpiece has a dimension comparable with the penetration depth δ_1 in the workpiece material, the electromagnetic field exists also for $y > h$ and it is necessary to develop a mathematical model of its distribution in a two-layer medium.

We indicate with the indexes *1*—the quantities in the first layer (metal workpiece), and *2*—the quantities of the second layer (the dielectric medium).

The distribution of the electromagnetic field in the system is described by the system of Maxwell's equations (2.1a) under the following conditions:

1. in the first layer $\dot{\bar{J}} \gg \frac{\partial \dot{\bar{D}}}{\partial t}$, in the second one $\dot{\bar{J}} \ll \frac{\partial \dot{\bar{D}}}{\partial t}$.
2. both layers are homogeneous, i.e. with constant physical properties: $\rho_1 = \text{const.}$, $\mu_{r1} = \text{const.}$, $\rho_2 = \infty$, $\varepsilon_{r2} = \text{const.}$, $\mu_{r2} = \text{const.}$
3. the inductor current is sinusoidal, and all parameters of the electromagnetic field are also sinusoidal

$$\overline{H} = \overline{H}_m \sin \omega t; \quad \overline{B} = \overline{B}_m \sin \omega t; \quad \overline{E} = \overline{E}_m \sin(\omega t + \varphi).$$

The Maxwell's equations (2.1a) for a two-layer system with sinusoidal time-varying electromagnetic field, can be written as follows in complex form:

- in the first layer

$$rot\, \dot{\overline{H}}_1 = \dot{\overline{J}}_1; \quad rot\, \dot{\overline{E}}_1 = -\mu_{r1}\mu_0 \frac{\partial \dot{\overline{H}}_1}{\partial t} \tag{2.53a}$$

- in the second layer

$$rot\, \dot{\overline{H}}_2 = \varepsilon_2 \frac{\partial \dot{\overline{E}}_2}{\partial t}; \quad rot\, \dot{\overline{E}}_2 = -\mu_{r2}\mu_0 \frac{\partial \dot{\overline{H}}_2}{\partial t} \tag{2.53b}$$

with: $\varepsilon_2 = \varepsilon_{r2}\varepsilon_0$—absolute dielectric constant of the second layer; ε_{r2}—relative dielectric constant of the second layer; $\varepsilon_0 = 8.842 \times 10^{-12}$ F/m; μ_{r1}, μ_{r2}—relative magnetic permeability of the first and second layer; $\mu_0 = 4\pi \cdot 10^{-7}$ H/m; $\rho_2 = \infty$—resistivity of the second layer, $\Omega \cdot m$.

The solution of system of Eq. (2.53a, b) is found with the following boundary conditions [10]:

$$\left.\begin{array}{ll} y = 0, & \dot{\overline{H}}_1 = \dot{\overline{H}}_e \\ y = h, & \dot{\overline{H}}_{1h} = \dot{\overline{H}}_{2h}, \dot{\overline{E}}_{1h} = \dot{\overline{E}}_{2h} \\ y = \infty, & \dot{\overline{H}}_2 = 0 \end{array}\right\}. \tag{2.54}$$

The solution for the first conducting layer can be written as follows:

$$\left.\begin{aligned}\dot{\bar{H}}_1 &= \dot{\bar{C}}_1 \cdot e^{-k_1 y} + \dot{\bar{C}}_2 \cdot e^{k_1 y}\\ \dot{\bar{E}}_1 &= 1_x \cdot k_1 \cdot \rho_1 \cdot \left(\dot{\bar{C}}_1 \cdot e^{-k_1 y} - \dot{\bar{C}}_2 \cdot e^{k_1 y}\right)\end{aligned}\right\} \tag{2.55}$$

with $k_1 = j\frac{\omega\mu_{r1}\mu_0}{\rho_1}$—damping coefficient of the electromagnetic field in the first layer;

For the second layer

$$\left.\begin{aligned}\dot{\bar{H}}_2 &= \dot{\bar{C}}_3 \cdot e^{-k_2 y} + \dot{\bar{C}}_4 \cdot e^{k_2 y}\\ \dot{\bar{E}}_2 &= 1_x \cdot \frac{k_2}{j\omega\varepsilon_{r2}\varepsilon_0} \cdot \left(\dot{\bar{C}}_3 \cdot e^{-k_2 y} - \dot{\bar{C}}_4 \cdot e^{k_2 y}\right)\end{aligned}\right\} \tag{2.56}$$

with $k_2 = j\omega\sqrt{\varepsilon_{r2}\varepsilon_0\mu_{r2}\mu_0}$—damping coefficient of the electromagnetic field in the dielectric medium.

The transformation of Eqs. (2.55) and (2.56) with the boundary conditions (2.54) gives the following expressions for the moduli of the magnetic and electric field vectors in the metallic layer:

$$\left.\begin{aligned}\dot{H}_1 &= \dot{H}_e \cdot \frac{sh\left[\frac{1+j}{\delta_1}(h-y)\right] + Q_\varepsilon \cdot ch\left[\frac{1+j}{\delta_1}(h-y)\right]}{sh\left[\frac{1+j}{\delta_1}h\right] + Q_\varepsilon \cdot ch\left[\frac{1+j}{\delta_1}h\right]}\\ \dot{E}_1 &= \dot{H}_e \cdot \frac{1+j}{\delta_1} \cdot \rho_1 \cdot \frac{ch\left[\frac{1+j}{\delta_1}(h-y)\right] + Q_\varepsilon sh\left[\frac{1+j}{\delta_1}(h-y)\right]}{sh\left[\frac{1+j}{\delta_1}h\right] + Q_\varepsilon \cdot ch\left[\frac{1+j}{\delta_1}h\right]}\end{aligned}\right\} \tag{2.57}$$

with: $Q_\varepsilon = \frac{(1+j)\rho_1}{\delta_1}\sqrt{\frac{\varepsilon_2}{\mu_2}}$.

In induction heating applications Q_ε is usually very small.

For example, if the first layer is constituted by a metal and the second one by air,

$$\sqrt{\frac{\varepsilon_2}{\mu_2}} = \sqrt{\frac{\varepsilon_0}{\mu_0}} = 2.65 \times 10^{-3}, 1/\text{Ohm}$$
$$Q_\varepsilon = (1+j) \cdot 5.27 \times 10^{-7} \cdot \sqrt{\rho_1\mu_{r1}f}.$$

and, for cold magnetic steel, with $\rho_1 = 1.6 \times 10^{-7}$ Ωm, $\mu_{r1} = 100$, and $f = 10^7$ Hz, the value of Q_ε is

$$Q_\varepsilon = (1+j) \cdot 5.27 \times 10^{-7}\sqrt{1.6 \times 10^{-7} \times 100 \times 10^7} = (1+j) \times 6.67 \times 10^{-6}.$$

In this case, in Eq. (2.57) the terms which include Q_ε, are negligible. We can assume $Q_\varepsilon = 0$.

Then, for induction heating we have:

$$\left.\begin{aligned} \dot{H}_1 &= \dot{H}_e \cdot \frac{sh\left[\frac{1+j}{\delta_1}(h-y)\right]}{sh\left[\frac{1+j}{\delta_1}h\right]} \\ \dot{E}_1 &= \dot{H}_e \cdot \frac{1+j}{\delta_1} \cdot \rho_1 \cdot \frac{ch\left[\frac{1+j}{\delta_1}(h-y)\right]}{sh\left[\frac{1+j}{\delta_1}h\right]} \end{aligned}\right\}. \tag{2.58}$$

Figure 2.12 gives the curves of the ratio $\left|\dot{H}_1/\dot{H}_e\right|$ of the magnetic field intensity inside the workpiece and at the surface facing the inductor, as a function of the relative thickness h/δ_1 and the relative coordinate y/h.

For values of $(h/\delta_1) \leq 1$ the diagram has a shape near a straight line. For $(h/\delta_1) \geq 3.0$, the curves practically are coincident with an exponential; this result is the same already obtained for the semi-infinite body (shown in the diagram by dotted lines).

Figure 2.13 gives the diagrams of the ratio $\left|\dot{E}_1/\dot{E}_e\right|$ of the electric field intensities and $\left|\dot{J}_1/\dot{J}_e\right|$ of the current densities inside the workpiece and at the surface, as a function of the relative thickness $h/\delta_1$and the relative coordinate y/h.

For $(h/\delta_1) \leq 0.5$ the distributions of $\dot{E}$ and $\dot{J}$ in the thickness of the workpiece are approximately uniform; for $(h/\delta_1) \geq 3.0$, $\left|\dot{E}_1/\dot{E}_e\right|$ and $\left|\dot{J}_1/\dot{J}_e\right|$ tend to the curves valid for a semi-infinite body (shown by dotted lines).

The electrical impedance per unit surface of the workpiece, called the *wave impedance*, is

$$\dot{z}_{e1} = R_{e1} + jX_{e1} = \frac{\dot{E}_1}{\dot{H}_1} = \frac{1+j}{\delta_1}\rho_1 \frac{ch\left[\frac{1+j}{\delta_1}(h-y)\right] \cdot sh\left[\frac{1+j}{\delta_1}h\right]}{sh\left[\frac{1+j}{\delta_1}h\right] \cdot sh\left[\frac{1+j}{\delta_1}(h-y)\right]} \tag{2.59}$$

The corresponding resistance and reactance are:

$$\left.\begin{aligned} R_{e1} &= \frac{\rho_1}{\delta_1} \cdot \frac{sh\frac{2h}{\delta_1} + \sin\frac{2h}{\delta_1}}{ch\frac{2h}{\delta_1} - \cos\frac{2h}{\delta_1}} = \frac{\rho_1}{\delta_1} \cdot F_{slab} \\ X_{e1} &= \frac{\rho_1}{\delta_1} \cdot \frac{sh\frac{2h}{\delta_1} - \sin\frac{2h}{\delta_1}}{ch\frac{2h}{\delta_1} - \cos\frac{2h}{\delta_1}} = \frac{\rho_1}{\delta_1} \cdot G_{slab} \end{aligned}\right\} \tag{2.60}$$

2.3.1.1 Electromagnetic Energy in the Workpiece

The electromagnetic wave carries energy, which is defined by the Poynting vector. In accordance with Eq. (2.19b), for sinusoidal quantities, the average value in a period of the energy flux vector has the form

Fig. 2.12 Relative distribution of magnetic field intensity in the thickness of a metallic sheet work-piece (*dashed line* curves for the semi-infinite body)

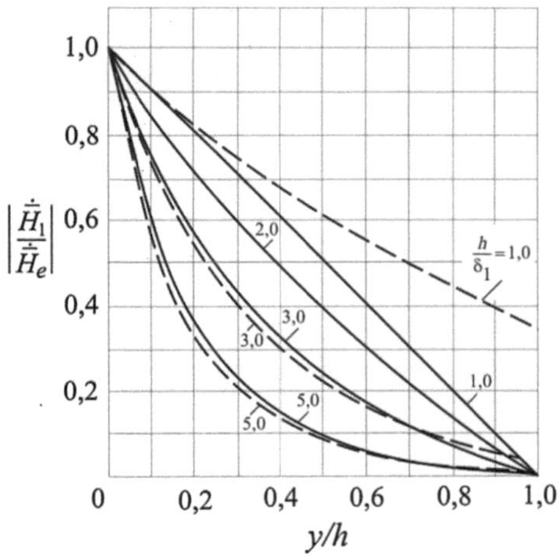

Fig. 2.13 Relative distribution electric field intensity and current density in the thickness of a metallic sheet workpiece (*dashed line* curves for the semi-infinite body)

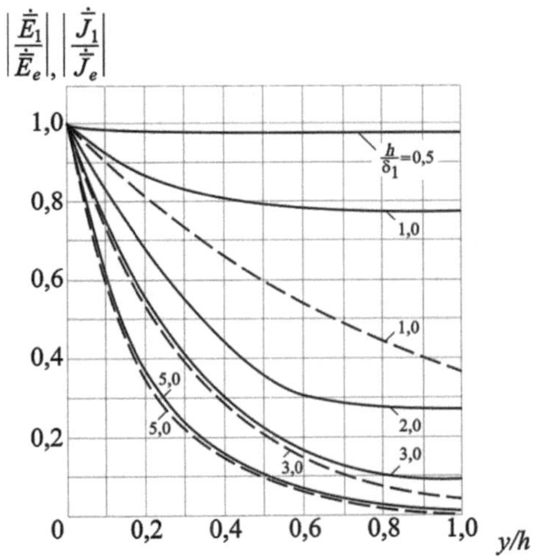

$$\overline{S} = \frac{1}{2}\left[\dot{\overline{E}}_1 \times \overline{H}_1^*\right], \tag{2.61a}$$

with $\overline{H}_1^*$ complex conjugate vector of $\dot{\overline{H}}_1$. By substituting in Eq. (2.61a) the expressions (2.58) of $\dot{\overline{H}}_1$ and $\dot{\overline{E}}_1$, we obtain

$$\dot{\bar{S}} = \frac{1}{2}\left\{[\bar{u}_z \times \bar{u}_x]\dot{H}_e \frac{\sqrt{2}\cdot e^{j\frac{\pi}{4}}}{\delta_1}\rho_1 \frac{ch\left[\frac{1+j}{\delta_1}(h-y)\right]}{sh\left[\frac{1+j}{\delta_1}h\right]} H_e^* \times \frac{sh\left[\frac{1+j}{\delta_1}(h-y)\right]}{sh\left[\frac{1+j}{\delta_1}h\right]}\right\} \tag{2.61b}$$

with $H_e = H_{em}\cdot e^{j\omega t}$, $H_e^* = H_{em}\cdot e^{-j\omega t}$.

After some transformations:

$$\dot{\bar{S}} = \bar{u}_y \cdot \rho_1 \cdot \frac{H_{em}^2}{2\delta_1}\left[\frac{sh\left(2\frac{h-y}{\delta_1}\right) + \sin\left(2\frac{h-y}{\delta_1}\right)}{ch\frac{2h}{\delta_1} - \cos\frac{2h}{\delta_1}} + j\frac{sh\left(2\frac{h-y}{\delta_1}\right) - \sin\left(2\frac{h-y}{\delta_1}\right)}{ch\frac{2h}{\delta_1} - \cos\frac{2h}{\delta_1}}\right] \tag{2.62a}$$

The flux of energy through the external surface of the workpiece ($y = 0$) is equal to

$$\dot{\bar{S}}_e = \bar{u}_y \cdot \rho_1 \cdot \frac{H_{em}^2}{2\delta_1}\left[\frac{sh\frac{2h}{\delta_1} + \sin\frac{2h}{\delta_1}}{ch\frac{2h}{\delta_1} - \cos\frac{2h}{\delta_1}} + j\frac{sh\frac{2h}{\delta_1} - \sin\frac{2h}{\delta_1}}{ch\frac{2h}{\delta_1} - \cos\frac{2h}{\delta_1}}\right] \tag{2.62b}$$

Then the modulus of the flux of electromagnetic energy through the surface of the workpiece, can be written as

$$\dot{S}_e = \rho_1 \cdot \frac{H_{em}^2}{2\delta_1}(F_{slab} + jG_{slab}) = S_{0e}(F_{slab} + jG_{slab}) \tag{2.63}$$

with:

$$\left.\begin{aligned} F_{slab} &= \frac{sh\frac{2h}{\delta_1} + \sin\frac{2h}{\delta_1}}{ch\frac{2h}{\delta_1} - \cos\frac{2h}{\delta_1}} \\ G_{slab} &= \frac{sh\frac{2h}{\delta_1} - \sin\frac{2h}{\delta_1}}{ch\frac{2}{\delta_1} - \cos\frac{2h}{\delta_1}} \end{aligned}\right\} \tag{2.64}$$

and S_{oe}—modulus of the flux of electromagnetic energy through the surface.

The coefficients F_{slab} and G_{slab} are those already used in Eq. (2.60) for the resistance and the reactance of the workpiece.

The analysis of Fig. 2.12 shows that the magnetic field on the side opposite to the one facing the inductor, is completely damped inside the workpiece. Hence the flux of energy $\dot{S}_e$ passing through the surface at $y = 0$, is equal to the total energy transferred to the workpiece.

Then introducing the parameters of the inductor and the Eq. (2.26), we obtain for the modulus of the energy flux through the surface

$$\dot{S}_e = 2 \times 10^{-3}(N_1 I)^2 \cdot \sqrt{\rho_1 \mu_{r1} f}(F_{slab} + jG_{slab}), \tag{2.65}$$

where: I—rms value of the inductor current; N_1—number of turns of the inductor per unit length.

Since the electromagnetic wave decays almost completely in the thickness of the metal body, then the real part of $\dot{S}_e$ gives the value of the active power dissipated in the metal, and the imaginary part—the reactive power in the volume of the body, per square meter of its surface, i.e.

$$p = 2 \times 10^{-3}(N_1 I)^2 \sqrt{\rho_1 \mu_{r1} f} \cdot F_{slab} \tag{2.66a}$$

$$q = 2 \times 10^{-3}(N_1 I)^2 \sqrt{\rho_1 \mu_{r1} f} \cdot G_{slab}. \tag{2.66b}$$

Figure 2.14 shows the diagrams of the function F_{slab}, of the active power dissipated in a plane metallic workpiece heated from one side, and the function G_{slab}, for the reactive power in the same heating conditions. The minimum value of F_{slab} occurs for $2h/\delta_1 = \pi$, where the value of F is 0.917.

From the analysis of the functions F_{slab} and G_{slab} we can draw the following conclusions:

1. F_{slab} decreases from ∞ (at $2h/\delta_1 = 0$) to values close to 1 for $2h/\delta_1 \geq 2.6$; G_{slab} increases from 0 (at $2h/\delta_1 = 0$) to 1 for $2h/\delta_1 \geq 2.6$. This means that for $2h/\delta_1 \geq 2.6$ the flat metal sheet shows the same behavior of a semi-infinite conducting body.
2. for $2h/\delta_1 = 3.14$ the active power in the flat metal workpiece is minimum, (with $F = 0.917$).
3. the condition $h/\delta_1 \geq 1.57$ is a guide for selecting the thickness of current carrying elements of different equipment (current supply bus-bars, thickness of

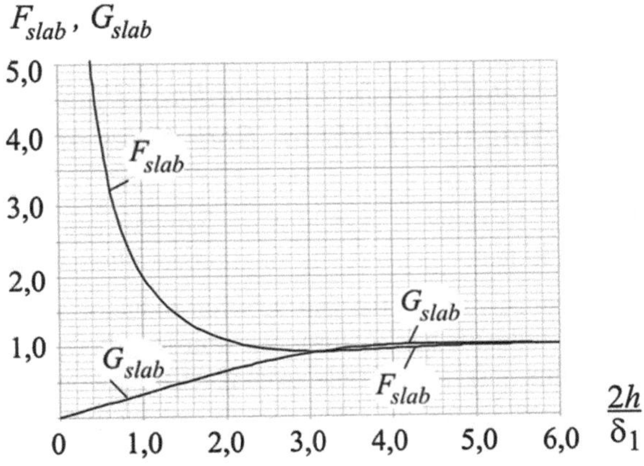

Fig. 2.14 Coefficients F_{slab} and G_{slab} for plane metallic workpieces heated from one side, as a function of the relative thickness

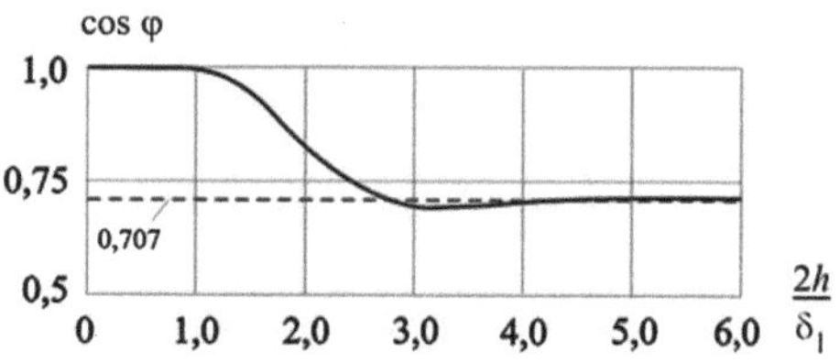

Fig. 2.15 Power factor of flat metal sheet as a function of the relative thickness

the walls of water-cooled tubes of inductors, connections to the heating workpiece) or electromagnetic shields.

4. The higher values of active power in the workpiece are induced at small values of h/δ_1. In practical calculations, for heating with high electrical efficiency it is sufficient to select the frequency in order to fulfill the condition $h/\delta_1 \leq 1.5$.

We can define now the resistance r_e and the reactance x_e per square meter of the surface of the flat workpiece by the use of Eq. (2.66a, b).

$$r_e = \frac{p}{(N_1 I)^2} = 2 \times 10^{-3} \cdot \sqrt{\rho_1 \mu_{r1} f} \cdot F_{slab} = \frac{\rho_1}{\delta_1} F_{slab}, \tag{2.67a}$$

$$x_e = \frac{q}{(N_1 I)^2} = 2 \times 10^{-3} \cdot \sqrt{\rho_1 \mu_{r1} f} \cdot G_{slab} = \frac{\rho_1}{\delta_1} G_{slab}. \tag{2.67b}$$

The power factor therefore is:

$$\cos \varphi = \frac{p}{\sqrt{p^2 + q^2}} = \frac{F_{slab}}{\sqrt{F_{slab}^2 + G_{slab}^2}}. \tag{2.68}$$

In Fig. 2.15 is given the diagram of the power factor as a function of the relative thickness $2h/\delta_1$.

For $2h/\delta_1 \geq 3$ the power factor is equal to 0.707, i.e. the same value of the semi-infinite body. For $2h/\delta_1 \rightarrow 0$ $\cos \varphi$ tends to 1.

2.3.2 Metal Slab in Longitudinal Flux Inductor (Exciting Magnetic Field on Both Sides of the Workpiece)

Figure 2.16 shows a sketch of the inductor and the metal workpiece.

The inductor generates a longitudinal alternating magnetic field, with magnetic field intensity $\dot{\overline{H}}_e$ in the gap between inductor and workpiece.

The inductor has N_1 turns per unit length.

In this case a plane electromagnetic wave is incident on both surfaces of the workpiece. Figure 2.17 specifies the geometrical dimensions of the workpiece.

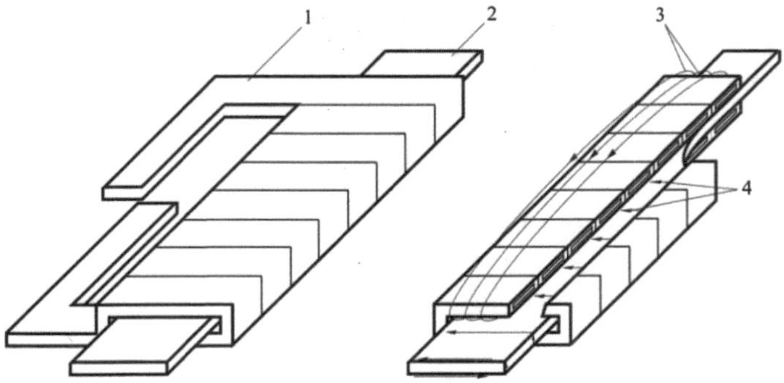

Fig. 2.16 Schematic of metal slab and inductor: *1* inductor; *2* workpiece; *3* lines of magnetic field; *4* direction of induced currents on the workpiece surface

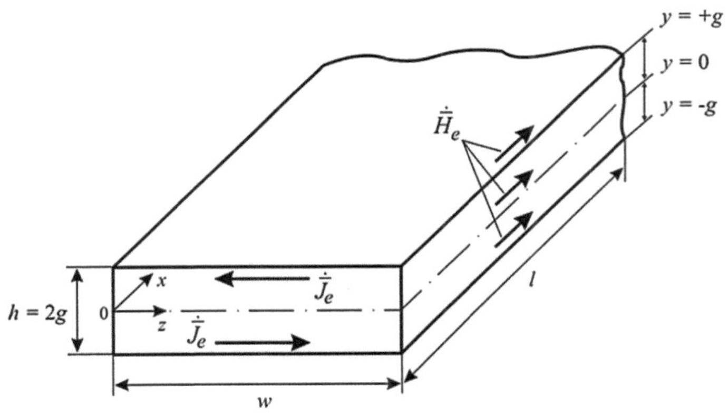

Fig. 2.17 Directions of H and J in the workpiece

The distributions of the magnetic and electric field intensities in this case are symmetrical on the opposite sides of the mean plane of the workpiece.

The coordinate system is chosen so that the origin of the axis *oy* is coincident with the plane of symmetry .

The distribution of the electromagnetic field in the workpiece is described by the Maxwell's equations (2.1a).

The solution of this equations system is found in this case with the conditions:

1. $\dot{\bar{J}} \gg \partial\dot{\bar{D}} / \partial t$ in the conducting workpiece
2. workpiece of homogeneous material with $\rho = \text{const}$, $\mu_r = \text{const}$.
3. sinusoidal electromagnetic quantities:

$$\overline{H} = \overline{H}_m \cdot \sin \omega t; \;\; \overline{B} = \overline{B}_m \cdot \sin \omega t; \;\; \overline{E} = \overline{E}_m \cdot \sin(\omega t + \varphi).$$

Maxwell's equations (2.1a) for this case take the form:

$$\frac{\partial^2 \dot{\overline{H}}}{\partial y^2} = k^2 \dot{\overline{H}}, \tag{2.69}$$

with $k^2 = j\frac{\omega\mu_0\mu_r}{\rho}$, and its solution is

$$\dot{\overline{H}} = \dot{\overline{C}}_1 \cdot e^{-ky} + \dot{\overline{C}}_2 \cdot e^{ky}. \tag{2.70}$$

Applying the boundary conditions

$$\begin{aligned} &for\ \ y = +g, \quad \dot{\overline{H}}_e = \dot{\overline{C}}_1 \cdot e^{+kg} + \dot{\overline{C}}_2 \cdot e^{-kg} \\ &for\ \ y = -g, \quad \dot{\overline{H}}_e = \dot{\overline{C}}_1 \cdot e^{-kg} + \dot{\overline{C}}_2 \cdot e^{+kg}, \end{aligned}$$

we obtain

$$\dot{\overline{C}}_1 = \dot{\overline{C}}_2 \quad ; \quad \dot{\overline{H}}_e = \dot{\overline{C}}_1\left(e^{-kg} + e^{kg}\right) = 2\dot{\overline{C}}_1 ch(kg)$$

and finally:

$$\dot{\overline{C}}_1 = \frac{\dot{\overline{H}}_e}{2ch(kg)}.$$

Then the module of the vector of the magnetic field intensity is

$$\dot{H} = \dot{H}_e \cdot \frac{ch(ky)}{ch(kg)},$$

or, taking into account that $k = (1+j)/\delta$,

$$\frac{\dot{H}}{\dot{H}_e} = \frac{ch\left[(1+j)\frac{y}{\delta}\right]}{ch\left[(1+j)\frac{g}{\delta}\right]}$$

The module of the ratio of the complex magnetic field intensities, i.e. the ratio of the current value of $\dot{H}$ and the value $\dot{H}_e$ at the surface, is

$$\left|\frac{\dot{H}}{\dot{H}_e}\right| = \sqrt{\frac{ch\left(2\frac{y}{\delta}\right) + \cos\left(2\frac{y}{\delta}\right)}{ch\left(2\frac{g}{\delta}\right) + \cos\left(2\frac{g}{\delta}\right)}}. \tag{2.71}$$

Figure 2.18 shows the curves of the module $\left|\dot{H}/\dot{H}_e\right|$ on half thickness of the workpiece, as a function of the relative thickness $(2g/\delta)$ and the relative coordinate y/g.

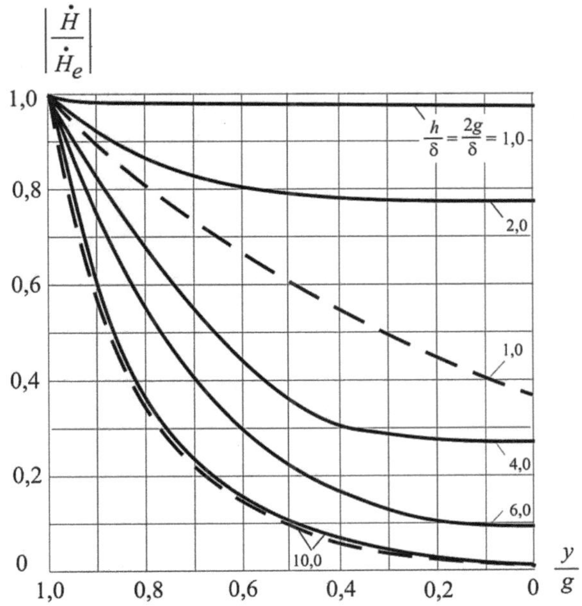

Fig. 2.18 Distribution of the magnetic field intensity in the thickness of a flat workpiece in longitudinal flux inductor (*dashed line* curves of the semi-infinite body)

For $(2g/\delta) \leq 1$, the magnetic field intensity is practically uniformly distributed on the slab thickness. With the increase of $(2g/\delta)$ the distribution tends to an exponential and for $(2g/\delta) \geq 10$ the distribution becomes coincident with that obtained for the semi-infinite homogeneous body (dashed curve).

We can calculate now the current density distribution in the thickness of the workpiece.

From Maxwell's equations (2.1a) we can obtain the module of the current density vector $\dot{J} = \frac{\dot{E}}{\rho} = \frac{d\dot{H}}{dy}$; by substituting in it $\dot{H}$ from Eq. (2.70) and taking into account that $\dot{C}_1 = \dot{C}_2$, we obtain:

$$\dot{J} = k\dot{C}_1\left(e^{ky} - e^{-ky}\right) = \dot{H}_e k \frac{\left(e^{ky} - e^{-ky}\right)}{2ch(kg)} = -\dot{H}_e k \frac{sh(ky)}{2ch(kg)}. \quad (2.72)$$

With the boundary conditions

$$\dot{J}_e = \dot{J}(+g) = -\dot{J}(-g), \quad \text{and} \quad \dot{J}(0) = 0,$$

we can write for the current density at the surface of the workpiece

$$\dot{J}_e = -\dot{H}_e k \frac{sh(kg)}{ch(kg)},$$

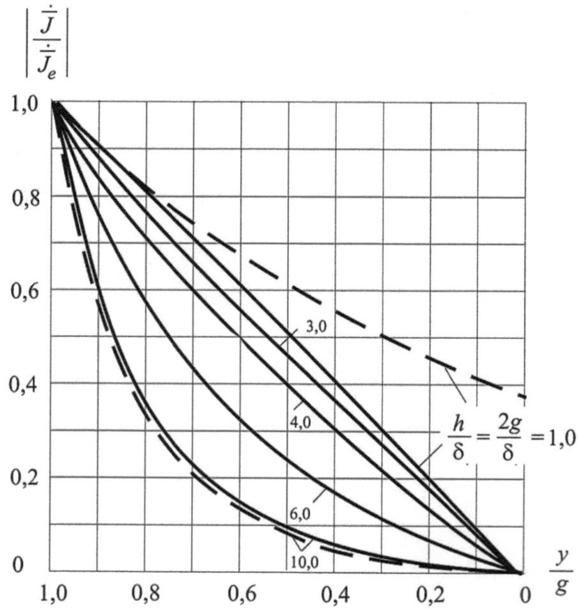

Fig. 2.19 Distribution of the current density in the thickness of a slab: *dashed line* curves for the semi-infinite body

Therefore

$$\dot{J}_e = \dot{J}_e \frac{sh(ky)}{sh(kg)} = \dot{J}_e \cdot \frac{sh[(1+j)y/\delta]}{sh[(1+j)g/\delta]}. \tag{2.72a}$$

The module of the ratio of the complex values of the current density (i.e. the current value $\dot{J}$ and the value at the surface $\dot{J}_e$) in the slab is therefore equal to

$$\left|\frac{\dot{J}}{\dot{J}_e}\right| = \sqrt{\frac{ch\left(\frac{2y}{\delta}\right) - \cos\left(\frac{2y}{\delta}\right)}{ch\left(\frac{2g}{\delta}\right) - \cos\left(\frac{2g}{\delta}\right)}}. \tag{2.73}$$

In Fig. 2.19 the distributions of $|\dot{J}/\dot{J}_e|$ in the half thickness of the slab are shown. On the axis of symmetry (at $y = 0$) the current density is zero.

For relative thickness $2g/\delta \leq 3$, the current density variation is practically linear from the maximum value J_e at the surface to zero on the axis.

When $2g/\delta$ increases, the variation of the current density tends to an exponential and, for $2g/\delta \geq 10$, the curve is the same obtained for the semi-infinite body.

In agreement with Eq. (2.5a) the current density is

$$\dot{\bar{J}} = \frac{\dot{\bar{E}}}{\rho} = -\bar{u}_z \frac{\partial \dot{H}}{\partial y} = -\bar{u}_z \dot{H}_e \frac{\partial}{\partial y}\left[\frac{ch(1+j)y/\delta}{ch(1+j)g/\delta}\right]$$
$$= -\bar{u}_z \dot{H}_e \cdot \frac{1+j}{\delta}\left[\frac{sh(1+j)y/\delta}{ch(1+j)g/\delta}\right]. \tag{2.74}$$

For $y = g$ the module of the vector of the current density becomes

$$\left|\dot{\bar{J}}_e\right| = \sqrt{2}\frac{H_e}{\delta} th\left[(1+j)\frac{g}{\delta}\right].$$

When the thickness of the slab is much greater than the penetration depth, i.e. when $g/\delta \to \infty$, the module of the current density at the surface $\left|\dot{\bar{J}}_e\right|$ tends to the value $\left(\sqrt{2}H_e\right)/\delta$, i.e. to the same value of the current density on the surface of a semi-infinite body.

In this case, the heated slab can be considered as constituted by two independent semi-infinite slabs, separated by the plane of symmetry.

This condition is verified with a precision of about 1 % for $2g/\delta \geq 5$. For example, a slab of steel C40 at 20 °C ($\mu_r = 100$, $\rho = 20 \times 10^{-8}$ Ωm) heated at 50 Hz, can be considered as a semi-infinite body when its thickness is 2g > 16 mm; if the slab is of stainless steel ($\mu_r = 1$, $\rho = 60 \times 10^{-8}$ Ωm) the same applies for 2g > 280 mm; if it is of copper ($\mu_r = 1$, $\rho = 1.75 \times 10^{-8}$ Ωm), for 2g > 85 mm.

2.3.2.1 Energy in a Slab with Double-Side Heating

The specific surface power absorbed by the slab can be determined applying the Poynting vector to the two surfaces of the slab

$$\dot{\bar{S}} = \left[\dot{\bar{E}} \times \bar{H}^*\right]_{y=g} = \bar{u}_y \rho \frac{H_e^2}{\delta}(1+j)\left\{\frac{sh\left[\frac{(1+j)g}{\delta}\right]}{ch\left[\frac{(1+j)g}{\delta}\right]}\right\}$$
$$= \bar{u}_y \rho \frac{H_e^2}{\delta}(1+j) th\left[\frac{(1+j)g}{\delta}\right] = \bar{u}_y \rho \frac{H_e^2}{\delta}(1+j)\left[\frac{sh\frac{2g}{\delta} + j\sin\frac{2g}{\delta}}{ch\frac{2g}{\delta} + \cos\frac{2g}{\delta}}\right] \tag{2.75}$$
$$= \bar{u}_y \rho \frac{H_e^2}{\delta}\left[\frac{sh\frac{2g}{\delta} - \sin\frac{2g}{\delta}}{ch\frac{2g}{\delta} + \cos\frac{2g}{\delta}} + j\frac{sh\frac{2g}{\delta} + \sin\frac{2g}{\delta}}{ch\frac{2g}{\delta} + \cos\frac{2g}{\delta}}\right]$$

Separating the real and imaginary terms, we can write the expressions of the active and reactive power

$$p = \rho \frac{H_e^2}{\delta}\left[\frac{sh\frac{2g}{\delta} - \sin\frac{2g}{\delta}}{ch\frac{2g}{\delta} + \cos\frac{2g}{\delta}}\right] = 2S_{0e} \cdot F_{2s}, \tag{2.76a}$$

$$q = \rho \frac{H_e^2}{\delta}\left[\frac{sh\frac{2g}{\delta} + \sin\frac{2g}{\delta}}{ch\frac{2g}{\delta} + \cos\frac{2g}{\delta}}\right] = 2S_{0e} \cdot G_{2s}, \tag{2.76b}$$

where $S_{0e} = H_e^2 \frac{\rho}{2\delta}$—density of the active (or reactive) component of the energy flow through the surface of a semi-infinite body (one-side heating of a flat metal workpiece); F_{2s} and G_{2s}—surface specific active and reactive power in the slab with induction heating on both sides (heating with longitudinal magnetic field with an inductor encircling the workpiece).

The power factor then results:

$$\cos\varphi = \frac{p}{\sqrt{p^2 + q^2}}.$$

Figure 2.20 gives the diagrams of F_{2s} and G_{2s} and cos φ as a function of the relative thickness of the slab.

From these curves it can be seen that the active power increases from 0 to a maximum equal to 0.417, when the relative thickness $2g/\delta$ increases from 0 to 2.25, while for a further increase of $2g/\delta$ to ∞, it decreases to 0.

The specific reactive power decreases from 1 to 0 when the relative thickness $2g/\delta$ varies from 0 to ∞.

The power factor has the minimum value $\cos\varphi = 0$, for $2g/\delta = 0$. When $2g/\delta$ increases, cos φ also increases, reaching the value 0.6 when the active power is

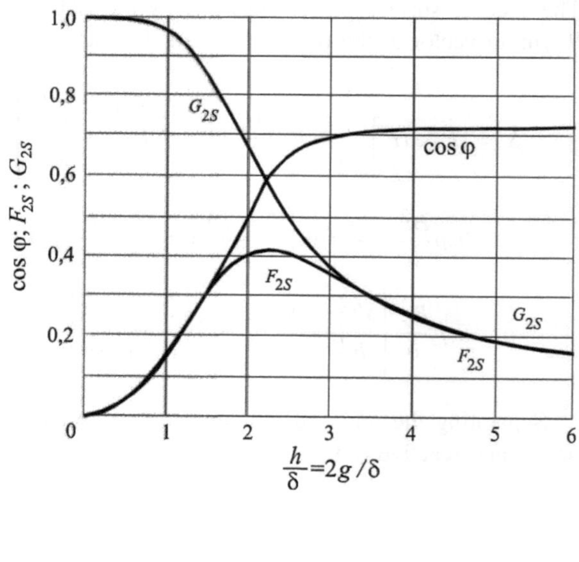

Fig. 2.20 Diagrams of specific active and reactive power and power factor in a flat slab as a function of the relative thickness

maximum (for $2g/\delta = 2.25$); with a further increase of $2g/\delta$, the power factor reaches—for $2g/\delta \geq 3.5$—the value 0.707, which is the same value in the semi-infinite body.

2.3.3 Direct Resistance Heating

In this paragraph we consider the distribution of the current density J in the thickness of a slab heated by direct resistance.

Figure 2.21 shows the scheme of the flow of a current in a workpiece of thickness $h = 2g$.

For the analysis of this problem we make the following hypotheses:

1. the thickness 2g of the slab is much greater of the penetration depth δ of the electromagnetic wave.
2. the width w of the workpiece is much greater than its thickness ($w \gg 2g$).
3. the workpiece is assumed to be homogeneous, i.e. its physical characteristics are constant (ρ = const., μ_r = const.).
4. in the workpiece flows a sinusoidal current.

In these hypotheses, the current density varies along the thickness of the workpiece and is a function only of the coordinate y. At the surface (for $y = \pm g$) the module of the current density is $\dot{J}_e$.

Considering the Ohm's law for the module of the current density vector, the Maxwell's equations (2.1a) give the following differential equation:

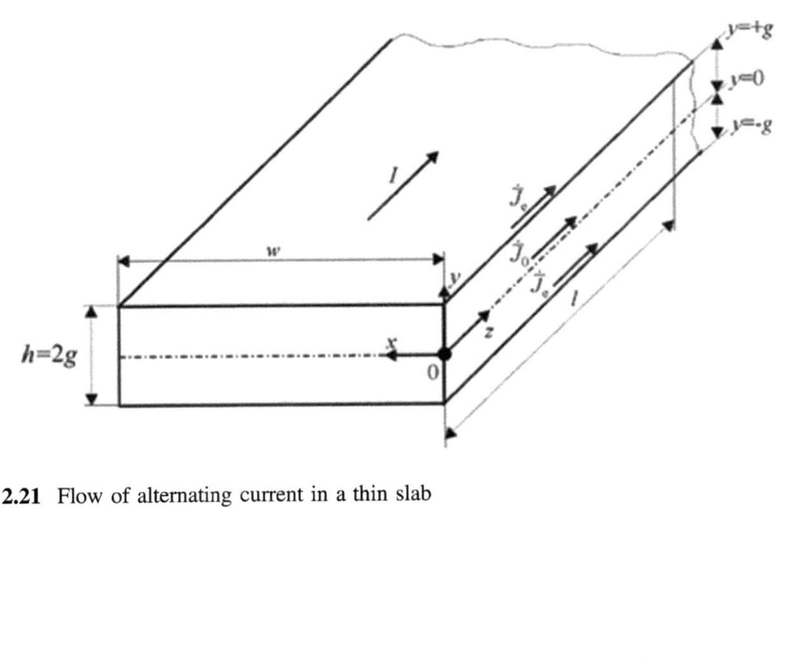

Fig. 2.21 Flow of alternating current in a thin slab

$$\frac{\partial^2 \dot{J}}{\partial y^2} = \frac{\mu}{\rho}\frac{\partial \dot{J}}{\partial t} = k^2 \dot{J}, \tag{2.77}$$

with $k^2 = j\frac{\omega\mu_0\mu_r}{\rho}$.

The solution of this equation has the form

$$\dot{J} = \dot{C}_1 e^{ky} + \dot{C}_2 e^{-ky}. \tag{2.78}$$

The integration constants $\dot{C}_1$ and $\dot{C}_2$ can be determined with the boundary conditions:

$$y = +g, \quad \dot{J}_e = \dot{C}_1 e^{kg} + \dot{C}_2 e^{-kg}, \tag{2.79}$$

$$y = -g, \quad \dot{J}_e = \dot{C}_1 e^{-kg} + \dot{C}_2 e^{kg}. \tag{2.80}$$

The simultaneous solution of Eqs. (2.79) and (2.80) gives for $\dot{C}_1$ and $\dot{C}_2$

$$\dot{C}_1 = \dot{C}_2 = \frac{\dot{J}_e}{2ch(kg)}.$$

Therefore, in a generic point of the workpiece is

$$\dot{J} = \frac{\dot{J}_e}{2ch(kg)}\left(e^{ky} + e^{-ky}\right) = \dot{J}_e \frac{ch(ky)}{ch(kg)} \tag{2.81}$$

or, being $k = (1 + j)/\delta$:

$$\frac{\dot{J}}{\dot{J}_e} = \frac{ch\left[(1+j)\frac{y}{\delta}\right]}{ch\left[(1+j)\frac{g}{\delta}\right]}. \tag{2.82}$$

The module of the current density along the thickness of the workpiece therefore becomes

$$\begin{aligned} \left|\frac{\dot{J}}{\dot{J}_e}\right| &= \left|\frac{ch\left[(1+j)\frac{y}{\delta}\right]}{ch\left[(1+j)\frac{g}{\delta}\right]}\right| = \left|\frac{ch\frac{y}{\delta}\cdot\cos\frac{y}{\delta} + jsh\frac{y}{\delta}\cdot\sin\frac{y}{\delta}}{ch\frac{g}{\delta}\cdot\cos\frac{g}{\delta} + jsh\frac{g}{\delta}\cdot\sin\frac{g}{\delta}}\right| \\ &= \sqrt{\frac{ch^2\frac{y}{\delta}\cdot\cos^2\frac{y}{\delta} + sh^2\frac{y}{\delta}\cdot\sin^2\frac{y}{\delta}}{ch^2\frac{g}{\delta}\cdot\cos^2\frac{g}{\delta} + sh^2\frac{g}{\delta}\cdot\sin^2\frac{g}{\delta}}} = \sqrt{\frac{ch2\frac{y}{\delta} + \cos 2\frac{y}{\delta}}{ch2\frac{g}{\delta} + \cos 2\frac{g}{\delta}}} \end{aligned} \tag{2.83}$$

Figure 2.22 shows the variation of the ratio $\left|\dot{J}/\dot{J}_e\right|$ as a function of the relative coordinate y/g, for different values of the ratio between the thickness 2g and the penetration depth δ.

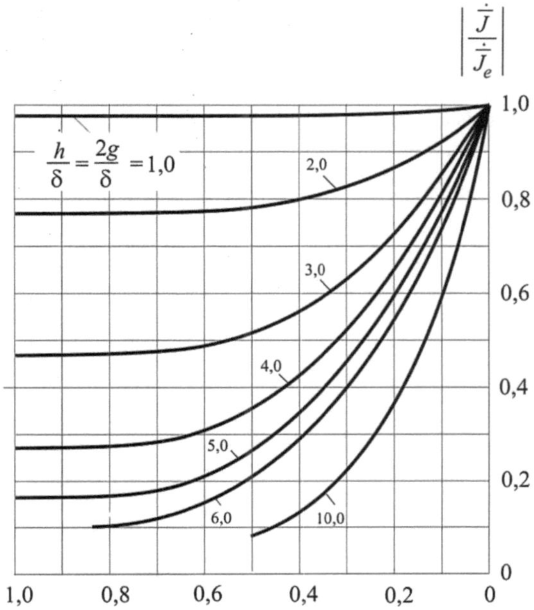

Fig. 2.22 Ratio $|\dot{J}/\dot{J}_e|$ of current densities inside the slab and at the surface, as a function of the relative coordinate y/g and the relative thickness 2g/δ

Since the magnetic field intensity is equal to:

$$\dot{H} = -\frac{\rho}{j\omega\mu_0\mu_r}\frac{\partial \dot{J}}{\partial y},$$

taking into account Eq. (2.81), we obtain

$$\frac{\dot{H}}{\dot{H}_e} = \frac{sh[(1+j)y/\delta]}{sh[(1+j)g/\delta]}$$

and

$$\left|\frac{\dot{H}}{\dot{H}_e}\right| = \sqrt{\frac{ch2\frac{y}{\delta} - \cos 2\frac{y}{\delta}}{ch2\frac{g}{\delta} - \cos 2\frac{g}{\delta}}}.$$

Figure 2.23 shows the module of the ratio of magnetic field intensity $\dot{H}$ in the slab and the value $\dot{H}_e$ at the surface, as a function of the relative thickness 2y/δ.

We can express now the current density $\dot{J}$ as a function of the total current $\dot{I}$ in the workpiece.

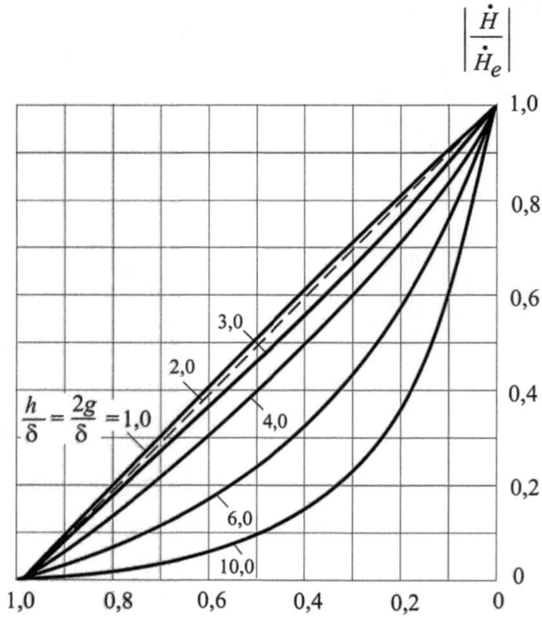

Fig. 2.23 Module of ratio of magnetic field intensity in the slab and at the surface, as a function of the relative coordinate y/g and the relative thickness 2g/δ

The total current in the workpiece is

$$\dot{I} = \int_{-g}^{g} w\dot{J}dy = \int_{-g}^{g} \frac{w\dot{J}_e ch(ky)}{ch(kg)} dy = \frac{w\dot{J}_e}{ch(kg)} \left[\frac{sh(ky)}{k}\right]_{-g}^{g}$$
$$= \frac{2\dot{J}_e \delta w}{(1+j)} th\left[(1+j)\frac{g}{\delta}\right] \tag{2.84}$$

Therefore

$$\dot{J}_e = \frac{\dot{I}}{2\delta w}(1+j) \cdot cth\left[(1+j)\frac{g}{\delta}\right]$$

and

$$\dot{J} = \frac{\dot{I}}{2\delta w}(1+j) \cdot \frac{ch\left[(1+j)\frac{y}{\delta}\right]}{sh\left[(1+j)\frac{g}{\delta}\right]}. \tag{2.85}$$

The impedance $\dot{Z}_e$, resistance R_e and internal reactance X_{ie} of the workpiece can be obtained from the voltage drop $\dot{U} = \rho\dot{J}_e\ell$, along the length ℓ of the workpiece. It results:

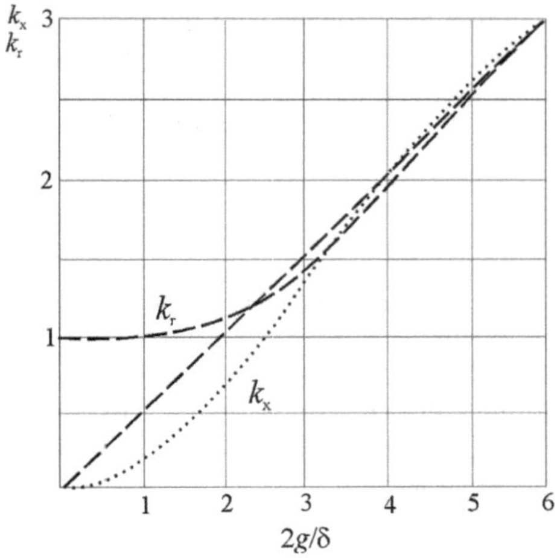

Fig. 2.24 Coefficients k_r and k_x of skin-effect in a slab, as a function of the ratio between thickness and penetration depth

$$\dot{Z}_e = \frac{\dot{U}}{\dot{I}} = R_e + jX_{ie} = \frac{\rho \dot{J}_e \ell (1+j)}{2\dot{J}_e \delta w th\left[(1+j)\frac{g}{\delta}\right]}$$
$$= \frac{\rho \ell}{2\delta w}\left[\frac{sh\frac{2g}{\delta} + \sin\frac{2g}{\delta}}{ch\frac{2g}{\delta} - \cos\frac{2g}{\delta}}\right] + j\frac{\rho \ell}{2\delta w}\left[\frac{sh\frac{2g}{\delta} - \sin\frac{2g}{\delta}}{ch\frac{2g}{\delta} - \cos\frac{2g}{\delta}}\right] \quad (2.86)$$

Denoting with $R_{dc} = (\rho\ell)/(2gw)$ the DC resistance of the workpiece, the skin-effect coefficients for the resistance and the internal reactance reactance are:

$$k_r = \frac{R_e}{R_{dc}} = \frac{1}{2}\left(\frac{2g}{\delta}\right)\left[\frac{sh\frac{2g}{\delta} + \sin\frac{2g}{\delta}}{ch\frac{2g}{\delta} - \cos\frac{2g}{\delta}}\right], \quad (2.87)$$

$$k_x = \frac{X_{ie}}{R_{dc}} = \frac{1}{2}\left(\frac{2g}{\delta}\right)\left[\frac{sh\frac{2g}{\delta} - \sin\frac{2g}{\delta}}{ch\frac{2g}{\delta} - \cos\frac{2g}{\delta}}\right]. \quad (2.88)$$

Figure 2.24 gives the diagrams of the coefficients k_r and k_x as a function of the ratio of the thickness 2g to the penetration depth δ.

When the ratio $2g/\delta$ is greater than 3, in Eqs. (2.87) and (2.88) the terms containing $\sin(2g/\delta)$ and $\cos(2g/\delta)$ become negligible in comparison with $sh(2g/\delta)$ and $ch(2g/\delta)$, and it is $sh(2g/\delta) \approx ch(2g/\delta)$. In this case the resistance R_e and the internal reactance X_{ie} have nearly the same values and the coefficients of skin-effect are practically equal, i.e. $k_r \approx k_x$.

It's interesting to compare the electrical characteristics of the workpiece with those of the semi-infinite slab.

To this end we calculate the power factor $\cos\varphi$ and the specific surface active p and reactive q power of the sheet workpiece.

From the knowledge of the resistance R_e and the module of the impedance $\dot{Z}$ of the workpiece, we can derive the equation of the power factor as a function of the ratio between the thickness 2g and the penetration depth δ

$$\cos\varphi = \frac{R_e}{\sqrt{R_e^2 + X_{ie}^2}} = \frac{sh\frac{2g}{\delta} + \sin\frac{2g}{\delta}}{\sqrt{2}\sqrt{sh^2\frac{2g}{\delta} + \sin^2\frac{2g}{\delta}}}. \tag{2.89}$$

Figure 2.25 gives the diagram of $\cos\varphi$ as a function of the ratio $2g/\delta$.

When $2g/\delta$ increases from 0 to 3, the power factor decreases from 1 to 0.707; for a further increase of the argument the power factor remains constant and equal to the value in a semi-infinite slab.

The specific surface power, developed in the workpiece, can be evaluated with the relationship:

$$\begin{aligned} 2\dot{S} &= \frac{\dot{U}\dot{I}^*}{\ell w} = \frac{\rho \dot{J}_e \ell\delta w 2J_e^*}{\ell w(1-j)} th\left[(1-j)\frac{g}{\delta}\right] \\ &= \rho\delta J_e e^{j\omega t} J_e e^{-j\omega t}(1+j)\frac{sh\left[(1-j)\frac{g}{\delta}\right]}{ch\left[(1-j)\frac{g}{\delta}\right]} \\ &= J_e^2\rho\delta\left[(1+j)\left(\frac{sh\frac{2g}{\delta} - j\sin\frac{2g}{\delta}}{ch\frac{2g}{\delta} + \cos\frac{2g}{\delta}}\right)\right] \\ &= J_e^2\rho\delta\left[\frac{sh\frac{2g}{\delta} + \sin\frac{2g}{\delta}}{ch\frac{2g}{\delta} + \cos\frac{2g}{\delta}} + j\frac{sh\frac{2g}{\delta} - \sin\frac{2g}{\delta}}{ch\frac{2g}{\delta} + \cos\frac{2g}{\delta}}\right] \end{aligned} \tag{2.90}$$

where

$J_e^* = J_e e^{-j\omega t}$ conjugate of complex current density at the surface of the workpiece;

$I^* = \frac{2}{(1-j)}\delta w J_e^* th\left[(1-j)\frac{g}{\delta}\right]$ conjugate of complex current.

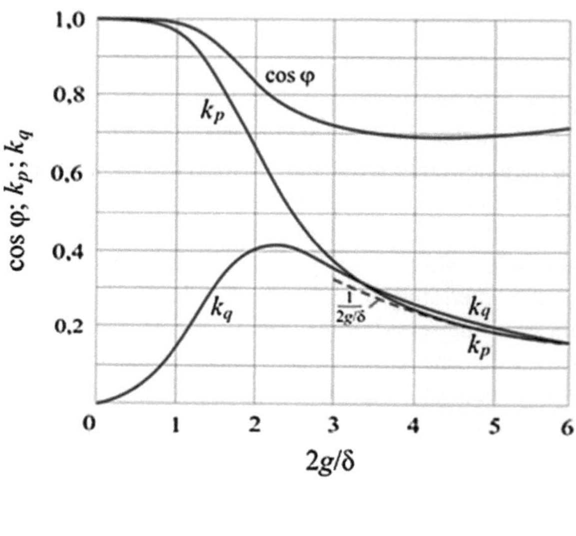

Fig. 2.25 Power factor and specific surface powers as a function of the relative thickness of the workpiece 2g/δ (active power—p; reactive—q)

Finally we obtain:

$$\dot{S} = p + jq = J_e^2 \rho \frac{\delta}{2} \left[\frac{sh\frac{2g}{\delta} + \sin\frac{2g}{\delta}}{ch\frac{2g}{\delta} + \cos\frac{2g}{\delta}} + j \frac{sh\frac{2g}{\delta} - \sin\frac{2g}{\delta}}{ch\frac{2g}{\delta} + \cos\frac{2g}{\delta}} \right] \\ = p_{dc}\left[k_p + jk_q\right] \tag{2.91}$$

with:

$k_p = p/p_{dc}$, $k_q = q/p_{dc}$ relative active and reactive specific surface power in the flat sheet workpiece;

$p_{dc} = J_e^2 \rho g$ specific surface power developed in a thin slab of thickness $2g$, when in it flows DC current of density J_e.

In Fig. 2.25 the values of k_p and k_q are shown as a function of the relative thickness of the workpiece $2g/\delta$.

The surface power p has the maximum value 1 for $2g/\delta = 0$; when the argument increases it decreases monotonically; the reactive power q has its minimum value equal to 0 for $2g/\delta = 0$, increases when the argument increases and reaches a maximum for $2g/\delta \approx 2.3$. For $2g/\delta > 2.3$ the function k_q starts to decrease, gradually becoming closer to the function k_p; starting from $2g/\delta \approx 2.3$, the two functions have a similar behaviour, decreasing with very close curves.

For arguments $2g/\delta > 4$ the two functions can be approximated with the function $1/(2g/\delta)$ (represented by the dotted line in Fig. 2.25), i.e. $k_p \approx k_q \approx \delta/2g$.

Therefore, for $2g/\delta \geq 4$, where $sh(2g/\delta) \gg \sin(2g/\delta)$, $ch(2g/\delta) \gg \cos(2g/\delta)$ and $sh(2g/\delta) \approx ch(2g/\delta)$, the active and reactive surface power are given by $p \approx q \approx J_e^2 \frac{\rho\delta}{2}$, the same as in semi-infinite slab.

2.4 Slab of Rectangular Cross-Section

2.4.1 Induction Heating

Figure 2.26 shows the schematic of inductor (2) and heated workpiece (1)—a metal slab with rectangular cross-section of infinite length. The inductor generates an alternating magnetic field, which in the gap between inductor and slab is given by the vector $\dot{\overline{H}}_e$.

The distributions of the magnetic and electric fields in the slab are described by Maxwell's equation (2.1a).

The solution of this system in a two-dimensional formulation is given in [5] with the following assumptions:

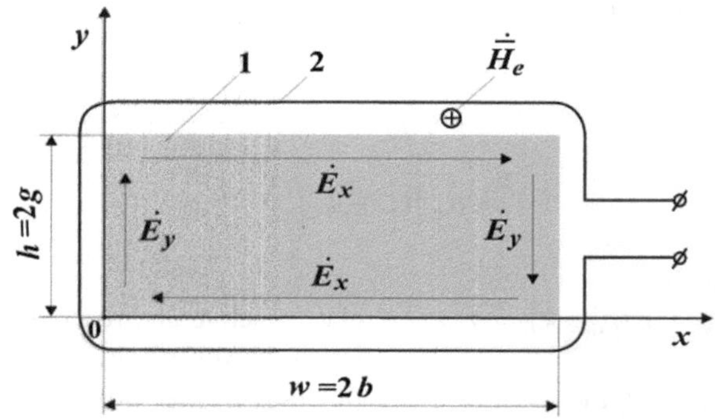

Fig. 2.26 Schematic of inductor and slab of rectangular cross-section: *1* workpiece; *2* inductor

1. slab made of homogeneous conductive material;

$$\dot{\bar{J}} \gg \frac{\partial \dot{\bar{D}}}{\partial t}$$

2. constant characteristics of slab material: $\rho = \text{const}$, $\mu_r = \text{const}$;
3. sinusoidal current in the inductor;
4. lengths of workpiece and inductor much greater than the dimensions of the slab cross-section

In the above hypotheses, after some transformations of system (2.1a), for the vector of the magnetic field intensity can be written:

$$\frac{\partial^2 \dot{\bar{H}}}{\partial x^2} + \frac{\partial^2 \dot{\bar{H}}}{\partial y^2} = k^2 \dot{\bar{H}}, \tag{2.92}$$

with $k^2 = j\frac{\omega\mu_r\mu_0}{\rho}$, and the components of the vector of the electric field intensity can be derived from the relations:

$$\dot{\bar{E}}_x = \rho \frac{\partial \dot{\bar{H}}}{\partial y} \quad \text{and} \quad \dot{\bar{E}}_y = -\rho \frac{\partial \dot{\bar{H}}}{\partial x}. \tag{2.93}$$

The boundary conditions are determined considering that the magnetic field intensity at the surface of the slab is:

$$\begin{aligned} \dot{\bar{H}} &= \dot{\bar{H}}_e \text{ for } x = 0,\ x = w \text{ and } 0 \le y \ge h \\ \dot{\bar{H}} &= \dot{\bar{H}}_e \text{ for } y = 0,\ y = h \text{ and } 0 \le x \ge w \end{aligned} \tag{2.94}$$

The solution of Eqs. (2.92) and (2.93) with the boundary conditions (2.94), gives for the magnetic field intensity the following equations [2, 7]:

$$\dot{H} = H_e\left\{1 - \frac{4}{\pi}k^2 \sum_{n=1,3,\ldots}^{\infty} \frac{1 - \frac{\cosh\left[\beta_n\left(\frac{h}{2}-y\right)\right]}{\cosh\left(\beta_n\frac{h}{2}\right)}}{n\beta_n^2} \sin\frac{n\pi x}{w}\right\} \tag{2.95a}$$

$$\dot{H} = H_e\left\{1 - \frac{4}{\pi}k^2 \sum_{m=1,3,\ldots}^{\infty} \frac{1 - \frac{\cosh\left[\alpha_m\left(\frac{w}{2}-x\right)\right]}{\cosh\left(\alpha_m\frac{w}{2}\right)}}{m\alpha_m^2} \sin\frac{m\pi y}{h}\right\} \tag{2.95b}$$

Since Eq. (2.95a, b) give the same distributions of $\dot{H}$ in the slab cross-section, they can be used indifferently for calculations.

Both equations describe the magnetic field intensity H by a Fourier series: the first one using a decomposition with sinusoidal functions along the coordinate x, the second one along y. The two forms of writing the solution for the magnetic field were needed in order to facilitate the derivation of the electric field intensity. In this way the component of the vector of the electric field intensity along x is obtained by differentiating the series (2.95a), while the component along y is obtained by differentiating the series (2.95b):

$$\begin{aligned} \dot{E}_x &= \rho\frac{\partial \dot{H}}{\partial y} = -H_e\rho\frac{4}{\pi}k^2\left\{\sum_{n=1,3,\ldots}^{\infty}\frac{1}{n\beta_n}\cdot\frac{\sinh\left[\beta_n\left(\frac{h}{2}-y\right)\right]}{\cosh\left(\beta_n\frac{h}{2}\right)}\sin\frac{n\pi x}{w}\right\} \\ \dot{E}_y &= -\rho\frac{\partial \dot{H}}{\partial x} = H_e\rho\frac{4}{\pi}k^2\left\{\sum_{m=1,3,\ldots}^{\infty}\frac{1}{m\alpha_m}\cdot\frac{\sinh\left[\alpha_m\left(\frac{w}{2}-x\right)\right]}{\cosh\left(\alpha_m\frac{w}{2}\right)}\sin\frac{m\pi y}{h}\right\} \end{aligned} \tag{2.96}$$

with $\beta_n^2 = \left(\frac{n\pi}{w}\right)^2 + k^2$; $\alpha_m^2 = \left(\frac{m\pi}{h}\right)^2 + k^2$.

In Fig. 2.27 the qualitative distributions of the relative electric field intensity in the cross-section of the slab are given in case of skin-effect weak $(h/\delta = 2)$ or strong $(h/\delta = 6)$ [6].

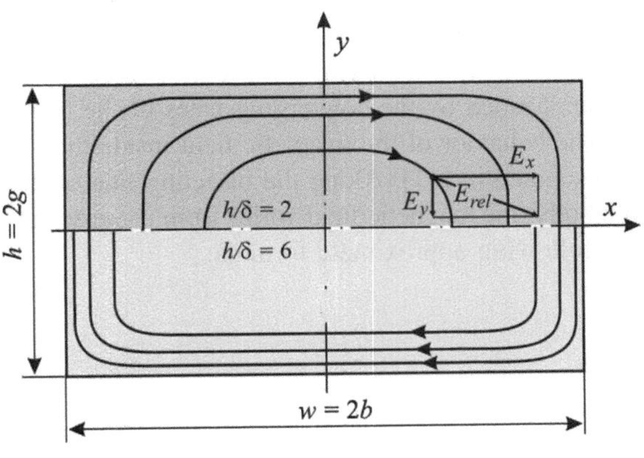

Fig. 2.27 Qualitative distributions of the electric field in the slab cross-section in case of skin-effect weak $(h/\delta = 2)$ or strong $(h/\delta = 6)$

The dimensionless value of the electric field intensity in a generic point of the cross-section can be determined with the relationship:

$$\mathrm{E_{rel}} = \frac{\delta}{H_e \rho} \sqrt{\dot{E}_x^2 + \dot{E}_y^2}.$$

The dimensionless distribution of the current density $\hat{J}$ can be obtained in a similar way.

The analysis of such distributions shows that on the corners at the coordinates $x = 0$, $y = 0$; $x = w$, $y = 0$; $x = w$, $y = h$; $x = 0$, $y = h$ (see Fig. 2.26), the current density is zero, while the maximum current density is located at the surface of the slab in the center of the wide side.

In case of strong skin effect (bottom part of Fig. 2.27), the current density is almost identical along the contour of the slab, except for the corners, where a decrease of the current density is observed for a length of (1.0–1.5) δ.

However, taking into account that in through heating the specific power in the slab is usually small, the zone of corners and the area near the narrow sides can be overheated, in comparison with the central region of the wide faces. This is due to the fact that, at low power densities, the temperature field in the slab is largely determined by the process of heat conduction. In fact, the heat is easier transmitted to the core of the slab from the wide faces than from corners and narrow faces (the thermal resistance from the wider face $R'_{th} = h/(\lambda w)$ is considerably lower than the thermal resistance $R''_{th} = w/(\lambda h)$ from the narrow faces). Overheating increases with the increase of the ratio w/h.

On the contrary, in case of through heating with weak skin effect and relatively low specific power, the region of weak current density covers the whole area of the narrow sides (upper part of Fig. 2.27), which leads to its under-heating. In this case, the lowest temperature of the cross section is reached in the corners. In the areas of the narrow sides and corners there is not enough power to provide the desired heating rate and to compensate for the flow of heat from this area towards the core and the heat losses from the body surface.

In paper [7] are given the equations of the fields H and E; they allow to evaluate the behavior of these functions within the slab, without the need to know the value of H_e on the surface of the body.

However this method is not accurate in computing E and H near the corners or angles formed by the converging faces of the heated body.

The behavior of the magnetic field inside the slab in the edge regions has been investigated in [11]. Along the bisecting line of the angle formed by two faces of the body, the magnetic field varies from its vertex towards the inside of the body by the following approximate formula:

$$H = H_e \cdot 2 \cdot e^{-(1-j)\frac{x\varphi_0}{2\delta}} \times \left[1 + \frac{1-j}{6} \cdot \frac{x}{\delta}\left(\frac{\varphi_0}{2}\right)^3 - \frac{1-j}{120} \cdot \frac{x}{\delta}\left(\frac{\varphi_0}{2}\right)^5 - \frac{j}{36} \cdot \left(\frac{\varphi_0}{2}\right)^6 + \cdots\right] \quad (2.97)$$

where x—distance from the vertex of the angle to the considered point along the bisector, m; φ_0—value of the angle in radians; δ—penetration depth, m.

From Eq. (2.97) it is evident that H varies along the bisector with an exponential law with an effective penetration depth $\delta_A = 2\delta/\varphi_0$.

For workpieces with the angle of converging faces $\varphi \leq \pi/2$, the magnetic field along the bisecting line decreases more slowly than along the normal to the sides of the angle, i.e. with $\delta_A = 2\delta/\varphi_0 > \delta$ (see Fig. 2.28a, b).

From Eq. (2.97) it is also clear that, when the module of H decreases of $e = 2.72$ times, the same skin-effect reached in the semi-infinite body at the depth δ, along the bisectrix φ_0, is obtained at the depth $\delta_A \approx 3.4 \cdot \delta/\varphi_0$.

In particular, in the case of a slab with rectangular cross-section, the effective penetration depth will be equal to $\delta_A \approx 2.16 \cdot \delta$.

For a body with angle $\varphi_0 \leq \pi/4$ it is $\delta_A \approx 4.3 \cdot \delta\delta_e$; for $\varphi_0 \leq \pi/8$ it is $\delta_A \approx 8.7 \cdot \delta$.

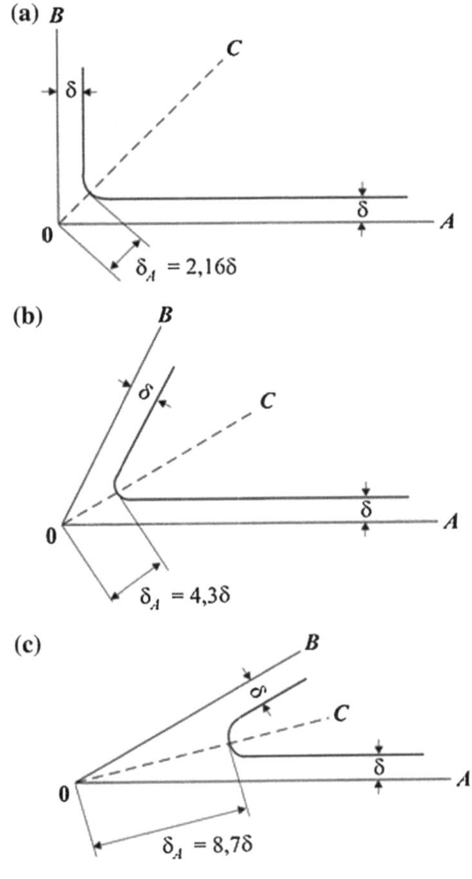

Fig. 2.28 Effective penetration depth δ_A of magnetic field in angle regions: **a** edges at 90°; **b** edges at 45° **c** edges at 22.5° (δ_A—effective penetration depth)

2.4.2 Energy in Slabs of Rectangular Cross-Section

The average energy absorbed in the slab per unit time is given by the flux of Poynting vector across the external surface S

$$\dot{\overline{S}} = \oint_s \frac{\dot{\overline{E}}_e \times \overline{H}_e^*}{2} d\overline{s}, \tag{2.98}$$

where $\dot{\overline{E}}_e$—vector of the electric field intensity; $\overline{H}_e^*$—conjugate of the vector of the magnetic field intensity at the surface of the slab.

After the introduction in Eq. (2.98) of the values $\dot{E}_{ex} = \dot{H}_e \cdot \rho$, $\dot{E}_{ey} = \dot{H}_e \cdot \rho$, and $\dot{H}_e$, with some mathematical manipulations the following solution is obtained for $\dot{S}$, which gives the active power P and the reactive power Q in the slab:

$$\dot{S} = P + jQ$$

with

$$P = H_e^2 \frac{\rho}{\delta}(h + w)\ell \cdot F_{ingot}\left(\frac{h}{\delta}; \frac{w}{h}\right), \tag{2.99}$$

$$Q = H_e^2 \frac{\rho}{\delta}(h + w)\ell \cdot G_{ingot}\left(\frac{h}{\delta}; \frac{w}{h}\right); \tag{2.100}$$

ℓ—length of an axial section of the infinite slab, m;

$$\begin{aligned} F_{ingot}\left(\frac{h}{\delta}; \frac{w}{h}\right) &= \frac{16}{\pi^2(h + w)} \\ &\times \left[\frac{h}{\delta}\sum_{m=1,3,\ldots}^{\infty} \frac{1}{m^2}\frac{r_m}{r_m^2 + s_m^2} \cdot \frac{\frac{s_m}{r_m}\sinh(r_m w) - \sin(s_m w)}{\cosh(r_m w) + \cos(s_m w)}\right. \\ &\left. + \frac{w}{\delta}\sum_{n=1,3,\ldots}^{\infty} \frac{1}{n^2}\frac{p_n}{p_n^2 + q_n^2} \cdot \frac{\frac{q_n}{p_n}\sinh(p_n h) - \sin(q_n h)}{\cosh(p_n h) + \cos(q_n h)}\right] \end{aligned} \tag{2.101}$$

$$\begin{aligned} G_{ingot}\left(\frac{h}{\delta}; \frac{w}{h}\right) &= \frac{16}{\pi^2(h + w)} \\ &\times \left[\frac{h}{\delta}\sum_{m=1,3,\ldots}^{\infty} \frac{1}{m^2}\frac{r_m}{r_m^2 + s_m^2} \cdot \frac{\sinh(r_m w) + \frac{s_m}{r_m}\sin(s_m w)}{\cosh(r_m w) + \cos(s_m w)}\right. \\ &\left. + \frac{w}{\delta}\sum_{n=1,3,\ldots}^{\infty} \frac{1}{n^2}\frac{p_n}{p_n^2 + q_n^2} \cdot \frac{\sinh(p_n h) + \frac{q_n}{p_n}\sin(q_n h)}{\cosh(p_n h) + \cos(q_n h)}\right] \end{aligned} \tag{2.102}$$

$$p_n = \frac{\pi n}{w}\sqrt{\frac{\sqrt{1+\frac{4}{(\pi n)^4}\left(\frac{w}{\delta}\right)^4}+1}{2}};$$

$$q_n = \frac{\pi n}{w}\sqrt{\frac{\sqrt{1+\frac{4}{(\pi n)^4}\left(\frac{w}{\delta}\right)^4}-1}{2}};$$

$$r_m = \frac{\pi m}{w}\sqrt{\frac{\sqrt{1+\frac{4}{(\pi m)^4}\left(\frac{h}{\delta}\right)^4}+1}{2}};$$

$$s_m = \frac{\pi m}{w}\sqrt{\frac{\sqrt{1+\frac{4}{(\pi m)^4}\left(\frac{h}{\delta}\right)^4}-1}{2}};$$

F_{ingot}, G_{ingot}—functions characterising active power and reactive power in the ingot.

The equivalent resistance R'_e and the reactance X'_e of the workpiece referred to the inductor clamps are defined by the relationships

$$P = R'_e I_1^2, \tag{2.103}$$

$$Q = X'_e I_1^2. \tag{2.104}$$

Since is $H_e = \left(\sqrt{2}N_1 I_1\right)/\ell$, then we can write:

$$R'_e = N_1^2 \rho \frac{2(h+w)}{\delta\ell} \cdot F_{ingot}\left(\frac{h}{\delta};\frac{w}{h}\right),$$

$$X'_e = N_1^2 \rho \frac{2(h+w)}{\delta\ell} \cdot G_{ingot}\left(\frac{h}{\delta};\frac{w}{h}\right).$$

Figures 2.29 and 2.30 give the diagrams of the functions F_{ingot} and G_{ingot}. The analysis of such functions shows the following:

- for values of $w/h \geq 1$, when h/δ tends to ∞:

$$\lim_{\frac{h}{\delta}\to\infty} F_{ingot}\left(\frac{h}{\delta};\frac{w}{h}\right) = 1 \quad \lim_{\frac{h}{\delta}\to\infty} G_{ingot}\left(\frac{h}{\delta};\frac{w}{h}\right) = 1$$

Hence, for a large increase of one linear dimension of the cross-section (e.g. the width w) in comparison with the penetration depth, the resistance and the reactance of the slab are the same of the semi-infinite body. In this case, it results:

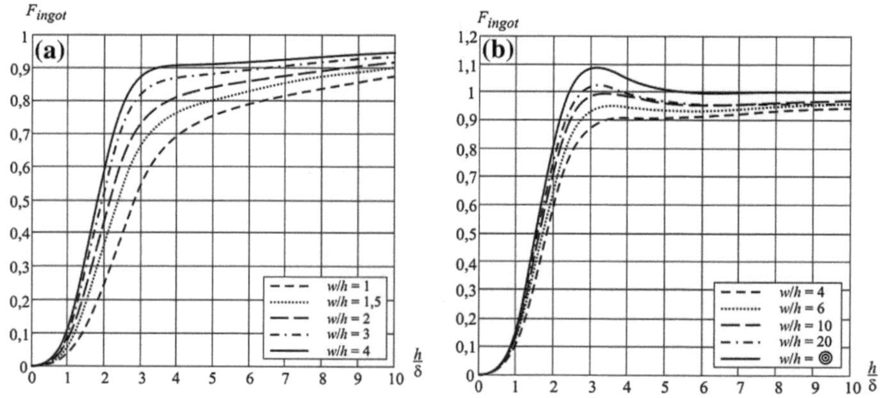

Fig. 2.29 **a** Diagrams of the function of active power F_{ingot}. **b** Diagrams of the function of active power F_{ingot}

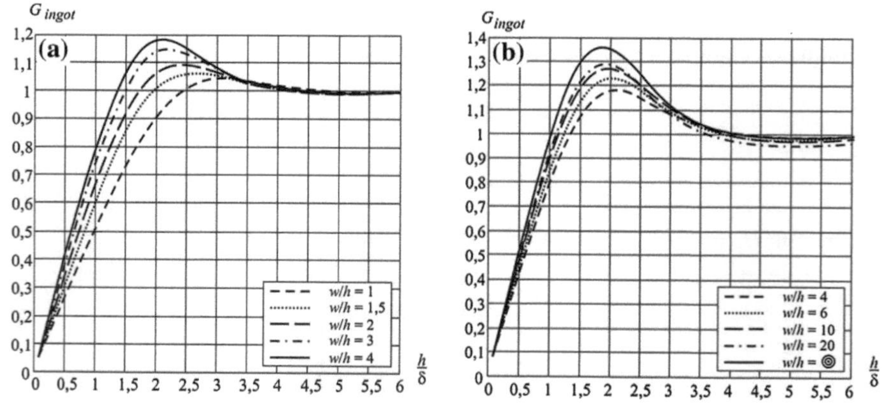

Fig. 2.30 **a** Diagrams of the function G_{ingot}, of reactive power. **b** Diagrams of the function G_{ingot}, of reactive power

$$\lim_{\frac{w}{\delta}\to\infty} F_{ingot}\left(\frac{h}{\delta};\frac{w}{h}\right) = \frac{\sinh\frac{h}{\delta} - \sin\frac{h}{\delta}}{\cosh\frac{h}{\delta} + \cos\frac{h}{\delta}};$$

$$\lim_{\frac{w}{\delta}\to\infty} G_{ingot}\left(\frac{h}{\delta};\frac{w}{h}\right) = \frac{\sinh\frac{h}{\delta} + \sin\frac{h}{\delta}}{\cosh\frac{h}{\delta} + \cos\frac{h}{\delta}}.$$

The equations for F_{ingot} and G_{ingot} are the same obtained for an infinite slab in the case of exciting field on one side (see Eq. 2.63).

- For practical calculations in case of strong skin effect, for $h/\delta \geq 5$, in [6] it is suggested to use the following approximation assuring an accuracy within 5 %.

$$F_{ingot}\left(\frac{h}{\delta};\frac{w}{h}\right) \approx 1 - \frac{2.5}{1+\frac{w}{h}}\cdot\frac{\delta}{h} \quad ; \quad G_{ingot}\left(\frac{h}{\delta};\frac{w}{h}\right) \approx 1.$$

It is interesting to analyse the distribution of the internal heating sources in the cross-section of a rectangular slab, given in the paper [13].

In contrast with paper [7], where the solutions for the slab were obtained in terms of Fourier series, in [13] the method of Galyorkin is used. This method provides an approximate solution, giving sufficient engineering accuracy.

Such distribution can be written as follows:

$$\begin{aligned} w(x,y) = \frac{\rho H_e^2}{\delta^2}\Bigg\{ & \left(p^2+q^2\right)\frac{\cosh(p-q)\frac{2x}{\delta} - \cos(p+q)\frac{2x}{\delta}}{\cosh(p-q)\frac{2b}{\delta} - \cos(p+q)\frac{2b}{\delta}} \\ & \times\left[1 + \frac{\cosh\frac{2y}{\delta}+\cos\frac{2y}{\delta}}{\cosh\frac{2g}{\delta}+\cos\frac{2g}{\delta}} \right. \\ & \left. - \frac{4\left(\cosh\frac{2y}{\delta}\cos\frac{2y}{\delta}\cosh\frac{2g}{\delta}\cos\frac{2g}{\delta} + \sinh\frac{2y}{\delta}\sin\frac{2y}{\delta}\sinh\frac{2g}{\delta}\sin\frac{2g}{\delta}\right)}{\cosh\frac{2g}{\delta}+\cos\frac{2g}{\delta}}\right] \\ & + \left(r^2+s^2\right)\frac{\cosh(r-s)\frac{2y}{\delta} - \cos(r+s)\frac{2y}{\delta}}{\cosh(r-s)\frac{2g}{\delta} + \cos(r+s)\frac{2g}{\delta}} \\ & \times\left[1 + \frac{\cosh\frac{2x}{\delta}+\cos\frac{2x}{\delta}}{\cosh\frac{2b}{\delta}+\cos\frac{2b}{\delta}} \right. \\ & \left. - \frac{4\left(\cosh\frac{2x}{\delta}\cos\frac{2x}{\delta}\cosh\frac{2b}{\delta}\cos\frac{2b}{\delta} + \sinh\frac{2x}{\delta}\sin\frac{2x}{\delta}\sinh\frac{2b}{\delta}\sin\frac{2b}{\delta}\right)}{\cosh\frac{2b}{\delta}+\cos\frac{2b}{\delta}}\right]\Bigg\} \end{aligned} \tag{2.105}$$

where the parameters p, q, r, s are given by the relationships:

$$r + js = \sqrt{\frac{2}{3 - \frac{b\dot{k}\tanh^2(b\dot{k})}{b\dot{k} - \tanh(b\dot{k})}}}, \tag{2.106}$$

$$p + jq = \sqrt{\frac{2}{3 - \frac{a\dot{k}\tanh^2(g\dot{k})}{a\dot{k} - \tanh(g\dot{k})}}}. \tag{2.107}$$

In Fig. 2.31 it is represented a family of curves of equal power density of the heating sources for large and small ratios of the thickness of the body to the penetration depth; as unit it is adopted the power density on the surface in the middle of the wide side of the cross section.

As seen from Fig. 2.31a, the curves follow the contour of the cross-section and are concentrated in a narrow surface layer. Increasing the aspect ratio of the cross-section, the distribution of power becomes even closer to the surface. This case

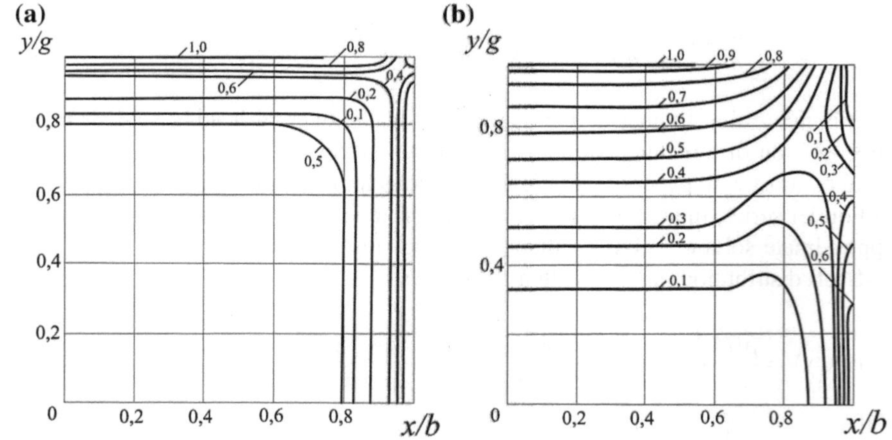

Fig. 2.31 Isolines of power density in slabs of rectangular cross-section (only one quarter of the cross-section is shown): **a** 2g/δ = 15, b/g = 1; **b** 2g/δ = 2, b/g = 5

occurs, for example during heating an aluminum slab of thickness 280 mm at industrial frequency.

The same dependence, corresponding for example to the case of heating a titanium slab of thickness of 200 mm at industrial frequency, is shown in Fig. 2.31b. In this case, the power density is not distributed in a uniform surface layer, but along narrow edges it is lower than along the large ones.

2.4.3 Direct Resistance Heating

In the study of the resistance-heating of workpieces with rectangular cross-section, a fundamental step is the solution of the electromagnetic problem, which allows to identify the main electromagnetic parameters, such as the current density distribution and the AC resistance and internal inductance.

In contrast with induction heating, where for the rectangular cross-section it was possible to obtain the analytical solution in a two-dimensional formulation from the knowledge of magnetic field at the surface of the workpiece [7], the derivation of an analytical solution for resistance heating is more difficult. A solution was proposed in paper [4], where analytical calculations are compared with numerical results.

However, the proposed method gives good results only for workpieces with square cross section.

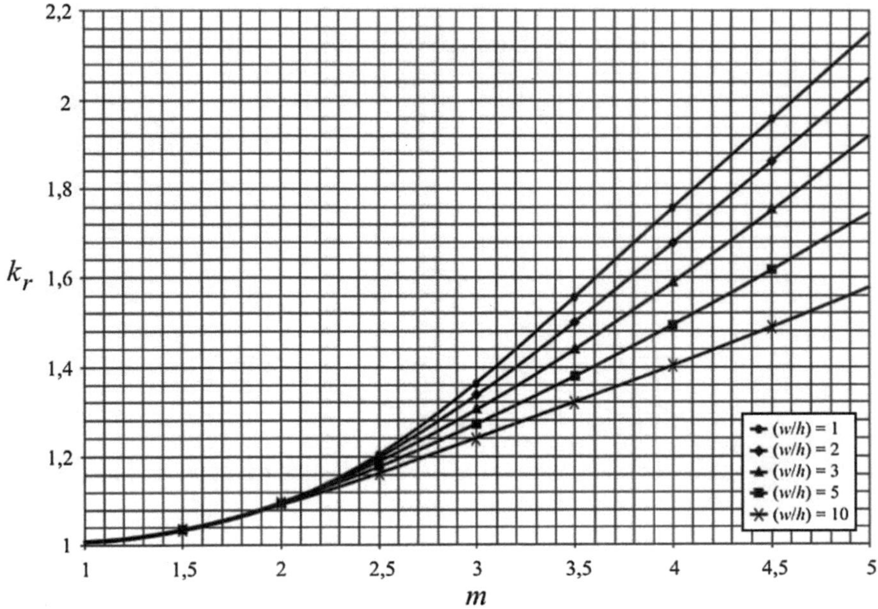

Fig. 2.32 Coefficient of AC resistance in a slab of rectangular cross-section as a function of w/h and m

Calculations made for aluminum and copper workpieces, have shown differences with numerical results of about 5 % for the current density and 1 % for the resistance. But when increasing the ratio *w/h*, this difference grows rapidly: to more than 10 % for the current density when $(w/h) > 1.5$, and for the resistance when $(w/h) > 3.5$.

Therefore this analytical method can not be recommended in the engineering practice for calculation of the resistance and the internal reactance in a wide range of geometrical parameters.

Therefore, in the following data obtained with numerical solutions are given [9].

Figures 2.32 and 2.33 give the diagrams representing k_r and k_x as a function of the ratio $m = \left(\sqrt{2}R_e\right)/\delta$, for different values of the relative width *w/h*, where $R_e = \sqrt{hw/\pi}$—is the equivalent radius of the slab cross-section.

The diagrams show that when w/h decreases k_r and k_x increase. These coefficients also increase with the increase of parameter *m*.

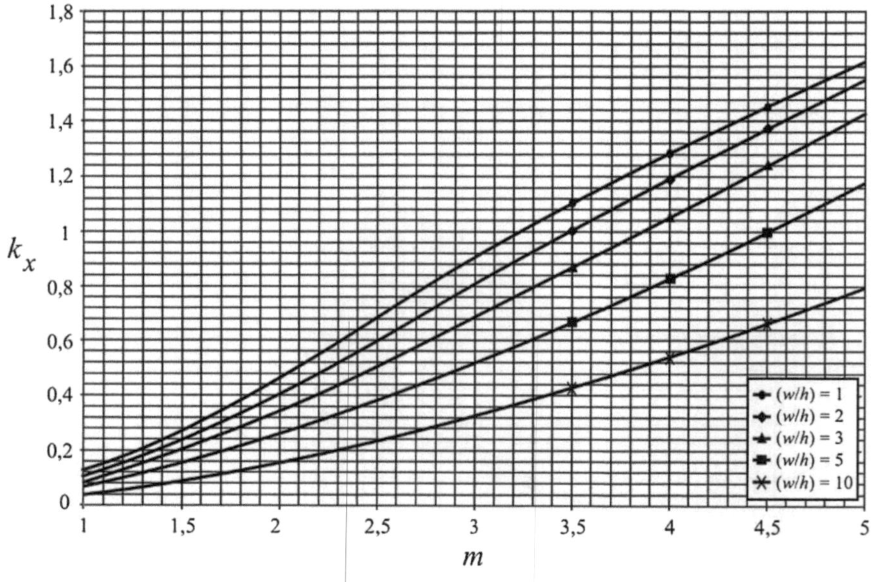

Fig. 2.33 Coefficient of AC internal reactance of a slab of rectangular cross-section as a function of w/h and m

2.5 Slab of Magnetic Material

2.5.1 Induction Heating

In case of strong skin effect in a slab made of ferromagnetic material, the functions F and G, here named as F_Φ and G_Φ can be calculated using the same formulas derived for the case of plane waves in a semi-infinite medium with variable μ_r:

$$F_\Phi \approx 1.32\left(1 - \frac{2.5}{1+\frac{w}{h}}\frac{\delta_e}{h}\right); \quad G_\Phi \approx 0.98. \tag{2.108}$$

These formulas in a ferromagnetic medium give an accuracy of 6 % for $h/\delta \geq 3$.

When skin effect is weak, like in the heating of ferromagnetic rods, rectangular sheets, strips workpieces at industrial frequency Eq. (2.108) gives fairly large errors.

For the general case, i.e. pronounced or weak skin effect, the functions F_Φ and G_Φ are given in Fig. 2.34 [6]. These plots have been obtained by solving the two-dimensional nonlinear problem by finite difference method. The calculations were performed for a strong magnetic external field, i.e. with $H_e \gg H_{cr}$.

The figures give also the values of F_Φ and G_Φ for slabs of square cross-section or infinite plates, with relative permeability μ_r = const. (dashed lines).

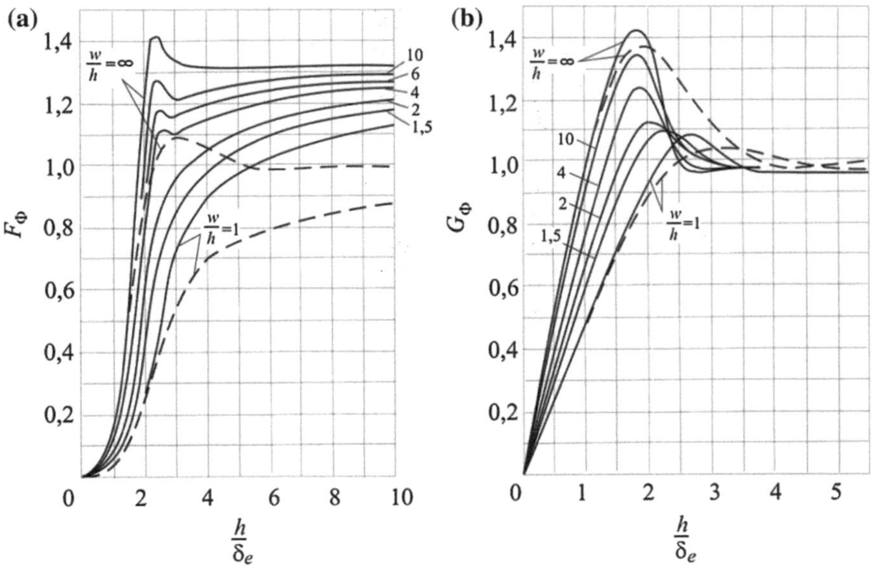

Fig. 2.34 Coefficients F_Φ and G_Φ of resistance (**a**) and internal reactance (**b**) in the induction heating of ferromagnetic workpieces of rectangular cross-section: *solid line* with $\mu_r = f(H)$; *dashed line* with μ_r = const

In case of weak skin effect ($h/\delta_e \le 2$—for square cross-section and $h/\delta_e \le 1.5$—for the plate) the diagrams of F_Φ and G_Φ with μ_r = var. and μ_r = const., are the same.

For $h/\delta_e \ge 2$ the variation of μ_r on the cross-section leads to an overall increase of F_Φ and a significant reduction of the width of the peaks of F_Φ and G_Φ.

The maximum of F_Φ is observed for $w/h = \infty$ in case of variable μ_r at $h/\delta_e \approx 2.5$. It decreases with the decrease of w/h and for $w/h = 4$ practically disappears.

2.5.2 Resistance Heating

This topic is dealt with in Sect. 4.1.

References

1. Davies, E. J. (1990). *Conduction and induction heating* (371 pp.). London: Peter Peregrinus Ltd.
2. Grinberg, G. A. (1948). *Selected problems of mathematical theory of electric and magnetic phenomena*. Moscow: USSR Academy of Science (in Russian).

3. Hegewaldt, F. (July/August 1961). *Induktives Oberflächenhärten* (pp. 434–456). Induktionsöfen: BBC Nachrichten (in German).
4. Inkin, A. I., Aliferov, A. I., & Blank, A. V. (2007). *Calculation of electrical parameters in resistance heating of workpieces with rectangular cross-section.* Elektrichestvo, *8*, 35–38 (in Russian).
5. Neiman, L. R. (1949). *Skin effect in ferromagnetic bodies* (190 pp.). Leningrad: Gostehnizdat (in Russian).
6. Nemkov, V., & Demidovich, V. B. (1988). *Theory and calculation of induction heating installations* (280 pp.). Leningrad: Energoatomiszdat. ISBN 5-283-04409-2 (in Russian).
7. Peisahovich, V. A. (1961). *Calculation of impedance of square and rectangular cross-section billets heated by induction, NITVH—industrial applications of high frequency currents* (Vol. 3, pp. 5–18). Leningrad: Mashinostroenie (in Russian).
8. Sarač, V. (1981). *Influence of current type and characteristics of power supply on the direct heating of steel billets* (20 pp.). Moscow: MEI (in Russian).
9. Silvester, P. (June 1967). *AC resistance and reactance of isolated rectangular conductors.* IEEE Transactions, Vol. PAS-86 (6), 770–774.
10. Vainberg, A. M. (1967). *Induction melting furnaces* (415 pp.). Moscow: Energia (in Russian).
11. Vologdin, V. P. (1947). *Surface induction hardening* (291 pp.). Moscow: Oborongiz (in Russian).
12. Zaimovskiĭ, A. S., & Čudnovskaja, Z. (1957). *Magnetic materials.* Moscow: GEI (ГЭИ) (in Russian).
13. Zimin, L. S. (1973). *Features of heating rectangular bodies, application of high-frequency currents in electroheat* (pp. 25–34). Leningrad: Mashinostroenie (in Russian).

Chapter 3
Electromagnetic Field in Cylindrical Bodies

Abstract The third chapter deals with the induction and direct resistance heating of cylindrical workpieces. For the induction heating cases, solenoidal inductors are considered, leading to an axisymmetric configuration. The chapter is subdivided into 4 sections. In Sect. 3.1, the induction and direct resistance heating are considered for solid cylindrical workpieces made of nonferrous metals. Analytical solutions of the electromagnetic problem give the distributions of the heating sources and main integral parameters, like equivalent resistance and reactance. The analytical solutions lead to quite complex formulas that can be easily evaluated thanks to dimensionless diagrams. In Sect. 3.2, analytical solutions are provided for nonlinear material, like ferromagnetic steel. The Sects. 3.3 and 3.4 deal with the heating of tubes, made of nonferrous and ferrous materials respectively. For induction heating cases, the use of internal or external inductors is considered. The main integral quantities can be easily evaluated with diagrams that are function of the dimensionless thickness and radii, internal and external, of the tube.

3.1 Solid Cylindrical Workpieces of Homogeneous Material

The heating of cylindrical workpieces can be analyzed in two different coordinate systems [1, 7–12]:

1. In induction heating, we consider the magnetic field intensity $\overline{H}$ directed along z, and electric field $\overline{E}$ tangential to the circumference of the cylinder, along the azimuthal coordinate φ (Fig. 3.1).
2. In resistance heating, we consider the electric field intensity $\overline{E}$ directed along the axis z of the cylinder and the magnetic field intensity $\overline{H}$ tangential to the circumference of the cylinder, i.e. directed along φ (Fig. 3.2).

S. Lupi et al., *Induction and Direct Resistance Heating*,
DOI 10.1007/978-3-319-03479-9_3

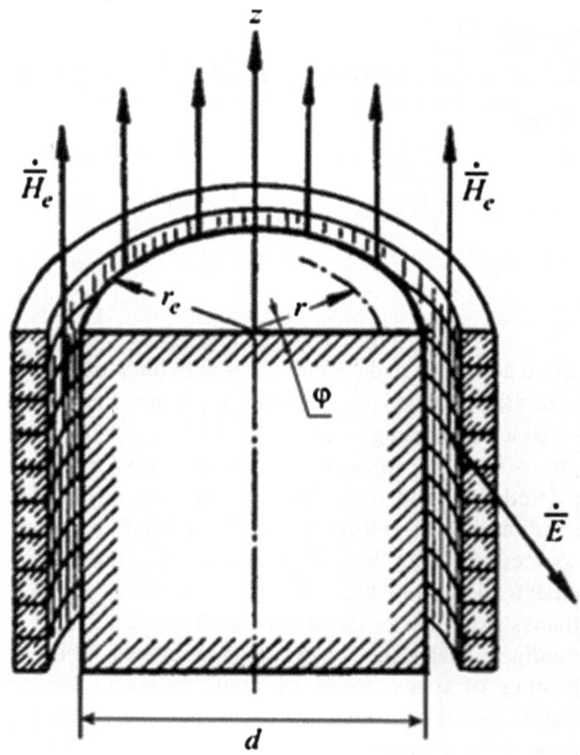

Fig. 3.1 Electric and magnetic field vectors in induction heating of cylindrical workpieces: *1* inductor; *2* workpiece

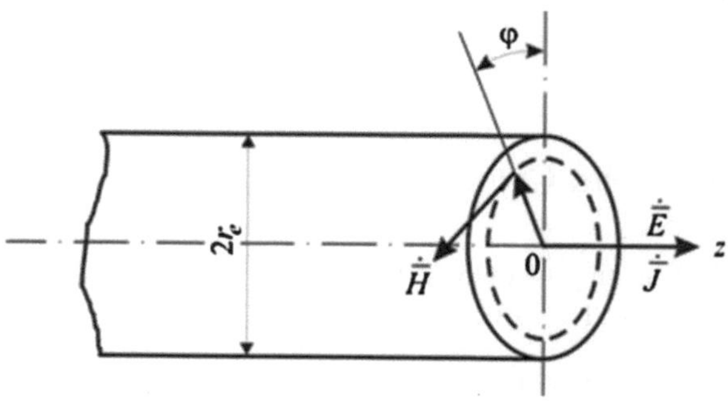

Fig. 3.2 Directions of magnetic and electric field vectors in resistance heating of a cylindrical body

3.1.1 Induction Heating of Cylindrical Workpieces

In induction heating applications different geometrical configurations are of interest:

(a) electromagnetic wave tangent to the external surface of a (solid or hollow) cylinder, like in:

- induction crucible furnaces for metal melting;
- through-heating of cylindrical workpieces for heat treatment or before mechanical hot forming;

- heating for the heat treatment of bearing rings;
- heating of tubes for pressure machining;
- heating of pipes for heat treatment;

(b) electromagnetic wave tangent to the inner surface of a hollow cylinder, e.g. in:

- hardening of the inner surface;
- heating of pipes for heat treatment (stress releasing);
- heating of molten metal in the channel of an induction channel furnace.

3.1.1.1 Distribution of Electromagnetic Field

The length of the inductor-load system (Fig. 3.1) is assumed infinite along the coordinate *z*. We will use the following notation: $\overline{H}_e, \overline{E}_e$- vectors of magnetic and electric fields intensity in the gap between the inductor and the heated body; $\overline{H}, \overline{E}$—the vectors inside the body.

As in the case of a plane electromagnetic wave, the equations describing the electromagnetic field in the body are Maxwell's equations (2.1).

We solve this equations system with the following assumptions:

1. The electromagnetic wave propagates in a conductive medium (the heated body), i.e.

$$\overline{J} > > \frac{\partial \overline{D}}{\partial t} \quad \text{and} \quad \frac{\partial \overline{D}}{\partial t} \approx 0.$$

2. The body is constituted by a homogeneous medium, with constant physical properties: $\rho = \text{const.}$, $\mu_r = \text{const.}$
3. The inductor current is sinusoidal and, as a consequence, also $\overline{H}, \overline{B}, \overline{E}$ are sinusoidal in time:

$$\overline{H} = \overline{H}_m \sin \omega t, \overline{B} = \overline{B}_m \sin \omega t, \overline{E} = \overline{E}_m \sin(\omega t + \varphi).$$

4. The sinusoidal time variation of vectors $\overline{H}, \overline{B}, \overline{E}$, allows us to express all quantities in the complex form:

$$\dot{\overline{H}} = \overline{H}_m \cdot e^{j\omega t}, \dot{\overline{B}} = \overline{B}_m \cdot e^{j\omega t}, \dot{\overline{E}} = \overline{E}_m \cdot e^{j(\omega t + \varphi)}.$$

With these hypotheses the system (2.1) of Maxwell's equations can be rearranged in the following form:

$$\nabla^2 \dot{\overline{H}} = j \frac{\omega \mu \dot{\overline{H}}}{\rho}, \tag{3.1}$$

with ∇^2 Laplacian operator, which in cylindrical coordinates is:

$$\nabla^2 = \frac{1}{r}\frac{\partial}{\partial r}\left(r\frac{\partial}{\partial r}\right) + \frac{1}{r^2}\frac{\partial^2}{\partial \varphi^2} + \frac{\partial^2}{\partial z^2}.$$

The magnetic field intensity $\dot{\overline{H}}$ is independent on the azimuthal coordinate φ, i.e. $\partial\dot{\overline{H}}/\partial\varphi = 0$; moreover it does not depend on z, $\partial\dot{\overline{H}}/\partial z = 0$. Therefore, Eq. (3.1) for the vector $\dot{\overline{H}}$ can be rewritten as follows:

$$\frac{\partial^2\dot{\overline{H}}}{\partial r^2} + \frac{1}{r}\frac{\partial\dot{\overline{H}}}{\partial r} - j\frac{\omega\mu\dot{\overline{H}}}{\rho} = 0. \quad (3.2)$$

Introducing the notations:

$$\begin{cases} \beta^2 = -j\frac{\omega\mu_0\mu}{\rho}r_e^2 = -jm^2; m = \frac{\sqrt{2}r_e}{\delta}; \xi = \frac{r}{r_e} \\ \delta = \sqrt{\frac{2\rho}{\omega\mu_0\mu_r}} \approx 503\sqrt{\frac{\rho}{\mu_r f}} - \text{penetration depth [m]} \end{cases}$$

we obtain:

$$\frac{\partial^2\dot{\overline{H}}}{\partial \xi^2} + \frac{1}{\xi}\cdot\frac{\partial\dot{\overline{H}}}{\partial \xi} + \beta^2\dot{\overline{H}} = 0.$$

This equation represents the Bessel differential equation of order zero. The general solution of this equation can be written in the form

$$\dot{\overline{H}} = \dot{\overline{C}}_1 \cdot J_0\left(\sqrt{-j}m\xi\right) + \dot{\overline{C}}_2 \cdot Y_0\left(\sqrt{-j}m\xi\right), \quad (3.3)$$

where $J_0(u)$ and $Y_0(u)$—are Bessel functions of order zero, of first and second type respectively, of the complex argument u; $\dot{\overline{C}}_1, \dot{\overline{C}}_2$—integration constants.
Since, from system (2.1) is

$$\dot{\overline{E}} = -u_\varphi \cdot \rho \cdot \frac{\partial\dot{H}}{\partial r} = -u_\varphi \cdot \frac{\rho}{R} \cdot \frac{\partial\dot{H}}{\partial \xi},$$

we can write:

$$\dot{\overline{E}} = u_\varphi \cdot \sqrt{-j} \cdot \frac{\sqrt{2}\rho}{\delta}\left[\dot{\overline{C}}_1 \cdot J_1\left(\sqrt{-j}m\xi\right) + \dot{\overline{C}}_2 \cdot Y_1\left(\sqrt{-j}m\xi\right)\right], \quad (3.4)$$

with: $J_0'(u) = \frac{\partial[J_0(u)]}{\partial u} = -J_1(u), Y_0'(u) = \frac{\partial[Y_0(u)]}{\partial u} = -Y_1(u)$, and $J_1(u), Y_1(u)$—modified Bessel functions of order 1 and first and second type, of argument u.

The constants $\dot{\overline{C}}_1$ and $\dot{\overline{C}}_2$ can be determined with the boundary conditions:

$$\begin{cases} \dot{\overline{H}} \neq \infty & for \quad \xi = 0 \\ \dot{\overline{H}} = \dot{\overline{H}}_e & for \quad \xi = 1 \end{cases}.$$

From the first one, being $Y_0(0) = -\infty$, it results $\dot{\overline{C}}_2 = 0$, while from the second one we obtain $\dot{\overline{C}}_1 = \dot{\overline{H}}/J_0(\sqrt{-j}m)$.

The results give the distribution of the complex vectors of the magnetic field, electric field and induced current density as a function of the nondimensional radius:

$$\dot{H} = \dot{H}_e \frac{J_0(\sqrt{-j}m\xi)}{J_0(\sqrt{-j}m)}, \tag{3.5}$$

$$\dot{E} = \sqrt{-j} \cdot \dot{H}_e \cdot \frac{\rho}{\delta} \cdot \sqrt{2} \cdot \frac{J_1(\sqrt{-j}m\xi)}{J_0(\sqrt{-j}m)}, \tag{3.6}$$

$$\dot{J} = \frac{\dot{E}}{\rho} = \sqrt{-j} \cdot \dot{H}_e \cdot \frac{\sqrt{2}}{\delta} \cdot \frac{J_1(\sqrt{-j}m\xi)}{J_0(\sqrt{-j}m)}. \tag{3.7}$$

The above solutions show that the moduli of the complex vectors of electric field intensity $\dot{E}$ and magnetic field intensity $\dot{H}$, depend not only on the relative radius $\xi = r/R$, but also on the parameter $m = \sqrt{2}R/\delta$. Moreover, the variations of the moduli $|\dot{E}|$ and $|\dot{H}|$ along the radius are not the same.

Figures 3.3 and 3.4 give the ratios between $|\dot{E}|, |\dot{J}|, |\dot{H}|$ and their values at the surface of the workpiece, as a function of the relative radius $\xi = r/R$ and the parameter $m = \sqrt{2}r_e/\delta$.

In all cases along the axis of the cylinder ($\xi = 0$), the moduli of $|\dot{E}|$ and $|\dot{J}|$ are equal to 0, while $|\dot{H}| \neq 0$. For low values of $m = \sqrt{2}r_e/\delta$, the modules of $|\dot{E}|$ and $|\dot{J}|$ decrease towards the axis almost linearly; for high values of m they decreases sharply near the surface.

The magnetic field intensity $|\dot{H}|$ for low values of m has a distribution nearly uniform, as shown by the curve of $|\dot{H}|$ for m = 1.

For $m \geq 10$ the curves of $|\dot{H}/\dot{H}_e|, |\dot{E}/\dot{E}_e|$ and $|\dot{J}/\dot{J}_e|$ are exponentials (dashed lines), as in the semi-infinite body.

The wave impedance of the cylinder is:

$$\begin{aligned} \dot{Z}_{e1} = R_{e1} + jX_{e1} = \frac{\dot{E}_e}{\dot{H}_e} &= \sqrt{-j} \cdot \frac{\rho}{\delta} \cdot \sqrt{2} \cdot \frac{J_1(\sqrt{j}m)}{J_0(\sqrt{j}m)} \\ &= \frac{\rho}{\delta} \cdot (F_{cyl} + jG_{cyl}), \end{aligned} \tag{3.8}$$

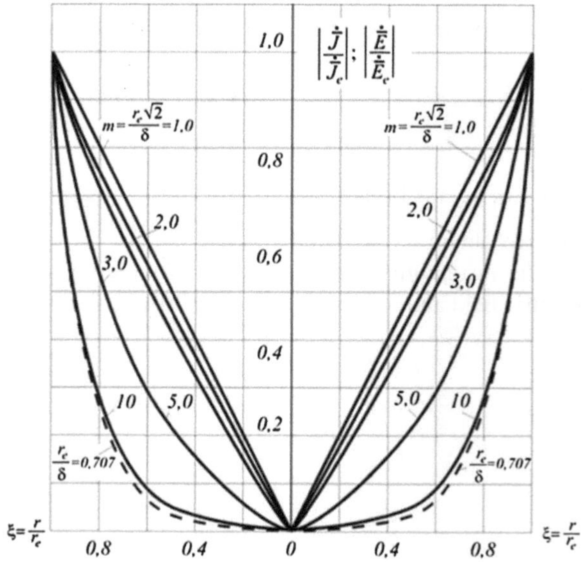

Fig. 3.3 Radial distributions in the cylinder of ratios $|\dot{E}/\dot{E}_e|$ of electric field intensities and $|\dot{J}/\dot{J}_e|$ of current densities [e—indicates surface values; *dotted lines* diagram of semi-infinite body]

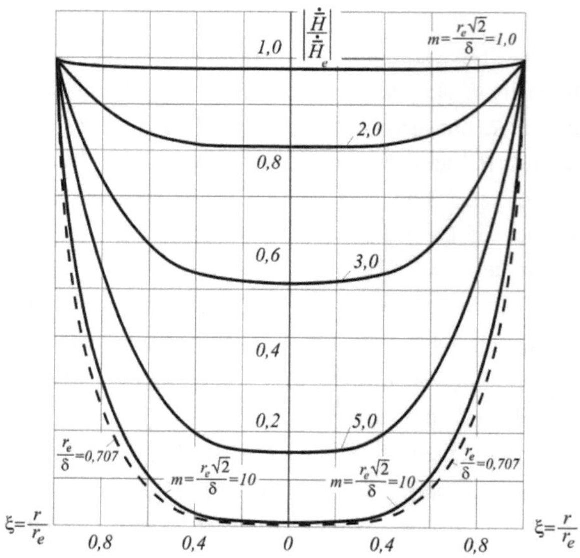

Fig. 3.4 Radial distributions in the cylinder of $|\dot{H}/\dot{H}_e|$ [e—indicates surface values; *dotted lines* diagram for the semi-infinite body]

with

$$F_{cyl} = \sqrt{2}\frac{ber'm \cdot ber\, m + bei'm \cdot bei\, m}{ber^2m + bei^2m}$$

$$G_{cyl} = \sqrt{2}\frac{bei'm \cdot ber\, m - ber'm \cdot bei\, m}{ber^2m + bei^2m}$$

$m = \sqrt{2}r_e/\delta$, *ber, bei, ber', bei'*—real and imaginary components of the Bessel functions of order zero given by the relationships

$$J_0\left(\sqrt{-j}u\right) = ber\,u + j\,bei\,u;$$
$$-\sqrt{-j}J_1\left(\sqrt{-j}u\right) = ber'u + j\,bei'u.$$

3.1.1.2 Energy in Solid Metallic Cylinder

The electromagnetic wave carries energy which can be evaluated by the Poynting's vector $\overline{S} = \left[\overline{E} \times \overline{H}\right]$.

If the modules of the electric and magnetic fields are sinusoidal in time, then the average value in a period of the energy flux vector has the form

$$\dot{\overline{S}} = \frac{1}{2}\left[\dot{\overline{E}} \times \dot{\overline{H}}^*\right], \tag{3.9}$$

with $\dot{\overline{H}}^*$—complex conjugate of the vector $\dot{\overline{H}}$.

Introducing in Eq. (3.9) the expressions (3.5) and (3.6) evaluated for $\xi = 1$, we obtain the complex vector of the flux of energy through the surface of a solid homogeneous conducting cylinder

$$\dot{\overline{S}}_e = -u_r\dot{H}_e^2\frac{\rho}{2\delta}\left(F_{cyl} + jG_{cyl}\right). \tag{3.10}$$

Since the total energy flowing through the external surface of the cylinder is completely released in it, then in order to determine the energy transferred to the cylinder it is sufficient to evaluate the flux through its surface.

Therefore, the flux of energy through the surface of the cylinder is:

$$\dot{S}_e = \dot{H}_e^2\frac{\rho}{2\delta}\left(F_{cyl} + jG_{cyl}\right),$$

with F_{cyl} and G_{cyl}—functions that characterize the active and reactive component of the energy in the solid cylinder.

In Fig. 3.5 the values of F_{cyl} and G_{cyl} are given as a function of $m = \sqrt{2}r_e/\delta$. For $m = \sqrt{2}r_e/\delta \geq 10$ is $F_{cyl} \approx G_{cyl} \approx 1.0$, and the values of the flux of energy $\dot{S}_e$ are the same obtained for a semi-infinite body.

The real and imaginary components of $\dot{S}_e$, give the active and reactive power flowing through the unit surface of the cylinder (W/m^2 and Var/m^2):

$$\begin{aligned} p &= H_e^2\frac{\rho}{\delta}F_{cyl} = 2 \cdot 10^{-3} \cdot (N_1 I)^2 \cdot \sqrt{\rho\mu_r f} \cdot F_{cyl}, \\ q &= H_e^2\frac{\rho}{\delta}G_{cyl} = 2 \cdot 10^{-3} \cdot (N_1 I)^2 \cdot \sqrt{\rho\mu_r f} \cdot G_{cyl}. \end{aligned} \tag{3.11}$$

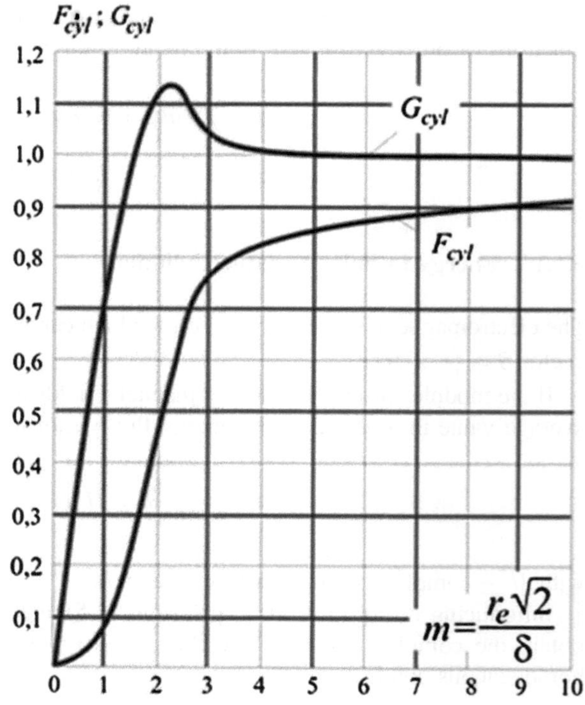

Fig. 3.5 Coefficients F_{cyl} and G_{cyl} as a function of $m = \sqrt{2}r_e/\delta$

In engineering calculations, instead of the power transmitted to the cylinder per unit surface, is often used the power per unit length of the cylinder, i.e.:

$$\begin{aligned} P_1 &= 2\pi r_e \cdot p = 6.3 \cdot d \cdot (N_1 I)^2 \cdot \sqrt{\rho \mu_r f} \cdot F_{cyl}, \\ Q_1 &= 2\pi r_e \cdot q = 6.3 \cdot d \cdot (N_1 I)^2 \cdot \sqrt{\rho \mu_r f} \cdot G_{cyl}, \end{aligned} \tag{3.11a}$$

with $d = 2r_e$—diameter of the cylinder, m.

Therefore the equivalent impedance of the cylindrical load of axial length ℓ seen at the inductor clamps is

$$\dot{Z}_{e1} = R_{e1} + jX_{e1} = N^2 \frac{\rho}{\delta} \left(F_{cyl} + jG_{cyl}\right) \frac{\pi d}{\ell} \tag{3.12}$$

The power factor is:

$$\cos\varphi = \frac{p_1}{\sqrt{p_1^2 + q_1^2}} = \frac{1}{\sqrt{1 + \left(\frac{G_{cyl}}{F_{cyl}}\right)^2}}. \tag{3.13}$$

Figure 3.6 shows the diagram of the power factor, as a function of $m = \sqrt{2}r_e/\delta$.

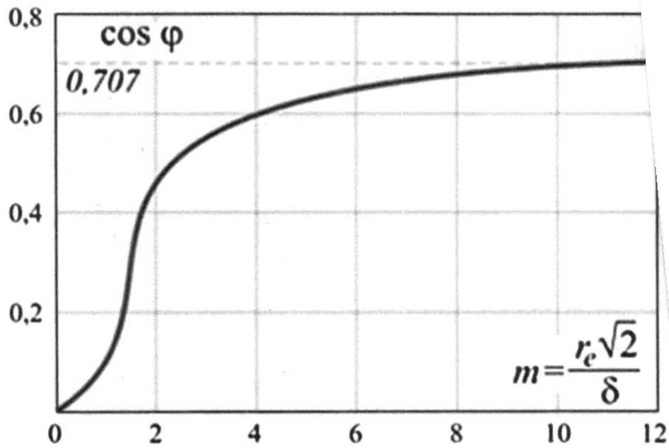

Fig. 3.6 Power factor in induction heating of solid cylindrical workpieces as a function of *m*

For values $m \geq 10$ the power factor tends to 0.707, like in the semi-infinite body. When the argument decreases, for $m \to 0$ the power factor tends to 0.

For small values of *m*, the induction heating is not effective, since the body becomes practically "*transparent*" to the electromagnetic wave.

3.1.2 Direct Resistance Heating

In the direct resistance heating of "long" straight cylindrical workpieces, electrical and magnetic quantities (current density, electric and magnetic field intensities, internal power sources) have one-dimensional radial distribution in the cross-section.

We consider now the flow of an alternating current in a metallic homogeneous cylindrical workpiece with the following assumptions:

1. The workpiece has infinite length. In this way the end effects, due to the uneven current density distribution near the contact system, are neglected. All field quantities, current density, electric and magnetic field intensity, are assumed to be dependent only on the radial position.
2. The same quantities are sinusoidal in time.
3. Relative permeability μ_r and resistivity ρ of the workpiece are constant.

In these hypotheses Maxwell's equations (2.1) can be written as

$$rot\,\dot{H} = \frac{\dot{E}}{\rho},\; rot\,\dot{E} = -j\,\omega\mu_r\mu_0\dot{H} \tag{3.14}$$

where $\dot{H}, \dot{E}$—complex magnetic field and electric field intensities, A/m and V/m; ρ—resistivity, Ohm·m; μ_r—relative magnetic permeability; $\mu_0 = 4\pi \cdot 10^{-7}$-magnetic constant of vacuum, H/m; $\omega = 2\pi f$—frequency pulsation of current, rad/s; *f*—frequency, Hz.

l coordinates system, Maxwell's equation can be rewritten as

$$\begin{cases} \dfrac{d\dot{H}}{dr} + \dfrac{1}{r}\dot{H} = \dfrac{\dot{E}}{\rho} \\ \dfrac{d\dot{E}}{dr} = j\omega\mu_r\mu_0\dot{H} \end{cases} \tag{3.15}$$

: second equation one can write

$$\frac{d\dot{H}}{dr} = \frac{1}{j\omega\mu_r\mu_0}\frac{d^2\dot{E}}{dr^2}.$$

ucing this equation in the first of (3.15), and taking into account that is we obtain

$$\frac{d^2\dot{J}}{dr^2} + \frac{1}{r}\frac{d\dot{J}}{dr} - j\frac{\omega\mu_r\mu_0}{\rho}\dot{J} = 0. \tag{3.16}$$

e introduce again the parameters

$$\xi = \frac{r}{r_e}; \quad k^2 = -j\frac{\omega\mu_r\mu_0}{\rho}r_e^2 = -jm^2; \quad m = \frac{\sqrt{2}r_e}{\delta}. \tag{3.17}$$

d Eq. (3.16) can be written in the form

$$\frac{d^2\dot{J}}{d\xi^2} + \frac{1}{\xi}\frac{d\dot{J}}{d\xi} + k^2\dot{J} = 0. \tag{3.18}$$

As known, the solution of Eq. (3.18) is

$$\dot{J} = \dot{C}_1 J_0(k\xi) + \dot{C}_2 Y_0(k\xi), \tag{3.19}$$

with J_0, Y_0—Bessel functions of order zero, of first and second kind.

The integration constants $\dot{C}_1$ and $\dot{C}_2$ can be determined with the boundary conditions

$$\begin{array}{l|l} \dot{J} \neq \infty & \text{for } \xi = 0, \\ \dot{J} = \dot{J}_e & \text{for } \xi = 1. \end{array}$$

Since it is $Y_0(0) = \infty$, it must be $\dot{C}_2 = 0$, while for $\xi = 1$ it is $C_1 = \dot{J}_e/\dot{J}_0(k)$, with $\dot{J}_e$—complex value of the current density at the surface of the cylinder.

Introducing the integration constants in Eq. (3.19), we obtain

$$\dot{J} = \dot{J}_e \frac{J_0(k\xi)}{J_0(k)} = \dot{J}_e \frac{J_0(\sqrt{-j}m\xi)}{J_0(\sqrt{-j}m)}. \qquad (3.20)$$

Since the argument of Eq. (3.20) is a complex value, on the basis of the relationship

$$J_0\left(\sqrt{-j}x\right) = ber\, x + j\, bei\, x$$

we can obtain the formulae for the complex current density and its modulus

$$\begin{cases} \dot{J} = \dot{J}_e \dfrac{ber(m\xi) + j\, bei(m\xi)}{ber(m) + j\, bei(m)} \\ \dfrac{|\dot{J}|}{|\dot{J}_e|} = \sqrt{\dfrac{ber^2(m\xi) + bei^2(m\xi)}{ber^2(m) + bei^2(m)}} \end{cases}. \qquad (3.21)$$

This formula can be written also in a different form, [1]:

$$\dot{J} = \dot{J}_e \frac{M(m\xi)e^{j\Phi(m\xi)}}{M(m)e^{j\Phi(m)}},$$

with

$$M(x) = \left(ber^2 x + bei^2 x\right)^{1/2},$$
$$\Phi(x) = arctg(bei\, x / ber\, x).$$

The above equations show that when the electromagnetic wave penetrates inside the cylinder, both module and phase angle of the current density vary.

Figure 3.7 illustrates the change of module and phase angle as a function of the distance x from the surface, for different values of m.

Figure 3.8 gives the diagrams of the relative current density along the radius for different values of m. For $m \leq 1$ the distribution of the current density is practically uniform, like in the DC case.

The total current flowing in the workpiece can be determined by the Ampere's law

$$\dot{I} = 2\pi r_e \dot{H}_e, \qquad (3.22)$$

with $\dot{H}_e$—magnetic field intensity at the surface of the cylinder, A/m.

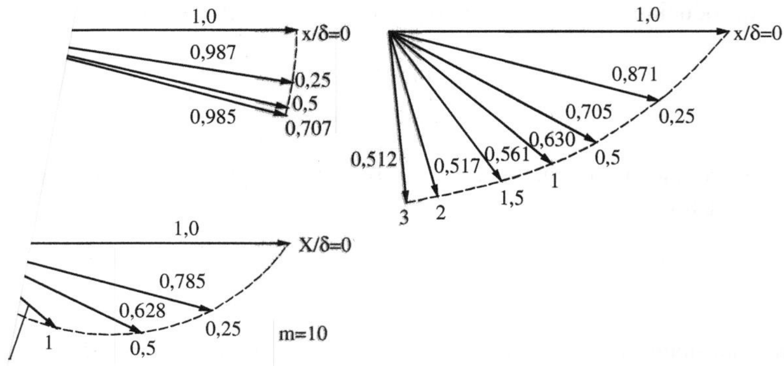

ctor representation of the variation of the current density as a function of the relative the workpiece for different values of $m = \sqrt{2}r_e/\delta$ (x-distance from the surface)

Normalized on along radius of n cylindrical ces, for different of m

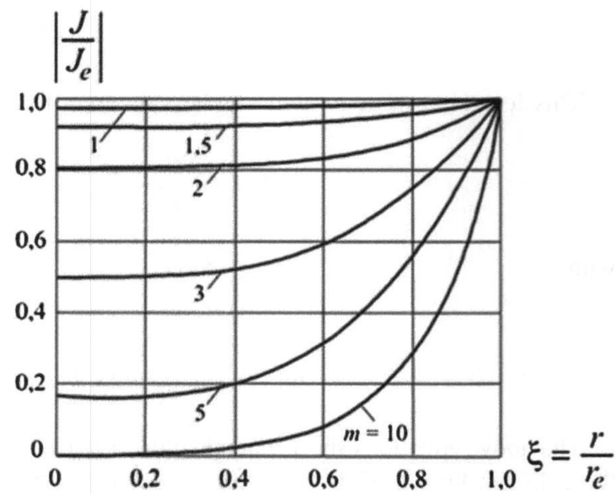

Since it is

$$J_0'(u) = \frac{dJ_0(u)}{du} = -J_1(u),$$

from Eq. (3.15) we obtain:

$$\dot{H} = j\sqrt{-j}\frac{\dot{J}_e}{\sqrt{2}}\delta\frac{J_1(\sqrt{-j}m\xi)}{J_0(\sqrt{-j}m)}. \tag{3.23}$$

Therefore, from Eq. (3.20) we have:

$$\dot{J}_e = \sqrt{-j}\frac{m}{2}\left(\frac{\dot{I}}{\pi R^2}\right)\frac{J_0(\sqrt{-j}m)}{J_1(\sqrt{-j}m)}. \tag{3.24}$$

and

$$\dot{J} = \sqrt{-j}\frac{m}{2}\left(\frac{\dot{I}}{\pi R^2}\right)\frac{J_0(\sqrt{-j}m\xi)}{J_1(\sqrt{-j}m)}. \tag{3.25}$$

The specific power per unit volume at the surface of the workpiece is $w_e = \rho|J_e|^2$. Therefore, from Eq. (3.25) we can write:

$$w = w_e\left(\frac{|\dot{J}|}{|\dot{J}_e|}\right)^2 = w_e\frac{ber^2(m\xi) + bei^2(m\xi)}{ber^2(m) + bei^2(m)}. \tag{3.26}$$

The relative distributions of heating sources along the radius are represented in Fig. 3.9.

Indicating with $w_{dc} = \rho\left(I/\pi r_e^2\right)^2$ the DC volume power density, from Eq. (3.24) it results

$$\frac{w_e}{w_{dc}} = \frac{m^2}{4}\frac{ber^2(m) + bei^2(m)}{ber'^2(m) + bei'^2(m)} = w^*(m). \tag{3.27}$$

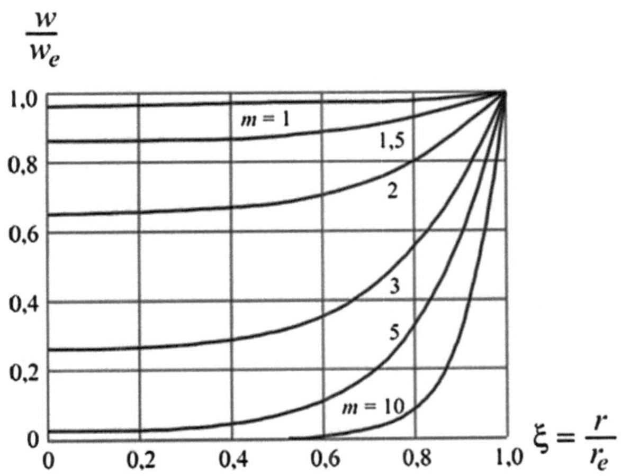

Fig. 3.9 Radial distribution of w/w_e in a cylindrical workpiece, as a function of the relative radius ξ and the parameter m

The average value in the workpiece cross-section of the power density distribution can be determined by:

$$
\begin{aligned}
w_{av} &= \frac{1}{\pi R^2} \int_{r=0}^{r=R} 2\pi R w(r) dr = 2 \int_{\xi=0}^{\xi=1} \xi w(\xi) d\xi \\
&= w_e \cdot \left\{ \frac{2(ber\, m\, bei' m - bei\, m\, ber' m)}{m(ber^2 m + bei^2 m)} \right\} = w_e \cdot k_r
\end{aligned}
\tag{3.28}
$$

The values of the functions $w^*(m)$ and k_r are given in Table 3.1.

Fundamental quantities are the equivalent impedance $\dot{Z}_{e1}$, the resistance R_{ac1} and the internal reactance X_{i1} of the workpiece per unit length.

This impedance can be evaluated as follows

$$
\dot{Z}_{e1} = R_{ac1} + jX_{i1} = \frac{\dot{E}_e}{\dot{I}} = R_{dc1}(k_r + jk_x), \tag{3.29}
$$

with: $\dot{E}_e$—electric field intensity at the surface of the workpiece; $R_{dc1} = \rho/(\pi r_e^2)$—DC resistance of unit length of the workpiece; k_r —skin-effect coefficient of the solid cylindrical workpiece:

$$
k_r = \frac{R_{ac1}}{R_{dc1}} = \frac{m(ber\, m\, bei' m - bei\, m\, ber' m)}{2(ber^2 m + bei^2 m)}; \tag{3.29a}
$$

k_x—skin-effect coefficient of the internal reactance:

$$
k_x = \frac{X_{i1}}{R_{dc1}} = \frac{m(ber\, m\, bei' m + bei\, m\, ber' m)}{2(ber^2 m + bei^2 m)} \tag{3.29b}
$$

The total active power in the workpiece per unit length, thus becomes

$$
P_1 = R_{ac1} I^2 = R_{dc1} I^2 k_r. \tag{3.30}
$$

The diagrams of the coefficients k_r and k_x as a function of m are shown in Fig. 3.10.

Table 3.1 Values of the functions $w^*(m)$ and k_r for different values of m

$m = \sqrt{2}R/\delta$	$w^*(m)$	k_r
1.0	1.026	1.004
1.5	1.129	1.025
2.0	1.394	1.079
3.0	2.641	1.318
5.0	7.191	2.042
10.0	26.821	3.809

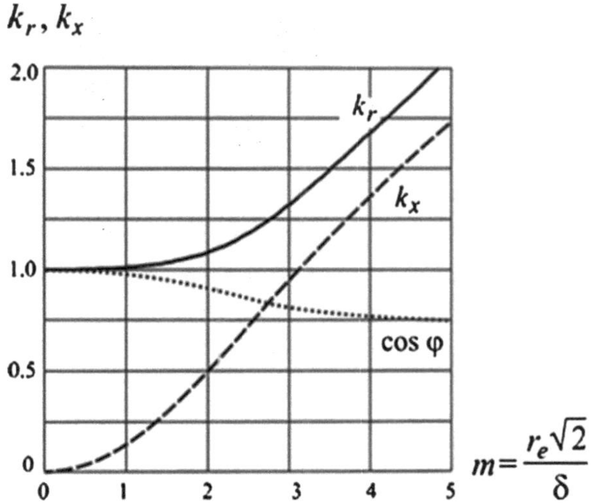

Fig. 3.10 Skin effect coefficients $k_r = R_{ac}/R_{dc}$ and $k_x = X_i/R_{dc}$ [R_{dc} − DC resistance of the workpiece]

In engineering calculations, k_r and k_x can be approximated as follows:

- for $m \ll 1$

$$\begin{cases} k_r \approx 1 + \frac{m^4}{192} \\ k_x \approx \frac{m^2}{8}\left(1 - \frac{m^4}{384}\right) \end{cases}, \tag{3.30}$$

- for $m > 3.5$

$$\begin{cases} k_r \approx \frac{m}{2\sqrt{2}} + \frac{1}{4} + \frac{3}{32}\left(\frac{\sqrt{2}}{m}\right) \\ k_x \approx \frac{m}{2\sqrt{2}} - \frac{3}{64}\left(\frac{2\sqrt{2}}{m}\right) - \frac{3}{16}\left(\frac{1}{m^2}\right). \end{cases} \tag{3.31}$$

For $m \ll 1$ is $R_{ac} \approx R_{dc}$ and $X_i \to 0$. For high values of m, $R_{ac} \approx X_i$ and the phase shift between applied voltage and current in the workpiece is 45°.

The power factor of the cylindrical workpiece then results:

$$\cos\varphi = \frac{R_{ac}}{\sqrt{R_{ac}^2 + X_i^2}} = \frac{1}{\sqrt{1 + \left(\frac{k_x}{k_r}\right)^2}}$$

In Fig. 3.10 the diagram of the power factor is shown as a function of m. Increasing m from 0 to 5, the power factor decreases from 1 to about 0.707; for a further increase of m, its value remains constant and equal to the power factor of the semi-infinite body.

3.2 Solid Cylindrical Workpieces of Ferromagnetic Material

3.2.1 Direct Resistance Heating

We consider a straight solid cylindrical workpiece heated by alternating current from ambient temperature up to the Curie point, making the following assumptions:

1. Electromagnetic quantities in the heated workpiece (current density, electric and magnetic field intensities) are sinusoidal.
2. Electrical resistivity constant and equal to the average value from the initial temperature and Curie point (ρ = const.).
3. Relative permeability variable with the magnetic field according to the realationship $\mu_r = KH^{(1-n)/n}$, with K and n—coefficients obtained for a particular ferromagnetic material from the characteristic $B = f(H)$. From the literature, the coefficient K is equal to 8130 for most ferromagnetic steels and n = 9.4—for structural steels, n = 9.26—for chromium-molybdenum steel. [2, 3].

In DRH installations for ferromagnetic workpieces, where $H_{me} > H_{cr}$, it is 4 < n < 20. Here H_{me} denotes the amplitude of the magnetic field intensity on the surface of the workpiece; H_{cr}, the value of the magnetic field corresponding to the maximum value of μ_r (see Fig. 2.6).

For cylindrical workpieces in [4] it is proposed to use the following law of variation of the relative permeability as a function of the workpiece radius

$$\dot{\mu}_r = \dot{\mu}_{re} \left(\frac{r}{r_e} \right)^{-2} \tag{3.32}$$

where $\dot{\mu}_{re} = \mu_{re} \cdot e^{-j\psi}$ -complex value of the relative magnetic permeability, dependent on the amplitude of the magnetic field intensity H_{me}; $\mu_{re} = K \cdot H_{me}^{(1-n)/n}$ -relative permeability on the workpiece surface; ψ—coefficient accounting for hysteresis losses, dependent for different ferromagnetic materials on H_{em}.

With the above assumptions the system of Maxwell's equations (2.1) takes the form [4]

$$\frac{d^2\dot{H}}{d\zeta^2} - \left(1 + j\dot{m}^2\right)\dot{H} = 0 \tag{3.33a}$$

$$\frac{d^2\dot{E}}{d\zeta^2} + 2\frac{d\dot{E}}{d\zeta} - j\dot{m}^2\dot{E} = 0 \tag{3.33b}$$

with $\dot{m} = \frac{\sqrt{2}r_e}{\delta_e} \cdot e^{-j\frac{\psi}{2}}$, $\zeta = \ln r$, r_e—external radius of the workpiece.

The solution of Eq. (3.33a) is:

$$\dot{H} = \dot{C}_1 \cdot e^{\sqrt{1+j\dot{m}^2\zeta}} + \dot{C}_2 \cdot e^{-\sqrt{1+j\dot{m}^2\zeta}}. \tag{3.34}$$

For $r = 0$ and $\zeta = -\infty$ the magnetic field intensity is nil and it must be $\dot{C}_2 = 0$, while for $r = r_e$ it is $\dot{H} = \dot{H}_e$ and $\dot{C}_1 = \frac{\dot{H}_{me}}{r_e^{\sqrt{1+j\dot{m}^2}}}$.

Therefore Eq. (3.34) becomes:

$$\dot{H} = \dot{H}_{me} \frac{r^{\sqrt{1+j\dot{m}^2}}}{r_e^{\sqrt{1+j\dot{m}^2}}}. \tag{3.35}$$

Since

$$\begin{aligned} \sqrt{1 + j\dot{m}^2} &= \sqrt{1 + j2k^2r_e^2(\cos\psi - j\,\sin\psi)} \\ &= \sqrt{1 + 2k^2r_e^2\sin\psi + j2k^2r_e^2\cos\psi} = \alpha_1' + j\alpha_1'', \end{aligned}$$

the roots of the characteristic equation are

$$\alpha_1' = \frac{\sqrt{\sqrt{1 + 4k^2r_e^2\sin\psi + 4k^4r_e^4} - 1 - 2k^2r_e^2\sin\psi}}{\sqrt{2}} \tag{3.36a}$$

$$\alpha_1'' = \frac{\sqrt{\sqrt{1 + 4k^2r_e^2\sin\psi + 4k^4r_e^4} - 1 - 2k^2r_e^2\sin\psi}}{\sqrt{2}} \tag{3.36b}$$

with $k = \sqrt{\frac{\omega\mu_e}{2\rho}} = \frac{1}{\delta_e}$.

The module of the magnetic field intensity therefore is:

$$|\dot{H}| = |\dot{H}_e| \left(\frac{r}{r_e}\right)^{-\alpha_1'},$$

the magnetic permeability becomes:

$$|\dot{\mu}_r| = |\dot{\mu}_{re}| \left(\frac{r}{r_e}\right)^{-2}$$

and:

$$\frac{|\dot{H}|}{|\dot{H}_e|} = \left(\frac{|\dot{\mu}_r|}{|\dot{\mu}_{re}|}\right)^{-\frac{\alpha_1'}{2}}. \tag{3.37a}$$

From the assumption made at the beginning regarding permeability, it follows that:

$$\frac{|\dot{H}|}{|\dot{H}_e|} = \left(\frac{|\dot{\mu}_r|}{|\dot{\mu}_{re}|}\right)^{-\frac{n}{n-1}} \tag{3.37b}$$

From Eq. (3.37a) and (3.37b) it results:

$$\alpha_1' = \frac{2n}{n-1} \tag{3.38}$$

In case of cylindrical configurations, in the hypothesis that magnetic permeability varies according to Eq. (3.32), the magnetic field intensity H can be zero only on the axis. Therefore, for given values of ψ and n (i.e., for a certain type of material), the solution (3.33a and 3.33b) is valid only for one specific value of the parameter $kr_e = r_e/\delta_e$.

The solutions of Eq. (3.33a) and (3.33b), give the magnetic and electric field intensities as a function of the radius. These relationships allow to calculate the equivalent resistance R_{ef}, internal reactance X_{if} and impedance $\dot{Z}_{ef}$ of the workpiece:

$$\dot{Z}_{ef} = R_{ef} + jX_{if} = \frac{\dot{U}}{\dot{I}} = \frac{\ell \dot{E}}{2\pi r_e^2}\left(\alpha_1' + 1 + j\alpha_1''\right) \tag{3.39}$$

The skin effect coefficients k_{rf}, k_{xf} and the relative impedance can be expressed in the form:

$$\left.\begin{aligned} k_{rf} &= \frac{R_{ef}}{R_{dc}} = \frac{\alpha_1'+1}{2} = \frac{3n-1}{2(n-1)} \\ k_{xf} &= \frac{X_{if}}{R_{dc}} = \frac{\alpha_1''}{2} \\ \frac{|\dot{Z}_{ef}|}{R_{dc}} &= \sqrt{\left[\frac{3n-1}{2(n-1)}\right]^2 + (\alpha_1'')^2} \end{aligned}\right\} \tag{3.40}$$

with $R_{dc} = \rho\ell/\pi r_e^2$—DC Resistance of the workpiece, Ohm; ℓ—length, m.

Table 3.2 shows the values of k_{rf} and k_{xf} for typical ferromagnetic materials.

On the contrary, the solution given in [4] for the semi-infinite ferromagnetic body taking into account hysteresis, has not this restriction, since the magnetic field intensity H may be zero at any distance from the surface.

This solution, dependent only on the body physical parameters and the frequency, is written in the form:

$$\dot{H} = \dot{H}_{me}\left(1 - \frac{y}{y_1}\right)^{\dot{\alpha}_1}$$

Table 3.2 Values of k_{rf} and k_{xf} for different materials

Material	n	k_{rf}	k_{xf}	
			$\psi = 0$	$\psi = 10\,°$
Ideal	∞	1.5		
Permalloy*	20.0	1.551	0. 930	0.850
Electrical Iron*	14.0	1.577		
Transformer Steel*	10.0	1.610	1.060	0.910
Construction Steel**	9.4	1.619		
Cr-Mo Steel**	9.26	1.621		
Cast Iron*	4.0	1.834	1.225	1.115

*From [4]; **From [5]

where:

$$\alpha_1 = \frac{2n}{n-1}\left(1 + j\sqrt{\frac{n+1}{2n} - \frac{(n-1)^2}{2n^2}k^2 y_1^2 \sin\psi}\right).$$

Considering that is $\alpha_1' = 2n/(n-1)$, in [4] ky_1 is given as a function of ψ and n:

$$ky_1 = \left(\frac{3n+1}{n-1}\right)\frac{\alpha_1}{2\cos\psi}\sqrt{\sqrt{1 - \left(\frac{n-1}{3n+1}\right)^2 \cos^2\psi} - \sin\psi}.$$

Therefore:

$$\alpha_1 = \alpha_1' + j\alpha_1'' = \frac{2n}{n-1} + j\frac{2n}{n-1}\sqrt{\frac{n+1}{2n} - \frac{(3n+1)^2}{8n^2}\frac{\sin\psi}{\cos\psi}\left(\sqrt{1 - \left(\frac{n-1}{3n+1}\right)^2\cos^2\psi} - \sin\psi\right)}.$$

In [4] Neiman has proved experimentally that for magnetic fields $H_{me} > 10^3$ A/m the angle ψ is in the range 0 ÷ 12°. On this basis here the cases $\psi = 0$ and $\psi = 10°$ have been considered.

The same theoretical and experimental studies have shown that in DRH heating the hysteresis in ferromagnetic materials reduces the value of k_{xf}. In fact, as shown in Table 3.2, for $\psi = 10°$ the value of k_{xf} is lower of about 8–14 % than the one for $\psi = 0°$.

Moreover, in [4] it is also suggested to use the following universal expressions, neglecting hysteresis losses, for the resistance, the internal inductance and the impedance of ferromagnetic workpieces:

$$R_{ef} = 1.4 \cdot \frac{\ell \rho}{\Pi} \cdot \sqrt{\frac{\omega \mu_e}{2\rho}} = 1.0 \cdot \frac{\ell}{\Pi} \cdot \sqrt{\omega \rho \mu_e}, \qquad (3.41)$$

$$X_{if} = 0.85 \cdot \frac{\ell \rho}{\Pi} \cdot \sqrt{\frac{\omega \mu_e}{2\rho}} = 0.6 \cdot \frac{\ell}{\Pi} \cdot \sqrt{\omega \rho \mu_e}, \qquad (3.42)$$

$$\dot{Z}_{ef} = \frac{\ell}{\Pi} \cdot \sqrt{\omega \rho \mu_e} \cdot (1 + j0.6) = 1.16 \cdot \frac{\ell}{\Pi} \cdot \sqrt{\omega \rho \mu_e} \cdot e^{j\phi}, \qquad (3.43)$$

with $\varphi = 31°$.

Then the skin effect coefficients can be written as follows: for $\chi \geq 2$

$$k_{rf} = \frac{R_{ef}}{R_{dc}} = 1.4 \cdot \frac{\rho \ell}{2\pi r_e} \cdot k \cdot \frac{\pi r_e^2}{\rho \ell} = 1.4 \cdot \frac{k r_e}{2} = 1.4 \cdot \chi; \qquad (3.44)$$

$$k_{xf} = \frac{X_{if}}{R_{dc}} = 0.6 \cdot 1.4 \cdot \chi = 0.84 \cdot \chi, \qquad (3.45)$$

for $0 \leq \chi \leq 2$

$$k_{rf} = 1 + 0.55\chi^2 - 0.025\chi^4, \qquad (3.46)$$

$$k_{xf} = 1.26\chi^2 - 0.42\chi^4 \qquad (3.47)$$

with $k = \sqrt{\frac{\omega \mu_e}{2\rho}} = \frac{1}{\delta_e}, \chi = k r_e / 2$ -for cylindrical workpieces; $\chi = kA/\Pi$- for workpieces with non-cylindrical cross section, with A and Π surface and perimeter of the cross section.

Figure 3.11 shows as a function of χ the diagrams of k_{rf} given by Eq. (2.157) and k_{xf} by Eq. (2.158) for ferromagnetic workpieces, taking into account the function $\mu_r = f(H)$. The same diagrams show also the curves of k_r and k_x, calculated with Eqs. (3.29a, 3.29b) for a semi-infinite body with μ_r = constant.

In Fig. 3.11 the highlighted points have been obtained with the data of Table 2.4 for ferromagnetic cylindrical workpieces: the values of $k_{r\Phi}$ are indicated by a triangle for n = 9.26; a square—for n = 9.4; a circle—for n = 10; a cross—for n → ∞. All these points lay very close to the curve of $k_{r\Phi}$ calculated for the semi-infinite ferromagnetic body. Therefore the diagrams of k_{rf} and k_{xf} can be used with sufficient accuracy also for ferromagnetic cylindrical workpieces.

Finally, in [4] it is proposed to use the graphics and, k_r and k_x of Fig. 3.11 also for straight workpieces with different cross-section shapes, when the magnetic field intensity H_{em} and the average value in a period of the Poynting vector $\overline{S}_{av}$ are the

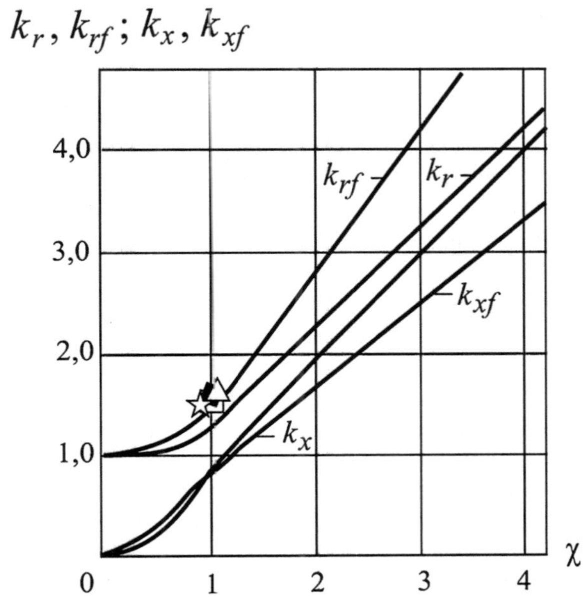

Fig. 3.11 Coefficients of skin effect for ferromagnetic (k_{rf} and k_{xf}) and nonmagnetic (k_r and k_x) workpieces [5]

same on all points of the surface. If this condition is fulfilled, the equivalent impedance $\dot{Z}_e$ of the straight workpiece is inversely proportional to the perimeter Π of the cross section. Such condition is satisfied in workpieces of circular cross section, rectangular cross section, or in equi-angular polygons.

The diagrams of k_r and k_x, given in Fig. 3.11 as a function of the argument $\chi = kA/\Pi$ (e.g. for cylinder: $A = \pi r_e^2, \Pi = 2\pi r_e$), are coincident with the curves for straight solid conductors of circular cross section with constant physical properties), calculated as a function of $\chi = kr_e/2$ (Fig. 3.10).

3.2.2 Induction Heating Solid Ferromagnetic Cylinder

The electromagnetic field in a solid cylinder with constant resistivity ρ is completely determined by the relative radius $m_0 = \sqrt{2}r_e/\delta_0$ (where δ_0-penetration depth with $\mu_r = 1$), the value of the magnetic field intensity H_m at any point, and the dependence $\mu_r = f(H)$.

In induction heating applications the magnetic field on the surface of the workpiece is usually $H_{me} > 10^3$ A/m and the relative permeability $\mu_r = f(H)$ first increases from the surface value $\mu_{re} = f(H_{me})$ towards the center of the cylinder, then it reaches a maximum and finally decreases (see Fig. 2.6b).

In this case, the dimensionless radius is $m_e = m_0\sqrt{\mu_{re}} > 3$ and it is fulfilled the condition $R_{ef} > X_{if}$ (with R_{ef}—resistance and X_{if}—internal reactance per unit length of the solid ferromagnetic cylinder).

However, in induction heating the condition $H_{me} > 5H_{cr}$ usually applies and the resistance and reactance R_{ef} and X_{if} are determined mainly by the outside layers of the cylinder, where the permeability μ_r diminishes with H.

Then, as shown by a large number of calculations, the functions F_{cf} and G_{cf}, characterizing active and reactive power (or resistance and internal reactance) in the solid ferromagnetic cylinder, do not depend on the absolute value of μ_e in a wide range of cases (for example, for carbon steels for $4 < \mu_e < 100$) and are determined only by the parameters m_e and $\alpha = (1-n)/n$ approximating the parabolic function $\mu_r = f(H)$.

The values of F_{cf} and G_{cf} as a function of m_e are given in Fig. 3.12 for three cases: magnetic permeability μ_r = const., permeability variable with the equation $\mu_r = K \cdot H^{\alpha}$, with $\alpha = -0.894$ (corresponding to the average magnetization curve of carbon steels) and $\alpha = -1$ (limiting case of constant magnetic induction for H > 0).

For $m_e \leq 1.5$, the three curves of the functions F_{cf} are coincident and the same occours for G_{cf}. This confirms the small change of μ_r in the cross section. For $m_e > 3$ the curves correspond to the approximate formulae

$$F_{cf} \approx \left(1 - \frac{1}{\sqrt{2}m_e}\right) \cdot f_1 \quad ; \quad G_{cf} \approx f_2 \tag{3.48}$$

where f_1 and f_2 are the coefficients for a semi-infinite ferromagnetic body excited by a plane wave.

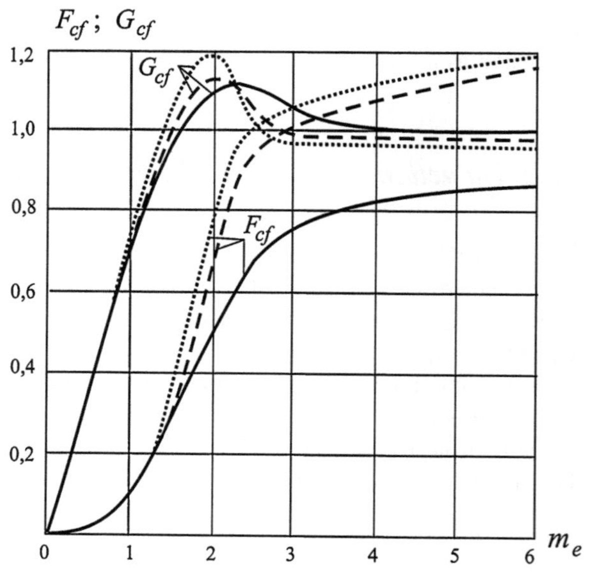

Fig. 3.12 Coefficients F_{cf} and G_{cf} of resistance and reactance of a ferromagnetic cylinder as a function of the relative radius m_e [*1* μ_r = const.; *2* μ_r = var. (with $\alpha = -0.894$); *3* μ_r = var. (with $\alpha = -1$)]

These coefficient are given in the bibliography as a function of α by the relationships [2]:

$$\left.\begin{aligned} f_1(n) &= \sqrt{2}(1+\alpha/2)^{-0.25}\cdot(2+\alpha/2)^{-0.5} \\ f_2(n) &= f_1(n)\cdot(1+\alpha/2)^{0.5} \end{aligned}\right\} \tag{3.48a}$$

with $\alpha = (1-n)/n$.

For $n \to \infty$ and $\alpha \to 1$ the values of the above functions are: $f_1 = 1.37$ and $f_2 = 0.97$ in accordance with the data of Table 2.2.

The selection of the frequency for heating ferromagnetic cylinders (e.g. for heat treatment of bars and wires) can be done with the condition $m_e \geq 2.8$, since for $m_e < 2.8$ the function F_{cf} decreases sharply, while the higher values of G_{cf} in the range $1.5 < m_e < 2.8$ would lead to a decrease of efficiency.

For carbon steels, in which α = −0.894, it is suggested to use the following relationships, which link the radius of the cylinder, the frequency of the current and the required surface density [2]:

$$\begin{aligned} R_{emin} &= 325\ p_0^{0.288}\ \rho^{0.356}\ f^{-0.644} \\ f_{min} &= 8\times 10^3\ r_e^{-1.55}\ \rho^{0.553}\ p_0^{0.447} \\ p_{0max} &= 1.910^{-9}\ r_e^{3.47}\ \rho^{-1.24}\ f^{2.24} \end{aligned} \tag{3.49}$$

where radius, frequency, resistivity and power are expressed in cm, hertz, ohm cm, W/cm², respectively.

Depending on the data of the problem, the design of a systems for induction heating of billets can be done with one of these formulas.

3.3 Hollow Cylindrical Workpieces of Homogeneous Material

3.3.1 Induction Heating with Internal and External Exciting Magnetic Field

We consider a cylindrical inductor-load system of infinite axial length with a hollow workpiece excited by an axial magnetic field, with intensities $\dot{H}_i$ and $\dot{H}_e$ at the internal and external surfaces (Fig. 3.13).

This is, for example, the case of a metallic tube heated with two inductors, one external and one internal.

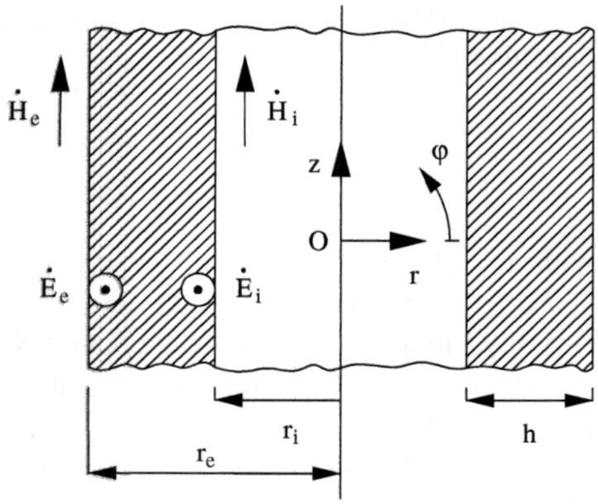

Fig. 3.13 Schematic of hollow cylindrical workpiece in axial magnetic field

We make reference to the Maxwell's equations:

- in the metal workpiece

$$\left.\begin{aligned} rot\dot{\bar{H}}_1 &= \frac{\dot{\bar{E}}_1}{\rho} \\ rot\dot{\bar{E}}_1 &= -\frac{\partial\dot{\bar{B}}_1}{\partial t} = -\mu_1\frac{\partial\dot{\bar{H}}_1}{\partial t} \end{aligned}\right\}$$

- in the internal cavity

$$\left.\begin{aligned} rot\ \dot{\bar{H}}_2 &= \varepsilon_2\frac{\partial\dot{\bar{E}}_2}{\partial t} \\ rot\ \dot{\bar{E}}_2 &= -\mu_2\frac{\partial\dot{\bar{H}}_2}{\partial t} \end{aligned}\right\}.$$

The general solution in the metal tube is analogous to the solution already given for the case of a solid metallic cylinder.

With the notations:

$$m_r = \frac{\sqrt{2}r}{\delta}, \quad m_i = \frac{\sqrt{2}r_i}{\delta}, \quad m_e = \frac{\sqrt{2}r_e}{\delta},$$

for the components of the field vectors we can write:

$$\left.\begin{aligned} \dot{H}_1 &= \dot{C}_1 J_0(\sqrt{-j}m_r) + \dot{C}_2 Y_0(\sqrt{-j}m_r) \\ \dot{E}_1 &= \sqrt{-j}\cdot\frac{\rho}{\delta}\cdot\sqrt{2}\cdot\left\{\dot{C}_1 J_1(\sqrt{j}m_r) + \dot{C}_2 Y_1(\sqrt{j}m_r)\right\} \end{aligned}\right\} \tag{3.50}$$

However, in this case $\dot{C}_2 \neq 0$, since in the metal of the hollow cylinder a reflected wave is present.

A general solution of the above equations can be written applying the boundary conditions:

$$\dot{H} = \dot{H}_e \text{ for } r = r_e, \text{ and } \dot{H} = \dot{H}_i \text{ for } r = r_i$$

It results:

$$\dot{C}_1 = \frac{\begin{vmatrix} \dot{H}_e Y_0(\sqrt{-j}m_e) \\ \dot{H}_i \quad Y_0(\sqrt{-j}m_i) \end{vmatrix}}{\mathrm{F_{oo}(m_i, m_e)}} = \frac{\dot{H}_e Y_0(\sqrt{-j}m_i) - \dot{H}_i Y_0(\sqrt{-j}m_e)}{\mathrm{F_{oo}(m_i, m_e)}}$$

$$\dot{C}_2 = \frac{\begin{vmatrix} J_0(\sqrt{-j}m_e)\dot{H}_e \\ J_0(\sqrt{-j}m_i)\dot{H}_i \end{vmatrix}}{\mathrm{F_{oo}(m_i, m_e)}} = -\frac{\dot{H}_e J_0(\sqrt{-j}m_i) - \dot{H}_i J_0(\sqrt{-j}m_e)}{\mathrm{F_{oo}(m_i, m_e)}}$$

with:

$$\mathrm{F_{oo}(x,y)} = J_0(\sqrt{-j}y)Y_0(\sqrt{-j}x) - Y_0(\sqrt{-j}y)J_0(\sqrt{-j}x)$$

By substitution of the constants $\dot{C}_1, \dot{C}_2$ into Eq. (3.50) we obtain:

$$\dot{H} = \dot{H}_e \frac{\mathrm{F_{oo}(m_i, m_r)}}{\mathrm{F_{oo}(m_i, m_e)}} - \dot{H}_i \frac{\mathrm{F_{oo}(m_e, m_r)}}{\mathrm{F_{oo}(m_i, m_e)}} \tag{3.51a}$$

$$\dot{E} = \sqrt{-2j}\frac{\rho}{\delta}\left[\dot{H}_e \frac{\mathrm{F_{o1}(m_i, m_r)}}{\mathrm{F_{oo}(m_i, m_e)}} - \dot{H}_i \frac{\mathrm{F_{o1}(m_e, m_r)}}{\mathrm{F_{oo}(m_i, m_e)}}\right] \tag{3.51b}$$

with:

$$F_{o1}(x, y) = J_1(\sqrt{-j}y)Y_0(\sqrt{-j}x) - Y_1(\sqrt{-j}y)J_0(\sqrt{-j}x).$$

3.3.2 Induction Heating with External Exciting Magnetic Field

In this case the hollow workpiece is surrounded by a coaxial solenoid inductor which produces in the gap between inductor and workpiece the exciting magnetic field intensity $\dot{H}_e$, and at the surface of the workpiece the induced electric field intensity $\dot{E}_e$ (see Fig. 3.14).

Equations 3.51a, 3.51b still apply and the magnetic field intensity $\dot{H}_e$ at the external surface of the tube is a given data, whereas the field $\dot{H}_i$ remains constant in the cavity.

Fig. 3.14 Schematic of a hollow cylindrical workpiece placed inside the inductor: *1* inductor; *2* hollow cylinder

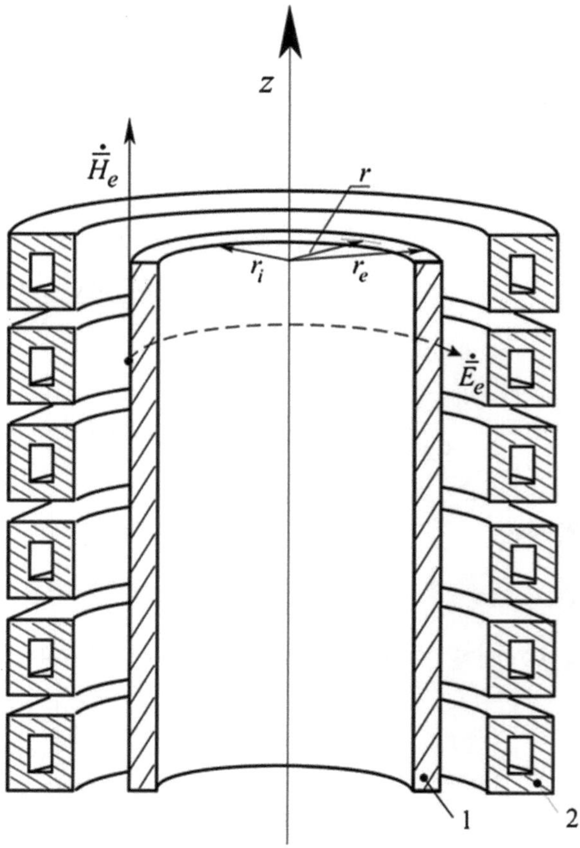

The unknown value $\overline{H}_i$ is determined by considering that the *e.m.f.* induced in an ideal circular conductor placed at the radius $r = r_i$ must be equal to the derivative of the flux linked with the same circuit.

Therefore, one can write:

$$\oint \dot{E} d\ell = 2\pi r_i \dot{E}_i = -j\omega\dot{\Phi}_i = -j\omega\mu_0 \dot{H}_i \pi r_i^2$$

or, for Eq. (3.51a):

$$\dot{E}_i = -j\frac{\omega\mu_0 r_i}{2}\dot{H}_i = \sqrt{-2j}\frac{\rho}{\delta}\left[\dot{H}_e\frac{\mathrm{F_{o1}(m_i, m_i)}}{\mathrm{F_{oo}(m_i, m_e)}} - \dot{H}_i\frac{\mathrm{F_{o1}(m_e, m_i)}}{\mathrm{F_{oo}(m_i, m_e)}}\right] \quad (3.52)$$

Considering that is:

$$F_{o1}(x,x) = \frac{2}{\sqrt{-j\pi x}},$$

from the previous relation, after some manipulations one obtains:

$$\frac{\dot{H}_e}{\dot{H}_i} = -j\frac{\pi m_i^2}{2\mu}\left[F_{00}(m_i,m_e) + F_{01}(m_e,m_i)\right].$$

Equations (3.51a) and (3.52) give the distributions of the magnetic field intensity $\dot{H}$ and the current density $\dot{J} = \dot{E}/\rho$ along the thickness of the tube wall.

Figures 3.15 show the normalized distributions of $|\dot{H}|/|\dot{H}_e|$ and $|\dot{E}|/|\dot{E}_e|$ inside the workpiece, for $m_i = 2, m_i = 5$ and $m_e = 7$.

It is worth to underline that in the workpiece cavity the magnetic field intensity remains constant and equal to the value at its internal surface, as illustrated in Fig. 3.16.

3.3.2.1 Energy in the Workpiece

The complex power $P_1 + jQ_1$ per unit length in the hollow cylinder can be determined by the Poynting's vector

$$\dot{\overline{S}} = \frac{1}{2}\left[\dot{\overline{E}} \times \overline{H}^*\right].$$

Indicating with $\dot{S}_e$ the flux of energy through the external surface, it is $P_1 + jQ_1 = \dot{S}_e \cdot 2\pi r_e$. The active component (W/m) can be written as follows

$$P_1 = (N_1 I)^2 \cdot \frac{\rho}{\delta} \cdot 2\pi r_e . F_{tube} \tag{3.53a}$$

and the reactive one (VAr/m), as

$$Q_1 = (N_1 I)^2 \cdot \frac{\rho}{\delta} \cdot 2\pi r_e . G_{tube} \tag{3.53b}$$

or:

$$\left.\begin{aligned} P_1 &= 6.24 \cdot 10^{-3} \cdot (N_1 I)^2 \cdot d \cdot \sqrt{\rho \mu_r f} \cdot F_{tube} \\ Q_1 &= 6.24 \cdot 10^{-3} \cdot (N_1 I)^2 \cdot d \cdot \sqrt{\rho \mu_r f} \cdot G_{tube} \end{aligned}\right\} \tag{3.54}$$

with $d = 2r_e$ and $F_{tube}\, G_{tube}$—coefficients containing Bessel functions of arguments $m_e = \left(\sqrt{2} r_e/\delta\right)$ and h/δ.

The values of these functions are given in the diagrams of Fig. 3.17.

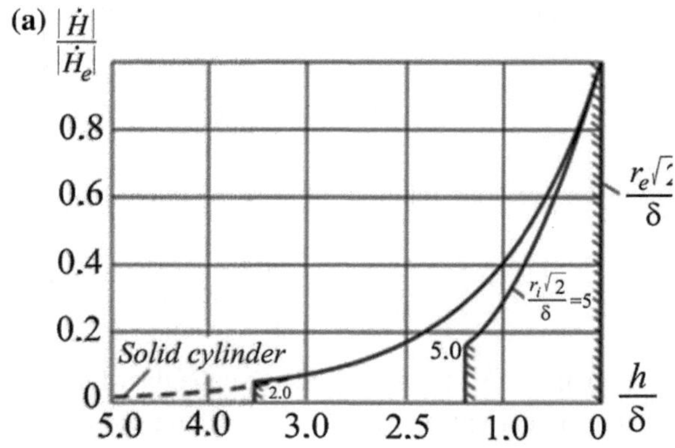

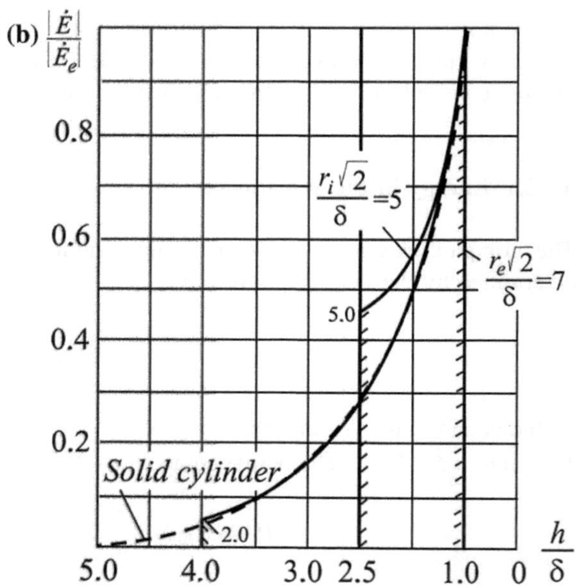

Fig. 3.15 **a** Distributions of magnetic field intensity in non-magnetic tubes, as a function of relative thickness of tube wall (h/δ), for $m_i = 2, m_i = 5$ and $m_e = 7$ (*dashed line*: *solid cylinder*), **b** Distributions of electric field intensity in non-magnetic tubes, as a function of relative thickness of tube wall (h/δ), for $m_i = 2$, $m_i = 5$ and $m_e = 7$ (*dashed lines*: *solid cylinder*)

A feature of the induction heating of hollow cylindrical workpieces with external inductor is the presence of peaks of the function F_{tube} in the range $h/\delta = 0.2 \div 0.5$. This means that the maximum active power induced in the hollow cylinder depends on the thickness of the cylinder wall and occurs for $h/\delta = 0.2 \div 0.5$.

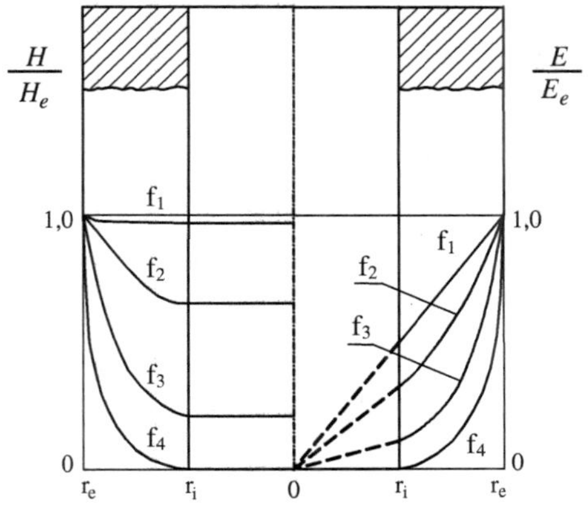

Fig. 3.16 Relative distributions of the magnetic and electric field intensity in tubes heated by external inductor, at different frequencies [2]

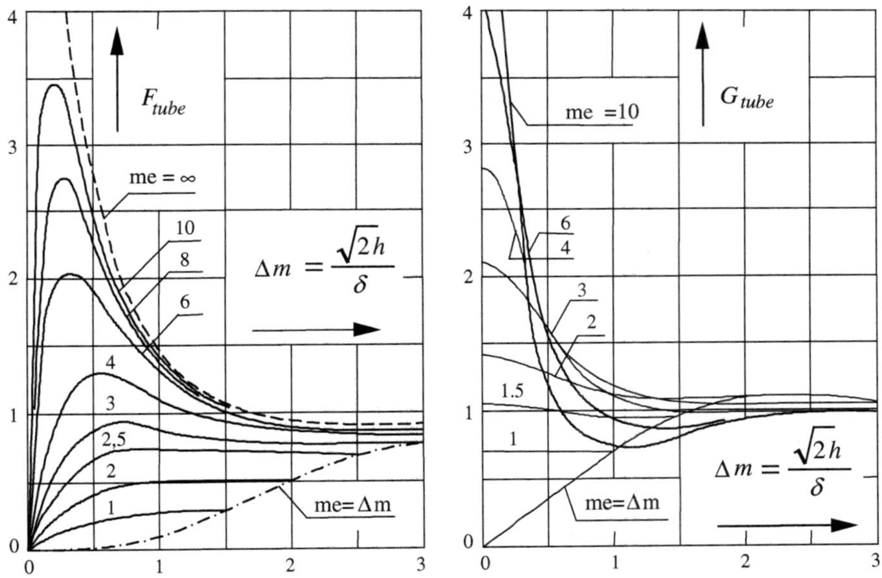

Fig. 3.17 Coefficient F_{tube} and G_{tube} as a function of $\sqrt{2}h/\delta$ and different values of $m_e = \sqrt{2}r_e/\delta$ [2]

The power factor cos φ is:

$$\cos\varphi = \frac{P_1}{\sqrt{P_1^2 + Q_1^2}} = \frac{1}{\sqrt{1 + \left(\frac{G_{tube}}{F_{tube}}\right)^2}}.$$

As shown in Fig. 3.18, also the curves of power factor have a maximum.

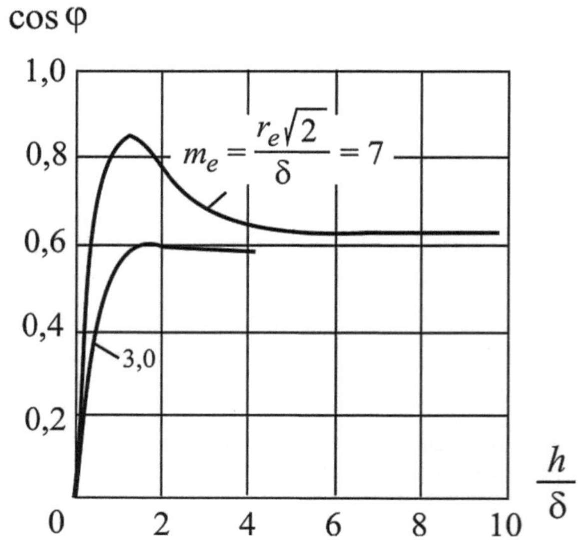

Fig. 3.18 Power factor in the heating of hollow cylindrical workpieces with external inductors

3.3.3 Induction Heating of Hollow Cylindrical Workpieces with Internal Inductors

Figure 3.19 represents the schematic of the inductor and workpiece system considered in this paragraph.The magnetic field intensity $\dot{H}_i$ is imposed on the internal surface of the workpiece by the exciting inductor whereas on the external surface it is $\dot{H}_e = 0$. This condition occurs since the configuration is of indefinite length.

Then Eqs. (3.51a, 3.51b) can be expressed in the form:

$$\frac{\dot{H}}{\dot{H}_i} = \frac{\mathrm{F_{oo}(m_r, m_e)}}{\mathrm{F_{oo}(m_i, m_e)}} = \frac{J_0(\sqrt{-j}m_e)Y_0(\sqrt{-j}m_r) - Y_0(\sqrt{-j}m_e)J_0(\sqrt{-j}m_r)}{J_0(\sqrt{-j}m_e)Y_0(\sqrt{-j}m_i) - Y_0(\sqrt{-j}m_e)J_0(\sqrt{-j}m_i)}$$

$$\begin{aligned}\frac{\dot{E}}{\dot{H}_i} &= \sqrt{-2j}\frac{\rho}{\delta}\left[\frac{\mathrm{F_{o1}(m_e, m_r)}}{\mathrm{F_{oo}(m_i, m_e)}}\right] \\ &= \sqrt{-2j}\frac{\rho}{\delta}\frac{J_1(\sqrt{-j}m_r)Y_0(\sqrt{-j}m_e) - Y_1(\sqrt{-j}m_r)J_0(\sqrt{-j}m_e)}{J_0(\sqrt{-j}m_e)Y_0(\sqrt{-j}m_i) - Y_0(\sqrt{-j}m_e)J_0(\sqrt{-j}m_i)}\end{aligned}$$

The functions $F_{00}(x, y)$ and $F_{01}(x, y)$ are again given by the relationships

$$\mathrm{F_{oo}(x,y)} = J_0(\sqrt{-j}y)Y_0(\sqrt{-j}x) - Y_0(\sqrt{-j}y)J_0(\sqrt{-j}x),$$
$$F_{o1}(x, y) = J_1(\sqrt{-j}y)Y_0(\sqrt{-j}x) - Y_1(\sqrt{-j}y)J_0(\sqrt{-j}x).$$

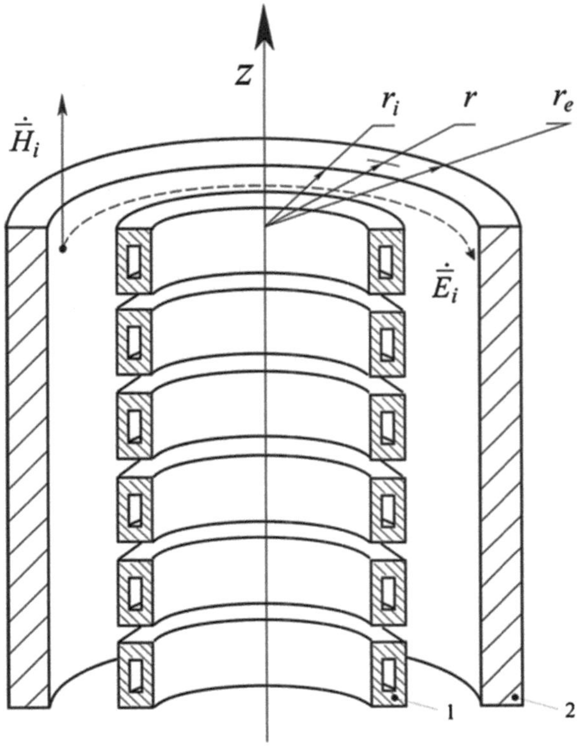

Fig. 3.19 Schematic of internal inductor and external cylindrical workpiece: *1* inductor; *2* workpiece

and, in addition, it has been taken into account that is

$$F_{00}(x,y) = -F_{0C}(y,x).$$

Figures 3.20 show the distributions of magnetic and electric field intensities in the thickness of the workpiece, for $m_e = \sqrt{2}r_e/\delta = 7$ and different values of $m_i = \sqrt{2}r_i/\delta$.

Final remarks.

1. In the heating of hollow cylindrical workpieces with internal inductors, the magnetic field intensity on the external surface of the cylinder is always nil, unlike the case of heating with external inductors, where on the internal surface can be H $\neq$ 0.
2. The electromagnetic wave diffusion in the workpiece excited from inside, is analogous to the one calculated for a flat plate.
3. The variation of E in the wall thickness is analogous to the distribution of E in the heating of a plate from one side.

Fig. 3.20 Distributions of relative values of $\dot{H}$ and $\dot{E}$ in the wall thickness of a tube heated by internal inductors (subscript (i) indicates values on the internal surface)

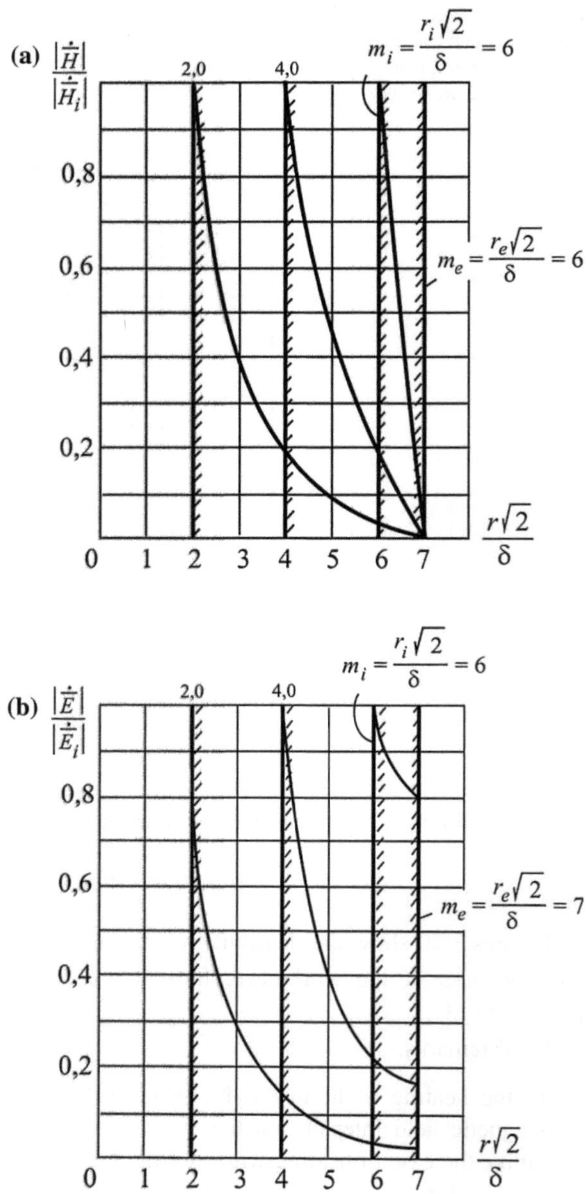

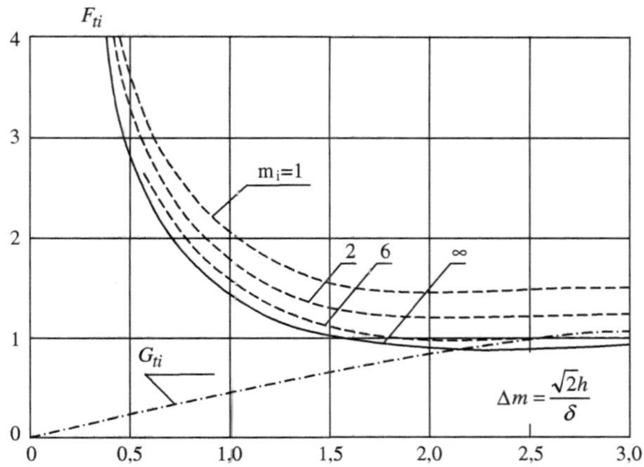

Fig. 3.21 Coefficients for the calculation of the active and reactive power in a tube heated from inside. $[m_i = \sqrt{2}r_i/\delta; \Delta m = \sqrt{2}h/\delta]$

3.3.3.1 Energy in Hollow Cylindrical Workpiece with Internal Exciting Magnetic Field

Active and reactive power per unit length are

$$\left.\begin{aligned} P_1 &= 6.24 \times 10^{-3} \cdot (N_1 I)^2 \cdot d_i \cdot \sqrt{\rho\mu_r f} \cdot F_{ti} \\ Q_1 &= 6.24 \times 10^{-3} \cdot (N_1 I)^2 \cdot d_i \cdot \sqrt{\rho\mu_r f} \cdot G_{ti} \end{aligned}\right\} \quad (3.55)$$

where $d_1 = 2r_i$ and the coefficients F_{ti}, G_{ti} are expressed by Bessel functions; their values are given in Fig. 3.21 as a function of $\sqrt{2h/\delta}$, for various values of m_i.

In particular, the curves show that the values of $F_{tube,i}$ increase when m_i decreases, i.e. when the curvature of the load increases.

For $m_i = \sqrt{2}r_i/\delta \geq 6$, with accuracy of 5 ÷ 10 %, the functions F_{ti} and G_{ti} are coincident with the corresponding functions F_{plate} and G_{plate} calculated for the plate $(m_i \to \infty)$, and it can be assumed $F_{ti} = F_{plate}$ and $G_{ti} = G_{plate}$.

The values of G_{ti} practically do not depend on m_i and grow linearly with the thickness h till $\Delta m \approx 2.5$. Above this value, it can be assumed $G_{ti} \approx 1$.

3.3.4 Resistance Heating of Cylindrical Hollow Workpieces

Equations (3.16) and (3.19) given for the direct resistance heating of a solid cylinder can be used also in this case with relevant boundary conditions:

$$\left.\begin{aligned} &\frac{\partial^2 \dot{J}}{\partial r^2} + \frac{1}{r}\frac{\partial \dot{J}}{\partial r} - j\frac{\omega\mu_0\mu_r}{\rho}\dot{J} = 0 \\ &\dot{J} = \dot{C}_1 J_0(\sqrt{-j}m_r) + \dot{C}_2 Y_0(\sqrt{-j}m_r) \end{aligned}\right\} \quad (3.56)$$

with $m_r = \sqrt{2}r/\delta$, J_0 and Y_0—Bessel function of order zero and first and second kind.

Moreover we make use of the following notation:

$$(m_e = \sqrt{2}r_e/\delta), (m_i = \sqrt{2}r_i/\delta = m_e\alpha, \alpha = \frac{r_i}{r_e})$$

The boundary conditions for determination of the integration constant $\dot{C}_1$ and $\dot{C}_2$ are:

$$\left.\begin{array}{r} \dot{J} = \dot{J}_e \quad for \quad r = r_e \\ \oint H\,d\Pi = 0 \quad for \quad r = r_i(\Pi = 2\pi r_i) \end{array}\right\} \tag{3.57}$$

From the second condition (3.57), at $r = r_i = \alpha r_e$ it results

$$\frac{\partial \dot{J}}{\partial r} = 0. \tag{3.58}$$

Then, applying the conditions (3.58) and the first of (3.57), from (3.56) one obtains:

$$\left.\begin{array}{l} \dot{J}_e = \dot{C}_1 J_0(\sqrt{-j}m_e) + \dot{C}_2 Y_0(\sqrt{-j}m_e) \\ 0 = \dot{C}_1 J_1(\sqrt{-j}m_e\alpha) + \dot{C}_2 Y_1(\sqrt{-j}m_e\alpha) \end{array}\right\}. \tag{3.59}$$

Solving the system (3.59), for the integration constants $\dot{C}_1$ and $\dot{C}_2$, we can write:

$$\dot{C}_1 = \frac{\begin{vmatrix} \dot{J}_e & Y_0(\sqrt{-j}m_e) \\ 0 & Y_1(\sqrt{-j}m_e\alpha) \end{vmatrix}}{\begin{vmatrix} \dot{J}_0(\sqrt{-j}m_e) & Y_0(\sqrt{-j}m_e) \\ \dot{J}_1(\sqrt{-j}m_e\alpha) & Y_1(\sqrt{-j}m_e\alpha) \end{vmatrix}} = -\dot{J}_e \cdot \frac{Y_1(\sqrt{-j}m_e\alpha)}{\mathrm{F_{o1}(m_e, m_e}\alpha)},$$

$$\dot{C}_2 = \frac{\begin{vmatrix} J_0(\sqrt{-j}m_e)\dot{J}_e \\ J_1(\sqrt{-j}m_e\alpha)0 \end{vmatrix}}{\begin{vmatrix} \dot{J}_0(\sqrt{-j}m_e) & Y_0(\sqrt{-j}m_e) \\ \dot{J}_1(\sqrt{-j}m_e\alpha) & Y_1(\sqrt{-j}m_e\alpha) \end{vmatrix}} = \dot{J}_e \cdot \frac{J_1(\sqrt{-j}m_e\alpha)}{\mathrm{F_{o1}(m_e, m_e}\alpha)},$$

with

$$F_{o1}(x, y) = J_1(\sqrt{-j}y)Y_0(\sqrt{-j}x) - Y_1(\sqrt{-j}y)J_0(\sqrt{-j}x).$$

The substitution of integration constants in Eq. (3.56), gives

$$\frac{\dot{J}}{\dot{J}_e} = \frac{J_o(\sqrt{-j}m_r)\mathbb{Y}_1(\sqrt{-j}m_e\alpha) - Y_o(\sqrt{-j}m_r)J_1(\sqrt{-j}m_e\alpha)}{J_o(\sqrt{-j}m_e)\mathbb{Y}_1(\sqrt{-j}m_e\alpha) - Y_o(\sqrt{-j}m_e)J_1(\sqrt{-j}m_e\alpha)} = \frac{F_{01}(m_r, m_e\alpha)}{F_{01}(m_e, m_e\alpha)} \tag{3.60}$$

Taking into account that from (3.15) it is

$$\dot{H} = \frac{1}{j\omega\mu_r\mu_0}\frac{\partial \dot{E}}{\partial r} = -j\frac{\delta^2}{2}\frac{\partial \dot{J}}{\partial r}$$

we can determine the law of variation of the magnetic field intensity in the thickness of the tube and evaluate its value at the external surface.

The magnetic field intensity on the external surface is obtained by Ampere's law $\dot{H}_e = \dot{I}/(2\pi r_e)$, and equating the values so obtained, finally we can write

$$\dot{J}_e = \frac{1}{j\sqrt{-j}} \cdot \frac{I}{\pi r_e^2} \cdot \frac{m_e}{2} \cdot \frac{F_{o1}(m_e, m_e\alpha)}{F_{11}(m_e, m_e\alpha)} \tag{3.61}$$

with

$$F_{11}(m_e, m_e\alpha) = J_1\left(\sqrt{-j}m_e\right)\mathbb{Y}_1\left(\sqrt{-j}m_e\alpha\right) - Y_1\left(\sqrt{-j}m_e\right)J_1\left(\sqrt{-j}m_e\alpha\right)$$

By substitution of Eq. (3.61) into (3.60), we can write

$$\frac{\dot{J}}{\frac{\dot{I}}{\pi r_e^2(1-\alpha^2)}} = \sqrt{-j}\cdot\left(1-\alpha^2\right)\cdot\frac{m}{2}\cdot\frac{F_{01}(m_r, m_e\alpha)}{F_{11}(m_e, m_e\alpha)}, \tag{3.62}$$

where: $\frac{\dot{I}}{\pi r_e^2(1-\alpha^2)}$ is the DC current density in the annular cross section of the workpiece.

The internal impedance per unit length can be evaluated as follows

$$\dot{Z}_{e1} = R_{e1} + jX_{i1} = \frac{\dot{E}_e}{\dot{I}} = \rho\frac{\dot{J}_e}{\dot{I}}. \tag{3.63}$$

By substituting in Eq. (3.63) $\dot{J}_e$ from (3.61), we obtain:

$$\frac{\dot{Z}_{e1}}{R_{dc1}} = \sqrt{-j}\cdot\left(1-\alpha^2\right)\cdot\frac{m}{2}\cdot\frac{F_{01}(m_e, m_e a)}{F_{11}(m_e, m_e\alpha)}, \tag{3.64}$$

where $R_{dc1} = \frac{\rho}{\pi r_e^2(1-\alpha^2)}$ is the DC resistance per unit length of the tubular workpiece.

Fig. 3.22 Coefficients of skin-effect k_r and k_x for resistance and internal reactance of non-magnetic tubes as a function of parameters m_e and α [7]

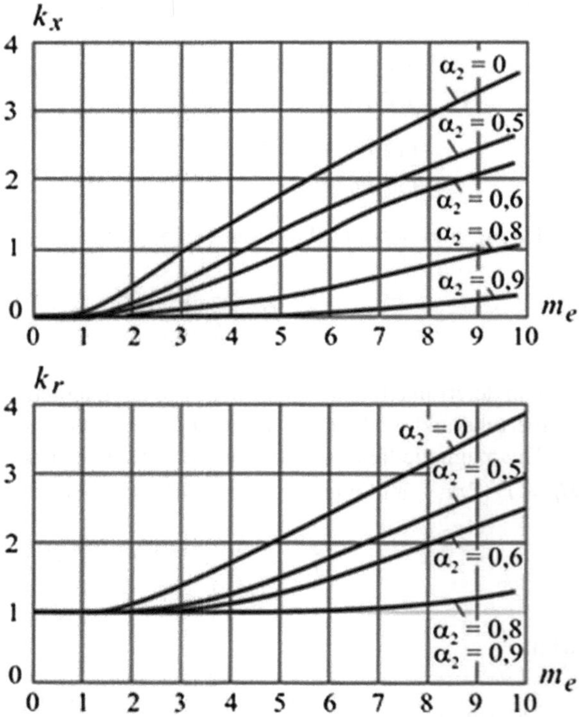

The real and imaginary parts of Eq. (3.64) give the coefficients of skin effect for the AC resistance $k_r = R_{ac}/R_{dc}$ and the internal reactance $k_x = X_i/R_{dc}$.

Figure 3.22 gives the values of these coefficients as a function of m_e and α.

3.4 Hollow Cylindrical Workpieces of Ferromagnetic Material

3.4.1 Induction Heating

In the induction heating of ferromagnetic workpieces skin effect is often very pronounced, and with sufficient accuracy for engineering purposes the calculation can be made with the equations given in Sect. 3.2. But when the penetration depth of the electromagnetic wave is comparable with the thickness of the workpiece, only numerical methods are available.

This may occur in the heating of ferromagnetic steel tubes. In this case the distribution of current and power density in the tube wall thickness cannot be given in an analytical form for a wide range of geometrical parameters and physical properties.

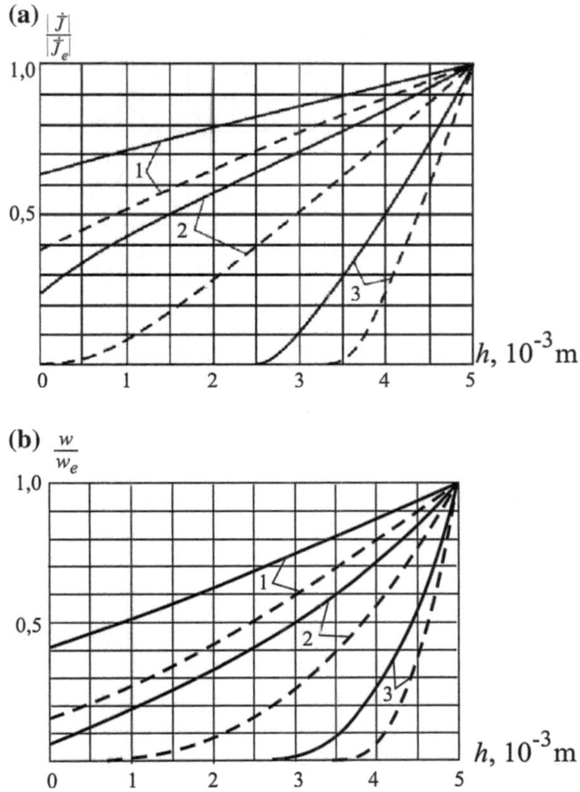

Fig. 3.23 **a** Distributions of relative current density J/J_e in the tube wall thickness [*Solid lines* $H_e = 3 \times 10^5$ A/m; *dotted lines* $H_e = 1 \times 10^5$ A/m; *1* f = 50 Hz; *2* f = 500 Hz; *3* f = 5000 Hz]. **b** Distributions of relative volume power density w/w_e in the tube wall thickness [*Solid lines* $H_e = 3 \times 10^5$ A/m; *dotted lines* $H_e = 1 \times 10^5$ A/m; *1* f = 50 Hz; *2* f = 500 Hz; *3* f = 5000 Hz]

The variations during heating of resistivity and relative magnetic permeability can affect results significantly.

A series of numerical simulations with the software package FLUX has been made for ferromagnetic steel tubes of infinite length, in the "*cold*" stage of the heating (when temperature does not exceeds 100 ℃).

The calculations have been done with the following data: tube external diameter $r_e = 0.1$ m, wall thickness $h = 0.005$ m, resistivity $\rho = 20 \times 10^{-8}$ ohm·m, relative permeability given by the equation $\mu_r = 8130H^{-0.894}$ as suggested in [2].

In Fig. 3.23 are given the distributions of the relative current density J/J_e and relative power density w/w_e in the tube wall thickness for different values of the magnetic field H_e at the surface of the tube and the frequency *f*.

The active and reactive power per unit surface of ferromagnetic tube are given by the equations:

$$p = H_e^2 \cdot \frac{\rho}{\delta_e} \cdot \sqrt{2} F_{tube,f}, q = H_e^2 \cdot \frac{\rho}{\delta_e} \cdot \sqrt{2} G_{tube,f}. \qquad (3.65)$$

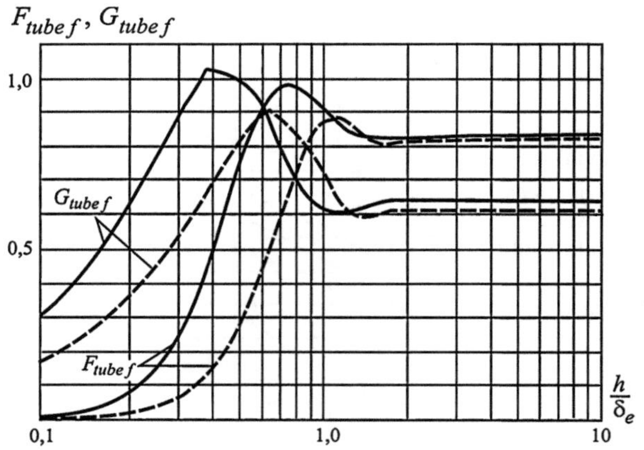

Fig. 3.24 Coefficients $F_{tube,f}$ and $G_{tube,f}$ of active and reactive power in a ferromagnetic tube of infinite axial length, as a function of the relative wall thickness (h/δ), for different magnetic field intensity H_e [*Solid lines* $H_e = 3 \times 10^5$ A/m; *dotted lines* $H_e = 1 \times 10^5$ A/m]

Figure 3.24 gives the coefficients $F_{tube,f}$ and $G_{tube,f}$, as a function of the relative thickness of the tube wall (h/δ_e), for different values of H_e.

These results show that the increase of H_e allows to obtain more uniform distributions of the current and power density in the wall thickness.

This can be explained by the fact that, increasing H_e, the relative magnetic permeability μ_r decreases and—as a consequence—the penetration depth δ_e increases.

On the contrary, with the increase of the frequency, the penetration depth δ_e decreases and a more uneven distributions of J and w in the tube thickness occur (Fig. 3.23).

Figure 3.24 shows that the diagrams of coefficients $F_{tube,f}$ and $G_{tube,f}$ of the active and reactive power as a function of (h/δ_e), have three characteristic regions.

In the first region the function monotonically increases up to a maximum. In the second one it decreases to a minimum. In the third one the curves increase again, approaching an asymptotic value.

Moreover, the graphs show the effect of the magnetic field intensity H_e on the values of $F_{tube,f}$ and $G_{tube,f}$. In particular, increasing H_e, the maxima are shifted towards smaller values of (h/δ_e).

3.4.2 Resistance Heating

We consider here the distribution of the electromagnetic field in a ferromagnetic tubular cylindrical workpiece of external radius r_e, internal radius r_i and infinite axial length.

The calculation of the electrical integral parameters, resistance R_{ef} and internal reactance X_{if}, has been done with numerical methods under the following assumptions:

1. Sinusoidal electromagnetic quantities in the workpiece (current, intensity of electric and magnetic fields);
2. Resistivity $\rho = const.$;
3. Relative magnetic permeability $\mu_r = f(H)$ given by a data sheet for the specific type of steel;
4. Workpiece of infinite length where the distributions of current density and magnetic field intensity vary only along the workpiece radius.

With the above assumptions Maxwell's equations give:

$$\left.\begin{aligned} &\frac{\partial^2 \dot{J}}{\partial r^2} + \frac{1}{r}\frac{\partial \dot{J}}{\partial r} - j\frac{\omega\mu_0\mu_r}{\rho}\dot{J} = 0 \\ &\dot{H} = -j\frac{\rho}{\omega\mu_0\mu_r}\frac{\partial \dot{J}}{\partial r} \end{aligned}\right\} \tag{3.66a}$$

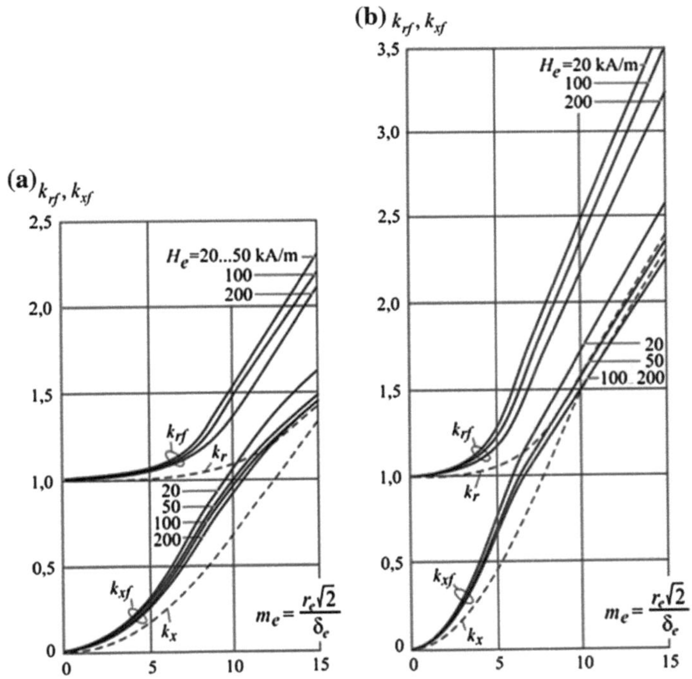

Fig. 3.25 Skin effect coefficients k_{rf} and k_{xf}, as a function of m_e and different values of the magnetic field intensity H_e. **a** $\alpha = r_i/r_e = 0.85$; **b** $\alpha = r_i/r_e = 0.75$; dashed lines: coefficients k_r, k_x with $\mu = \mu_e = const.$ [6]

and can be solved with the boundary conditions:

$$\left.\begin{aligned} \dot{H} &= \dot{H}_e = \tfrac{\dot{I}}{2\pi r_e} \quad for \quad r = r_e; \\ \dot{H} &= 0, \qquad\qquad for \quad r = r_i \end{aligned}\right\} \tag{3.66b}$$

The solution of the equation system (3.66a) and (3.66b) has been done by means of Finite Element Method, and the evaluation of the resistance, internal reactance and total impedance of the workpiece per unit length, can be done with the relationship:

$$\dot{Z}_{ef1} = \frac{\dot{E}_e}{\dot{I}} = \rho \frac{\dot{J}_e}{\dot{I}} = R_{ef1} + jX_{if1} = R_{dc1}\left(k_{rf} + jk_{xf}\right) \tag{3.67}$$

where R_{dc1} is is the DC resistance per unit length of the tubular workpiece.

Figure 3.25 gives for two values of the ratio $\alpha = r_i/r_e$, the coefficients $k_{rf} = R_{ef}/R_{dc}$ and $k_{xf} = X_{if}/R_{dc}$ as a function of $m_e = \sqrt{2}r_e/\delta_e$, and different values of the magnetic field intensity H_e.

The diagrams show that:

- when H_e increases, the coefficients k_{rf} and k_{xf} decrease;
- in the range $H_e = (20 \div 200) \times 10^3$ A/m and $m_e \leq 15$, the coefficients k_{rf} and k_{xf} of the ferromagnetic tubular workpiece are higher than the corresponding values k_r and k_x, calculated with $\mu_r = \mu_e =$ const.;
- when the ratio $\alpha = r_i/r_e$ increases, the coefficients k_{rf} and k_{xf} decrease.

References

1. Slukhotskii, E., Nemkov, V. et al. (1981). *Induction heating installations* (328 pp.). Leningrad: Energoisdat (in Russian).
2. Nemkov, V., & Demidovich, V. B. (1988). *Theory and calculation of induction heating installations* (280 pp.). Leningrad: Energoatomiszdat. ISBN 5-283-04409-2.
3. Zaimovskiĭ, A. S., & Čudnovskaja (1957). *Magnetic materials*. Moscow: GEI (ГЭИ) (in Russian).
4. Neiman, L. R. (1949). *Skin effect in ferromagnetic bodies* (90 pp.). Leningrad: Gostehnizdat (in Russian).
5. Aliferov, A., & Lupi, S. (2004). *Direct resistance heating of metals* (223 pp.). Novosibirsk: NGTU. ISBN 5-7782-0475-2.
6. Lupi, S., & Nunes, M. F. (1990). *Riscaldamento dei metalli mediante conduzione diretta di corrente* (86 pp.). Padova: CLEUP. ISBN 88-7178-338-7 (in Italian).
7. Davies, E. J. (1990). *Conduction and induction heating* (385 pp.). London: Peter Peregrinus Ltd. ISBN 0 86341 174 6.
8. EDF (1996). *Induction, conduction électrique dans l'industrie* (780 pp.). Paris: CFE DOPEE85.

9. Rudnev, V., Loveless, D., Cook, R., & Black, M. (2003). *Handbook of induction heating* (777 pp.). New York: Marcel Dekker Inc. ISBN 0-8247-0848-2.
10. Lupi, S. (2005). *Elettrotermia* (453 pp.). Padua: Libreria Progetto (in Italian).
11. Aliferov, A., & Lupi, S. (2010). *Induction and direct resistance heating of metals* (411 pp.). Novosibirsk: NGTU. ISBN 978-5-7782-1622-8 (in Russian).
12. Rudnev, V., & Totten, G.E., (Eds.). (2014). *ASM Handbook. Volume 4C—induction heating and heat treatment* (820 pp.). ASM international, Materials Park, OH (USA). ISBN-10 1-62708-012-0.

Chapter 4
Special Problems in Induction and Resistance Heating

Abstract This chapter deals with some specific applications of induction and direct resistance heating, that have been investigated also by the authors of this book. When available, analytical solutions for these 'special' applications are provided. Section 4.1 is about the resistance heating of ferromagnetic workpieces of rectangular cross-section, Sect. 4.2 describes the direct resistance heating of bended conductors. The transverse flux heating is a technology that is applied for treating thin slabs and stripes with high electrical efficiency and is the topic of the Sect. 4.3. Planar coils, often named pancake inductors, are described in the Sect. 4.4, while the effects due to the finite length of inductor are discussed in Sect. 4.5. Induction hardening of workpieces of complex geometry is one of the most important applications of induction heating, single and dual frequency processes for contour hardening are presented in Sect. 4.6. The innovative process of through heating by rotating a nonferrous billet inside a DC field produced by superconductive coils is detailed in Sect. 4.7, while in Sect. 4.8 the induction heating is produced by rotation of permanent magnets. Section 4.9 deals with the induction heating of hollow workpieces heated by internal inductors and Sect. 4.10 describes the forces acting in induction heating applications.

4.1 Resistance Heating of Ferromagnetic Workpieces of Rectangular Cross-section

In the design of installations for Direct Resistance Heating (DRH) of ferromagnetic workpieces of rectangular cross-section it is required to know the variations of the main process parameters. In particular, the supply current during heating, the value of the total impedance of the installation, the transient temperature field and temperature differential in the cross-section of the workpiece, and the process time depending on the secondary voltage of the supply transformer.

Typically, the supply of DRH installations is made using power transformers fed at industrial frequency, whose secondary voltage is selected in accordance with the

S. Lupi et al., *Induction and Direct Resistance Heating*,
DOI 10.1007/978-3-319-03479-9_4

value of the current flowing in the workpiece. This current in turn depends on the impedance of the high-current circuit of the installation, the resistance and internal reactance of the workpiece, which are temperature dependent, and the resistance of the contact system [1, 2].

Moreover, the short heating times usually adopted (typically from several tens to few hundreds of seconds), must be consistent with the process technological requirements on the temperature distribution over the cross-section of the workpiece.

A DRH installation usually consists of the following elements (Fig. 4.1): a high-power single-phase transformer (4); a high current secondary circuit (3), which comprises fixed bus-bars and short flexible cables (for adapting the supply to different workpiece lengths); the contact systems (2) and the heated workpiece (1).

Figure 4.2 shows a simplified equivalent circuit of the installation, in which the resistances of secondary winding, bus-bars, cables and contact systems have been neglected, due to their relative small value.

The external reactance X_{ext} includes the inductances of transformer windings, bus-bars and cables [3–5]. In this scheme X_{iw} and R_w represent respectively the internal reactance and the resistance of the workpiece, which during the heating process undergo significant variations [6–8].

In Ref. [4] are given experimental results about the dependence of the external reactance X_{ext} from the area of the surface of the circuit comprising the transformer secondary winding, secondary bus-bars and cables and the workpiece, when they lay in the same plane.

The value of X_{ext} is one of the main factors influencing the production capacity of the installation.

The influence of the external reactance X_{ext} on the performance of the installation will be analyzed in the following.

From one point of view, high values of X_{ext} limit too much the current flowing in the workpiece, and hence the productivity of the installation. On the other hand, the

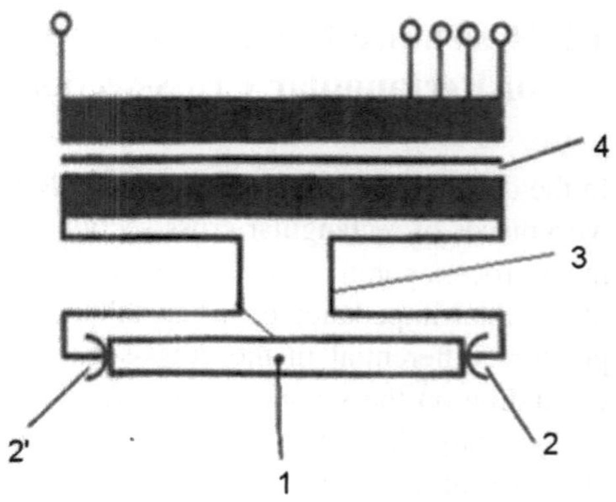

Fig. 4.1 Schematic of a DRH installation

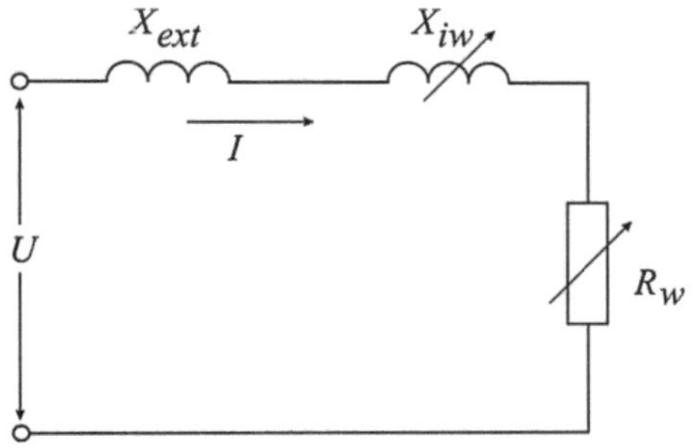

Fig. 4.2 Equivalent circuit of a DRH installation

reactance X_{ext} increases with the length of the workpiece, but at the same time the greater is the length, the greater becomes the productivity of the installation.

A number of Refs. [4, 7] provide formulas for the calculation of X_{ext} for rectangular conductors with different layouts, either with uniform distribution of current density or considering skin and proximity effects. These formulas and diagrams are only useful in the preliminary design of the system.

It is normal practice in the layout of DRH systems, to limit the external reactance by placing the supply conductors as close as possible to the workpiece. Therefore, for a more accurate evaluation of X_{ext} it must be taken into account also the redistribution of the current in the workpiece during the heating transient.

This can be done only through numerical simulations.

Some information about the dependence of the external reactance X_{ext} of a system of two parallel conductors of rectangular cross-section, as a function of the distance between them are given in Fig. 4.3 [9]. The data refers to a system of two parallel conductors made of non-magnetic material (copper or aluminum), of unit length and cross-section 20 × 120 mm. The results are in good agreement with experimental data [4].

A number of studies [10, 11] has developed analytical models for the calculation of the resistance and the reactance of ferromagnetic or nonmagnetic workpieces of rectangular cross-section. But these results are sufficiently accurate results only for ratios between the height and width of the rectangular cross-section greater than 10.

A comprehensive analysis of thermal, electrical and energetic parameters of DRH installations for heating ferromagnetic workpieces requires the solution of a coupled transient electromagnetic and thermal problem, with a high number of variables. Its solution is possible only through numerical methods; here the finite element method has been implemented in the commercial software FLUX 2D [12].

To take into account the influence of the external inductance X_{ext} on the electrical and energetic parameters of the installation, the electromagnetic-thermal problem has been solved coupled with the external circuit elements [13].

The electromagnetic process has been solved on the basis of the following differential equations, describing for the vector magnetic potential:

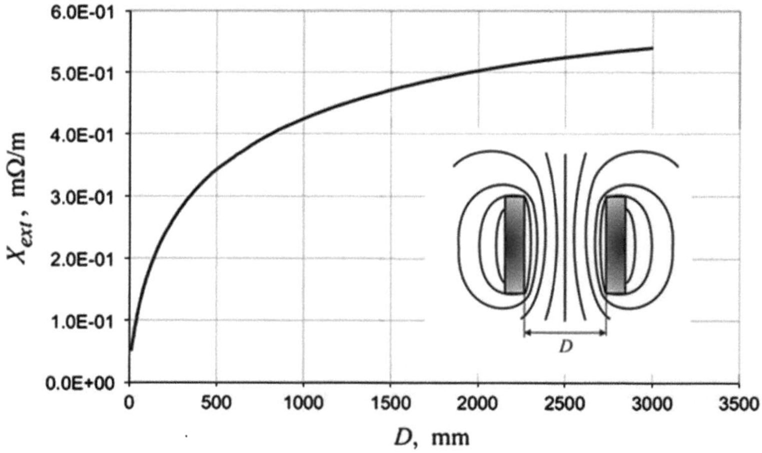

Fig. 4.3 External reactance X_{ext}, of one meter length of a system of two parallel conductors with rectangular cross-section (20 × 120 mm) of aluminium or copper, as a function of the distance D between them

$$\nabla \times \frac{1}{\mu} \nabla \times \dot{A}_z + j\frac{\omega}{\rho} \dot{A}_z = \dot{J}_z^{im}, \tag{4.1}$$

where μ—magnetic permeability; ω—pulsation; ρ—electrical resistivity; $\dot{A}_z, \dot{J}_z^{im}$—magnetic vector potential and impressed current density normal to the 2D domain.

The total current density $\dot{J}_z$ in the workpiece can be obtained from:

$$\dot{J}_z = \dot{J}_z^{ind} + \dot{J}_z^{im} = -\frac{1}{\rho}\left(j\omega \dot{A}_z + \nabla \dot{U}\right), \tag{4.2}$$

with $\dot{J}_z^{ind}$—induced current density; U—voltage applied to the workpiece ends from the external source.

In "long" workpieces, the voltage U varies only along the axial direction z with a linear law, so that the gradient of U along z is constant, i.e.

$$\dot{J}_z^{im} = -\frac{1}{\rho} \nabla \dot{U} = -\frac{1}{\rho}\frac{d\dot{U}}{dz} = \text{const.} \tag{4.3}$$

The solution for J_z allows the calculation of the distribution of the internal heating sources in the workpiece

$$w = \rho \dot{J} \times J^*. \tag{4.4}$$

The power of the internal heating sources, given by Eq. (4.4), is used to solve the thermal problem in the workpiece through the differential equation of heat conduction

$$c\gamma\frac{\partial\vartheta}{\partial t} = div(\lambda \cdot grad\vartheta) + w = \lambda\left(\frac{\partial^2\vartheta}{\partial x^2} + \frac{\partial^2\vartheta}{\partial y^2} + \frac{\partial^2\vartheta}{\partial z^2}\right) + w \tag{4.5}$$

with ϑ—temperature of slab cross-section, °C; t—time, s; λ—thermal conductivity, W/(m·K); γ—density, kg/m^3; c—specific heat of slab material, J/(kg·K).

The boundary conditions of the thermal problem are defined by the heat flux Φ_{th} transmitted by radiation and convection from the workpiece surface to the ambient

$$\Phi_{th} = \alpha_c(\vartheta - \vartheta_a) + k\varepsilon\left(T^4 - T_a^4\right),$$

with α_c—coefficient of heat transmission by convection, W/(m^2·K); k—Stefan-Boltzman constant, W/(m^2·K^4); ε—emissivity factor of the surface; $\vartheta, \vartheta_a, T, T_a$—Celsius and absolute temperatures of the surface and the surrounding medium, °C, K.

The physical properties of the slab are not linear, since thermal conductivity λ, electrical resistivity ρ and specific heat c depend on local temperature, while the relative permeability μ_r depends on temperature and magnetic field intensity. Therefore, the coupled transient electromagnetic and thermal problem must be solved iteratively.

The results of numerical solutions have been compared with the experimental data given in [5]. Figures 4.4 and 4.5 show this comparison for the case of resistance heating of ferromagnetic steel workpieces, 100 × 100 mm cross-section, 4.4 m length. The experimental data, shown in Figs. 4.4a and 4.5a, are in good agreement with the numerical results of Figs. 4.4b and 4.5b [13].

In particular, Fig. 4.4 shows the temperature transient at points 2, 3, 6 of the workpiece cross-section, indicated in the figure. The results are for supply voltages U = 95 and 130 V and value of the external reactance $X_{ext} = 1$ mΩ/m.

Figure 4.5a gives the experimental values of the temperature differential over the cross-section, as a function of the surface temperature, for different heating times (97, 220 and 460 s respectively) which give the same final thermal conditions.

Figure 4.5b shows the variations of the maximum temperature difference in the cross-section as a function of the heating time (final surface temperature 1,200 °C), with supply voltage of 95 and 130 V, corresponding to heating times of 210 and 100 s, respectively.

As shown by these data, the experimental and calculated curves agree very well within the technical accuracy of the experiment.

Moreover, the diagrams of Figs. 4.4 and 4.5 illustrate the characteristics of the heating transients in the resistance heating of ferromagnetic straight workpieces with rectangular cross-section.

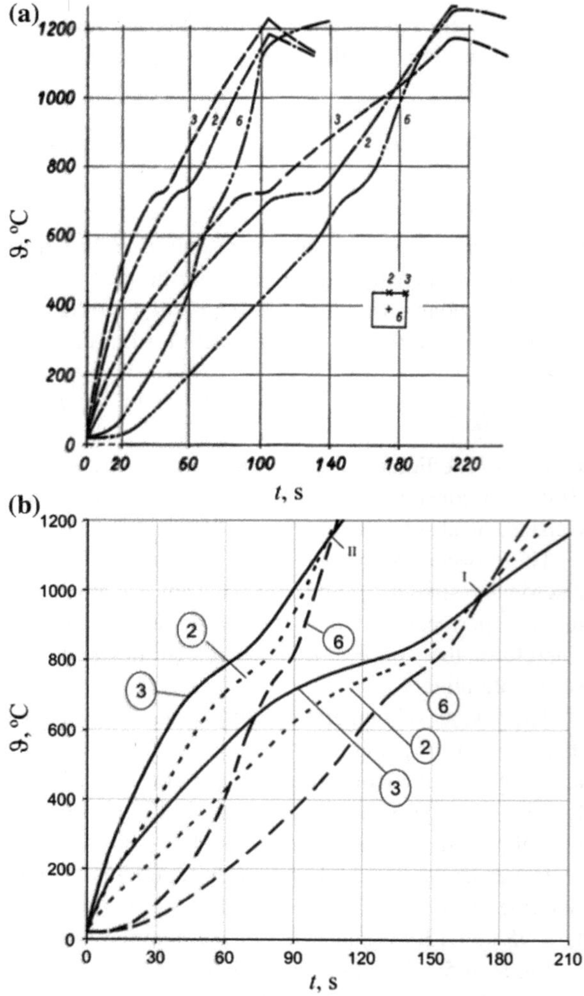

Fig. 4.4 Transient temperature distributions at surface and axis of a ferromagnetic workpiece, 4.4 m long, 100 × 100 mm cross-section **a** experimental data from [5]; **b** results calculated with $X_{ext} = 1\,\mathrm{m}\Omega/\mathrm{m}$; $I - U = 95$ V, $II - U = 130$ V

At the initial stage of heating, up to the temperature of magnetic transformation (Curie point at about 750–770 °C), the surface temperature is higher than the temperature of the axis. In this stage, the temperature differential in the cross-section reaches a maximum and then, with further heating, it reduces.

When reaching a certain temperature, dependent on the heating rate, the temperatures of the center and surface are the same. With further heating the temperature of the center begins to be higher than the surface temperature.

Such behavior can be explained as follows. Below the Curie point the workpiece is magnetic; because of the pronounced skin effect the power of the internal heating sources is significantly high in the external layers. Moreover, the heating sources are more concentrated in the corner regions (point 3 of the cross-section) than in the

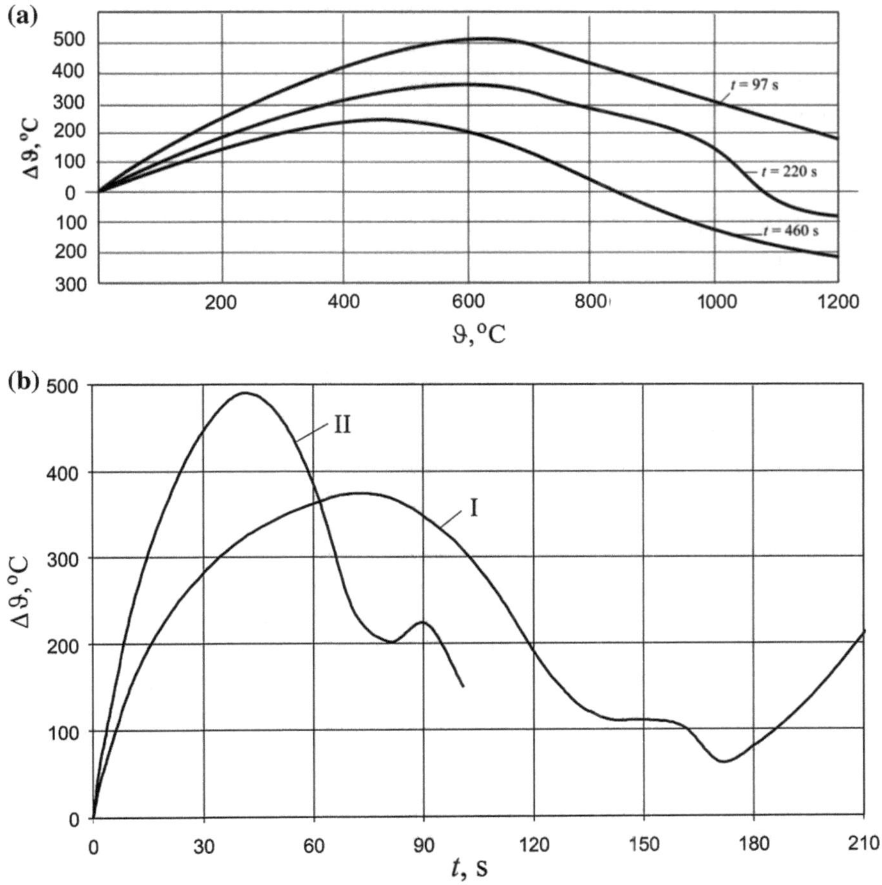

Fig. 4.5 Temperature differential in the workpiece cross-section during heating: **a** as a function of maximum surface temperature (experimental data from [5]); **b** as function of the heating time (calculations with $X_{ext} = 1\ \text{m}\Omega/\text{m}$); $I - U = 95$ V; $II - U = 130$ V

center of the side surface (point 2). Therefore, at the initial stage of the heating transient the temperature of point 3 is higher than that of point 2.

After passing through the Curie point the power in the surface layers decreases (due to the increase of the penetration depth). In this stage the heat losses by radiation and convection from the surface of the workpiece will substantially grow and reduces the rate of heating of surface layers. At the same time, the heating rate of the central region will increase due to the increase of the penetration depth, in comparison with its value in the initial stage of the heating.

It should be noted that above Curie point, the heating rate of the corner region (point 3) is lower than that of point 2 due to the more intensive losses.

The diagrams of Figs. 4.4 and 4.5 also show that the increase of the supply voltage increases the heating rate and reduces the heating time required for reaching the same final temperature. At the same time they show that the temperature differential in the cross-section of the workpiece increases in the first stage up to a maximum, and then it reduces till the intersection the temperature curves of the center and the surface.

4.1.1 *Influence of Power Supply Circuit on Heating Transient*

The influence of the power supply circuit is studied through a parametric analysis of design elements of the heating system: geometric dimensions of slab cross-section, applied voltage U, reactance X_{ext} of the external circuit.

The distributions of current density and specific power in a ferromagnetic workpiece depend on size and shape of cross-section, current flowing intensity and temperature. But, the current flowing in the workpiece is a complex function of voltage U, external reactance X_{ext} and impedance Z_w of the workpiece which, in turn, varies with the current and temperature values.

Numerical modeling allows to obtain precise values of the above quantities during the heating process, taking into account the changes of electrical, magnetic and thermal material properties in the cross-section of the workpiece.

Figure 4.6 shows some examples of numerical simulations made for ferromagnetic steel bars, 10 m length, 50 × 50 mm cross-section, with supply voltage $U = 100$ V and reactance $X_{ext} = 1$ and 2.5 mΩ/m, which are typical values for this kind of installations [9, 13].

The results show that for workpieces of relative small cross-section (less than 50 × 50 mm) the temperature differential $\Delta\vartheta$ in the cross-section is relatively small below Curie point. Above Curie point a redistribution of the temperature pattern occurs: the temperature of point 3, which up to Curie point was the highest, considerably reduces above Curie point; the contrary holds for the temperature of point 6. Moreover, with the increase of temperature above the magnetic transformation point, the differential $\Delta\vartheta$ increases and becomes significant, exceeding the value occurring below Curie point.

The Curie transition has a strong influence also on the electrical (Z_e, R_e, X_e) and energetic (P, cos φ, η_{th}) parameters of the installation, as can be seen from the diagrams of Fig. 4.6b–d.

Figure 4.7 gives other results for the heating of ferromagnetic workpieces, 10 m length, cross-section 20 × 20 mm and 50 × 50 mm, with different supply voltages. Figure 4.7a gives the time required to achieve the final temperature of 1,200 °C, as a function of the external reactance X_{ext}, for different values of the supply voltage U. Figure 4.7b shows the variation of the final temperature differential $\Delta\vartheta$ in the cross-section, as a function of the same parameters.

The diagrams show that the influence of the external reactance X_{ext} is less important for the small cross-sections (20 × 20 mm) than for the larger one.

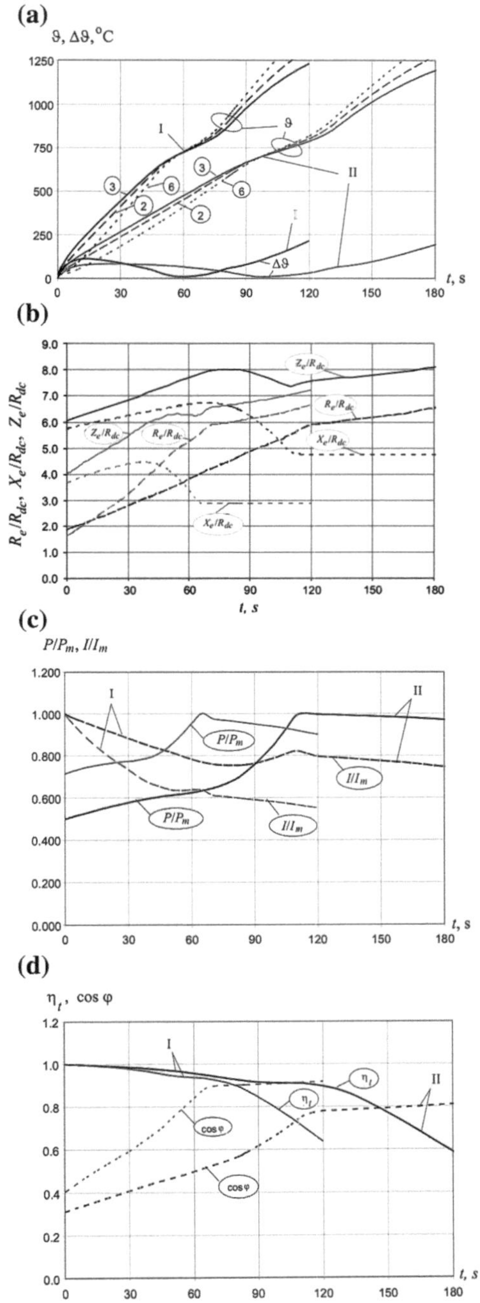

Fig. 4.6 Variations of thermal and electrical parameters of heating process: **a** temperature distributions at point 2, 3, 6 (of Fig. 4.4) and temperature differential in the cross-section; **b** relative values of resistance, reactance and total impedance, referred to the DC resistance R_{dc} at the start of heating; **c** relative values of current and active power, referred to their maximum values during the process; **d** power factor and thermal efficiency; workpiece: 50 × 50 mm cross-section, 10 m length, U = 150 V, R_{dc} = 0.8 mΩ/m; **I**—$X_{ext} = 1\,\mathrm{m}\,\Omega/\mathrm{m}$, I_m = 31,194 A, P_m = 1,758 MW; **II**—$X_{ext} = 2.5\,\mathrm{m}\,\Omega/\mathrm{m}$: I_m = 20,708 A, P_m = 1,288 MW

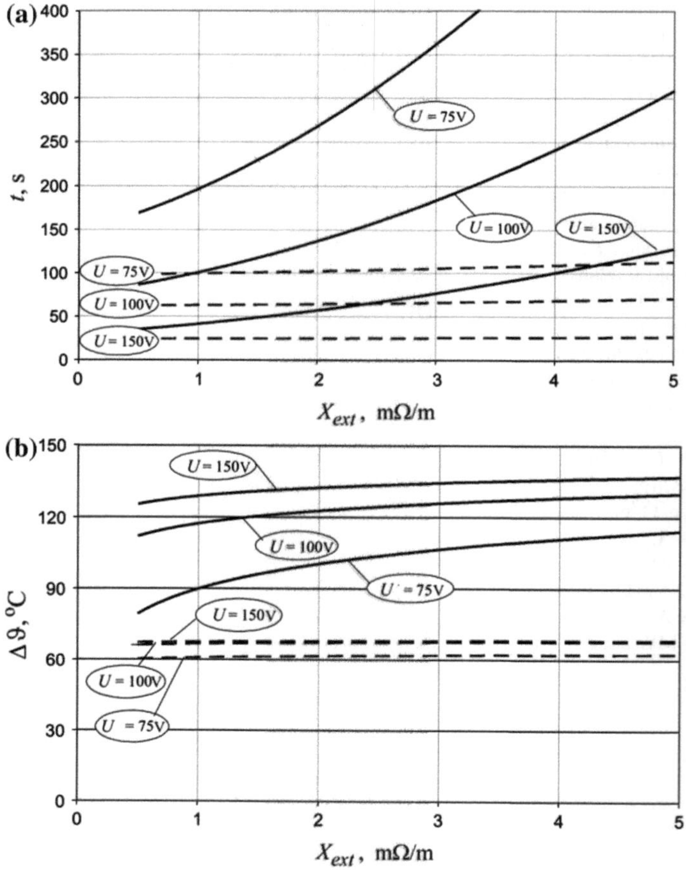

Fig. 4.7 Heating times of ferromagnetic steel bars up to 1,200 °C (**a**), and final temperature differential (**b**) as a function of the external reactance X_{ext} for different supply voltages U (*solid lines* cross-section 50 × 50 mm; *Dashed lines* cross-section 20 × 20 mm; length 10 m)

For workpieces of small cross-section (up to 20 × 20 mm), in the case here investigated the temperature differential usually required at the end of heating (typically lower than 100 °C) can be obtained with voltages U up to 150 V.

As shown in Fig. 4.8, which refers to the heating of ferromagnetic workpieces, 10 m long, 70 × 70 mm and 100 × 100 mm, a different situation occurs. In these cases it is possible to obtain at the end of heating temperature differentials which satisfy the technological requirements only by choosing heating transients, in which the intersection of temperatures of points 2–3–6 occurs as close as possible to the final heating temperature.

In Ref. [14], this criterion has been applied for equalization of the temperature field in workpieces of large cross-section by a suitable choice of the heating rate, in order to achieve the intersection of the curves of center and surface temperatures at the end of heating.

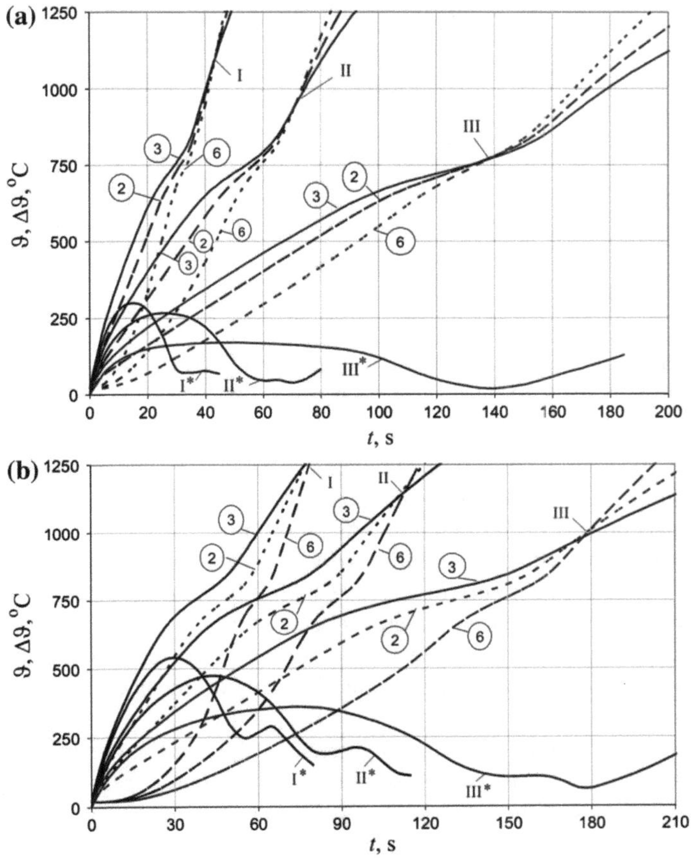

Fig. 4.8 Transient temperature of points 2, 3, 6 (Fig. 4.4) and temperature differential in the workpiece during the heating process ($X_{ext} = 1\ \text{m}\Omega/\text{m}$; supply voltage: U = 100–150–200 V; workpiece: l = 10 m; cross-section: **a** 70 × 70 mm; **b** 100 × 100 mm)

However, it should be noted that when the curves of temperatures of points 2, 3, 6 (of Fig. 4.4) intersect, the temperature differential in the cross-section of the workpiece is not necessarily equal to zero, because the maximum temperature may arise in an internal point, different from the positions 2, 3, 6.

This situation is confirmed by Fig. 4.9, which shows the graphs of maximum and minimum temperatures as well as the maximum temperature differential in the heated cross-section, as a function of time.

In the same figure data are given which show that, at constant voltage, an increase of the external reactance X_{ext} leads to a reduction of the maximum current I_m, of the maximum active power P_m delivered to the workpiece and a reduction of the average power factor cos φ_m of the installation.

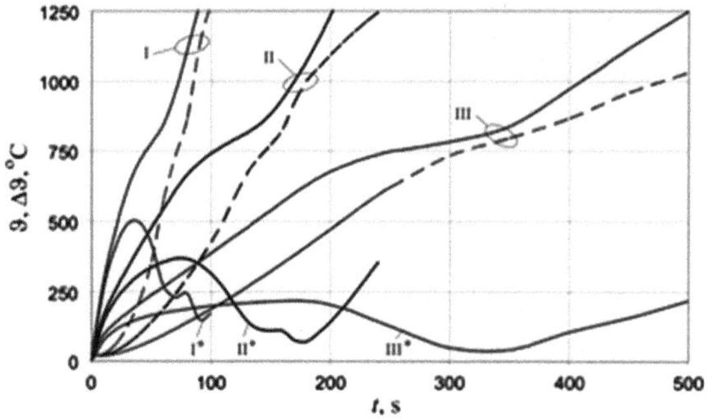

X_{ext}, mOhm	I_m, kA	P_m, MW	$\cos\varphi_m$
1,0	88.17	9,171	0,394
2,5	56.33	3,968	0,263
5,0	35.03	1,534	0,182

Fig. 4.9 Temperatures ϑ_{max} (*solid lines*), ϑ_{min} (*dashed lines*) and ($\vartheta_{max} - \vartheta_{min}$) in the cross-section during heating (*bar*: 100 × 100 mm, 10 m length; supply voltage: 225 V; X_{ext}=1–2.5–5 m Ω)

Figure 4.10 show the influence of external reactance X_{ext} and applied voltage U on the main process parameters in the heating of ferromagnetic workpieces, 10 m long, cross-sections 70 × 70 mm and 100 × 100 mm.

Figure 4.10a, b give for the above workpieces the heating times required for reaching a surface temperature of 1,200 °C; Fig. 4.10c, d give the corresponding final temperature differential in the cross-section.

It is interesting also to analyze the variations of the electrical and energetic parameters with the applied voltage *U* and the value of the external inductance X_{ext} (see Table 4.1).

The increase of voltage *U*, providing greater values of the maximum current in the heated workpiece, produces lower values of the relative magnetic permeability μ_r and higher values of the internal reactance of the heated workpiece (in the stage below the Curie point). As a consequence, with the increase of *U* also the average power factor cos φ will be lower.

Conclusions

In this paragraph the analysis of heating transients in the DRH of ferromagnetic straight workpieces with rectangular cross-section and the main design parameters of the installation are presented. The results have been obtained by a numerical model developed using the commercial software FLUX [12].

The comparison of numerical results with experimental data given in the literature has shown that the model gives a fairly accurate representation of the behavior of the electrical, power and thermal characteristics of the installation.

The provided diagrams allow the designer to predict heating times and final temperature differentials in the cross-section, as a function of the installation external reactance and the supply voltage, for a wide range of sizes of the heated workpieces.

Fig. 4.10 Parameters for heating magnetic steel bars to 1,200 °C as a function of the reactance X_{ext}, for different supply voltages U (workpiece: length: 10 m—**a**, **c** cross-section: 70 × 70; **b**, **d**: 100 × 100 mm; diagrams **a**, **b**: heating times; **c**, **d**: final temperature differential)

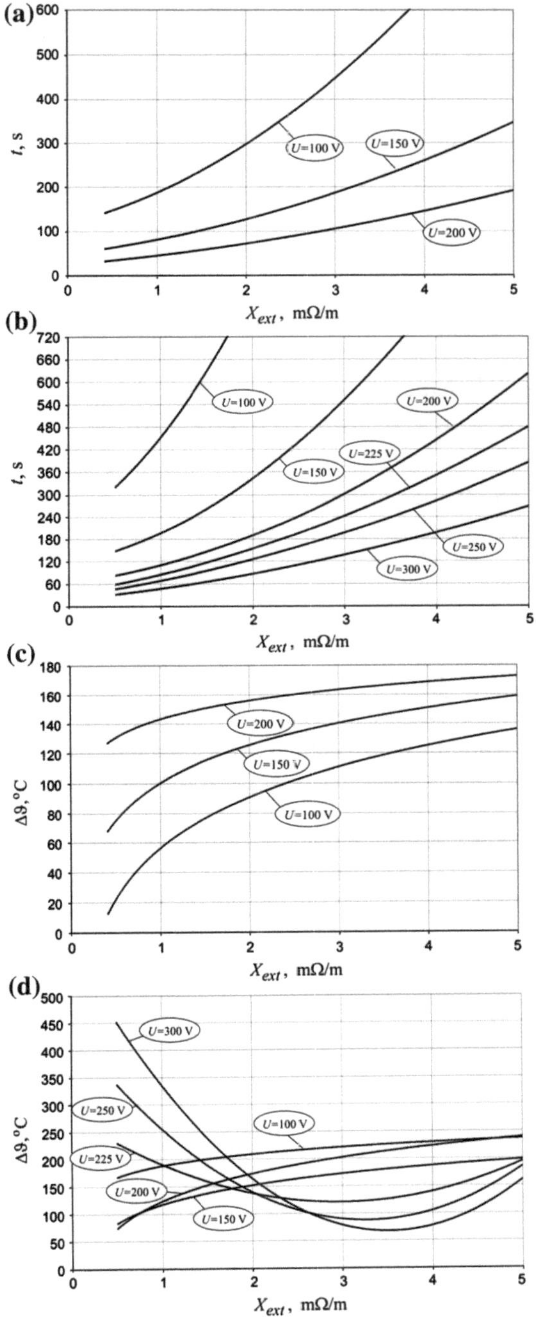

Table 4.1 Maximum current, maximum active power and average cos φ during heating

Supply voltage, V	cross-section, mm					
	70 × 70			100 × 100		
	I_m, kA	P_m, MW	cos φ_{av}	I_m, kA	P_m, MW	cos φ_{av}
100	33.94	2.161	0.564	35.59	1.817	0.420
150	53.68	4.862	0.538	57.72	4.100	0.395
200	73.72	8.642	0.522	77.67	7.272	0.373

4.2 Curved Conductors with Cylindrical Cross-Section

4.2.1 Electromagnetic Processes in Toroidal Conductors of Circular Cross-section

Toroidal conductors of circular cross-sections are widely used in electrical and electro-technological devices. Parts of toroid conductors with bending angles from 90° to 360° can be used as elements of electrical equipment in melting and induction heating units (e.g. single-turn inductors). Moreover, there are a number of direct resistance-heating systems where curved workpieces with cylindrical cross-section are heated.

It is therefore important to evaluate not only the distributions of current and power densities, but also the resistance and reactance of this type of conductors, as a function of their geometrical parameters.

We first consider a single toroid conductor of circular cross-section shown in Fig. 4.11.

Several authors have presented analytical solutions for the current density distribution in the cross-section of a toroid conductor [8, 15–18].

These solutions are valid under the following assumptions: (a) electromagnetic wave concentrated in a thin surface layer, (b) non-magnetic conductor material, (c) constant resistivity of the conductor, and (d) electric and magnetic fields with exponential distribution along radius of the conductor cross-section.

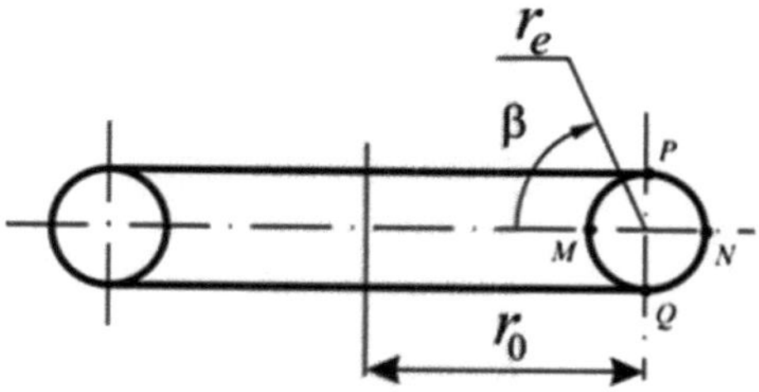

Fig. 4.11 Geometry of a toroidal conductor of circular cross-section: r_0, r_e—radii of torus and cross-section; β—angular coordinate along cross-section circumference

In Ref. [16], A. Fok, gives the following expression for the current density distribution:

$$J(\beta) = \frac{c}{4\pi^3} \cdot \frac{A}{a^2} \left[2\left(\frac{1}{\alpha} - \cos(w(\beta))\right)\right]^{1.5} \times \left[\frac{1}{f_0} - 2\sum_{n=1}^{\infty} \frac{\cos(n \cdot w(\beta))}{4n^2 - 1} \cdot \frac{1}{f_n}\right] \tag{4.6}$$

with c—velocity of light in vacuum; *I*—r.m.s. value of the current flowing in the toroid conductor;

$A = \frac{I \cdot \pi a}{c} \cdot \left(\frac{g_0}{f_0} - 2\sum_{n=1}^{\infty} \frac{1}{4n^2-1} \cdot \frac{g_n}{f_n}\right)^{-1}$—magnetic vector potential;

$a = \sqrt{r_0^2 - r_e^2}; \alpha = r_e/r_0; w(\beta) = \arccos\left[\frac{\cos(\pi - \beta) \cdot \frac{1}{\alpha} + 1}{\cos(\pi - \beta) + \frac{1}{\alpha}}\right];$

$f_n = \frac{1}{\pi} \cdot \int\limits_{-v_0}^{+v_0} \sqrt{2[ch(v_0) - ch(t)]} \cdot ch(nt)dt$—circular function;

$g_n = \frac{1}{2\pi} \cdot \int\limits_{0}^{\pi} \sqrt{2[ch(v_0) - \cos(t)]} \cdot ch(nt)dt$—circular function;

$ch(v_0) = 1/\alpha$.

Equation (4.6) is recommended by the author for all values of α; but for $\alpha \le 0.2$ it can be re-arranged in the simpler form:

$$J(\beta) = \frac{c}{4\pi^3} \cdot \frac{A}{a^2} \cdot \frac{(2 \cdot ch(v_0))^{1.5}}{f_0} \times \left[1 - (v_0 - 0.5 + 2\ln 2) \cdot \frac{\cos(w(\beta))}{ch(v_0)}\right]. \tag{4.7}$$

In Ref. [18], V. Vologdin gives another simple equation, obtained by modifying formulae (4.7), which is recommended in engineering calculations for $\alpha \le 0.1$:

$$J(\beta) = J_{\delta e} \frac{1 - c_1 \cdot \left[1 - \frac{1-\alpha^2}{1-\alpha\cos(\beta)}\right]}{1 - c_1 \cdot \left(1 - \sqrt{1 - \alpha^2}\right)} \tag{4.8}$$

where $J_{\delta e} = \frac{I \cdot \sqrt{2}}{2\pi\delta_e r_e}$—current density in a straight conductor of circular cross-section, averaged in a layer of thickness equal to the penetration depth δ in the conductor material;

$$c_1 = \ln 4 - 0.5 + \ln\frac{1 + \sqrt{1 - a^2}}{a}.$$

In Ref. [8], G. Šneerson gives a solution analogous to Eq. (4.6), but transformed introducing Legendre functions:

$$\begin{aligned} J(\beta) = &- \frac{I \cdot L}{8\sqrt{2}\pi^3 \cdot \mu_0 r_e^2}\left[\frac{\frac{1}{\alpha} - \cos(\psi(\beta))}{\frac{1}{\alpha} - 1}\right]^{\frac{3}{2}} \\ &\times \left[\frac{1}{P^1_{-1/2}\left(\frac{1}{\alpha}\right)} + 2\sum_{n=1}^{\infty}\frac{\cos(n\psi(\beta))}{P^1_{-1/2}\left(\frac{1}{\alpha}\right)}\right] \end{aligned} \tag{4.9a}$$

with:
P^1_v, Q^1_v—associated Legendre functions of first and second kind,

$$\begin{aligned} L = &\frac{\mu_0 r_e \pi^2}{2}\sqrt{\left(\frac{1}{\alpha}\right)^2 - 1} \\ &\times \left[\frac{Q^1_{-1/2}\left(\frac{1}{\alpha}\right)}{P^1_{-1/2}\left(\frac{1}{\alpha}\right)} - 2\sum_{n=1}^{\infty}\frac{1}{4n^2 - 1}\cdot\frac{Q^1_{n-1/2}\left(\frac{1}{\alpha}\right)}{P^1_{n-1/2}\left(\frac{1}{\alpha}\right)}\right]^{-1} \end{aligned} \tag{4.9b}$$

—inductance of the toroid conductor.

In paper, Tayler [17] has proposed a solution using circular functions, but converted to another form, which also includes Legendre functions:

$$J(\beta) = \frac{c}{4\pi a}\sqrt{\frac{1}{\alpha} - \cos(w(\beta))} \cdot \sum_{n=0}^{\infty}\left[C_n \cdot \cos(n\psi(\beta))\right], \tag{4.10}$$

where:

$$C_0 = \frac{1}{4}\cdot\frac{P^0_{-1/2}\left(\frac{1}{\alpha}\right)}{P^1_{-1/2}\left(\frac{1}{\alpha}\right)} - \left[\frac{3}{16\frac{1}{\alpha}} + \frac{15}{512\left(\frac{1}{\alpha}\right)^2}\right]\cdot\frac{P^0_{-1/2}\left(\frac{1}{\alpha}\right)}{P^1_{-1/2}\left(\frac{1}{\alpha}\right)}$$

$$C_n = \frac{1}{2}\left(n-\frac{1}{2}\right)\left(n-\frac{3}{2}\right)\frac{B_{n-1}}{\left(\frac{1}{\alpha}\right)^{n-1}}\cdot\frac{P^0_{n-3/2}\left(\frac{1}{\alpha}\right)}{P^1_{n-3/2}\left(\frac{1}{\alpha}\right)} - n\left(n+\frac{1}{2}\right)$$
$$\cdot\frac{B_n}{\left(\frac{1}{\alpha}\right)^{n-1}}\cdot\frac{P^0_{n-1/2}\left(\frac{1}{\alpha}\right)}{P^1_{n-1/2}\left(\frac{1}{\alpha}\right)} + \frac{1}{2}\left(n+\frac{1}{2}\right)\frac{B_n}{\left(\frac{1}{\alpha}\right)^{n}}\cdot\frac{P^0_{n+1/2}\left(\frac{1}{\alpha}\right)}{P^1_{n-1/2}\left(\frac{1}{\alpha}\right)}$$
$$+\frac{1}{2}\left(n+\frac{3}{2}\right)\left(n+\frac{1}{2}\right)\frac{B_{n+1}}{\left(\frac{1}{\alpha}\right)^{n+1}}\cdot\frac{P^0_{n+1/2}\left(\frac{1}{\alpha}\right)}{P^1_{n+1/2}\left(\frac{1}{\alpha}\right)}$$

$$B_0 = 1; B_1 = -\frac{1}{2} - \frac{5}{64\cdot\left(\frac{1}{\alpha}\right)^2}; B_2 = -\frac{1}{16} - \frac{3}{128^2_0};\ldots$$

$$B_n = \frac{2n-3}{4n}B_{n-1}.$$

Current Density Distributions

In order to assess the applicability of the above solutions to engineering tasks, the current density distributions in the cross-section of a toroid conductor, calculated with Eqs. (4.6)–(4.10), have been compared with the ones computed with the numerical software FLUX [15].

The results of calculations, made with $m = \sqrt{2}r_e/\delta_e = 15$ and $\alpha = 0.05; 0.125; 0.2; 0.25$, are presented in Fig. 4.12 using the relative units $J^* = J(\beta)/J(\pi)$.

Figure 4.13 shows the current density distributions along the the axes of symmetry *PQ* and *MN* of the cross-section of the toroid conductor (see Fig. 4.11). The distribution of the current density is normalized with reference to the value at point M, i.e. it is given the ratio $\overline{J} = J/J_M$.

The current density distributions have been evaluated along two axes that have their origin in the points M for the line *MN* and P for the line *QP* (Fig. 4.11); moreover, the abscissa is normalised to the conductor diameter by the ratio $\overline{x} = x/2r_e$.

Resistance and Reactance of Toroid Conductors

As shown by the previous results, in the examined range of the parameter α the most accurate solution for the current density is given by Eqs. (4.6) and (4.9a).

The current density distribution is important for the evaluation of the inductance of the current carrying conductor; its analytical expression is given in Eq. (4.9b).

The inductance values given by this equation have been compared with the ones obtained by numerical simulations done with the software FLUX [12].

To the authors' knowledge there are not in the literature analytical solutions for the calculation of the resistance of toroidal conductors. Therefore, the analysis of the behaviour of the resistance has been done only by numerical calculations.

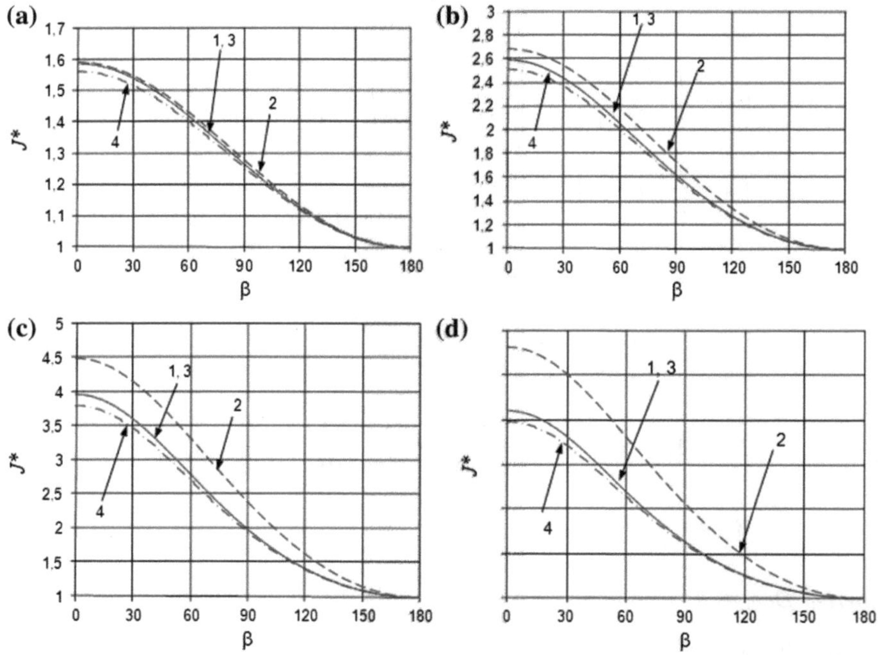

Fig. 4.12 Comparison of analytical and numerical calculations: *1* Eq. (4.10); *2* Eqs. (4.7) and (4.8); *3* Eqs. (4.6) and (4.9); *4* with flux (m = 15; $j^* = j(\beta)/j(\pi)$. **a** α = 0.05, **b** α = 0.125, **c** α = 0.2, **d** α = 0.25)

Figure 4.14 shows the diagrams of L_a and L_n, the inductance values obtained with the analytical and numerical methods respectively.

Figure 4.15 shows, for different values of *m* and *α*, the differences ε_{an} between analytical and numerical results of inductance, calculated with the formula:

$$\varepsilon_{an} = (L_a - L_n)/L_a \cdot 100\,\%.$$

In Fig. 4.15 is highlighted the region in which the difference ε_{an} between numerical and analytical results is lower than 5 %. This region corresponds to the following relationship, linking the parameters *m* and *α*:

$$\alpha \leq 0.032\,\mathrm{m} - 0.089 \qquad (4.11)$$

As regards current density, Eqs. (4.7) and (4.8) give almost identical values.

A. Fock suggests to use Eq. (4.7) for values of $\alpha \leq 0.2$. For $\alpha = 0.2$ and $m = 15$, the difference between the results given by Eqs. (4.7) and (4.8) and the numerical values obtained with FLUX, reaches 23.45 %.

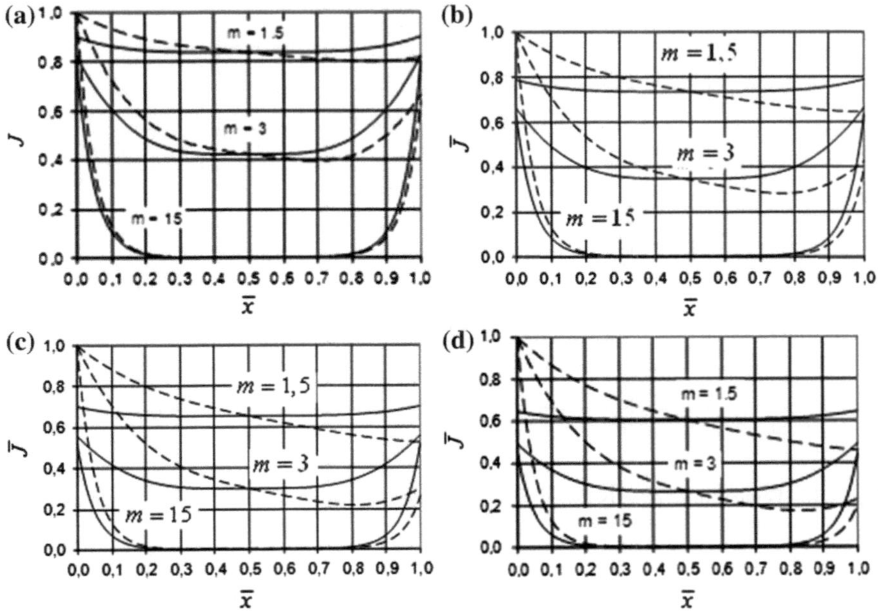

Fig. 4.13 Current density distributions along symmetry axes of the cross-section, normalised to the value of current density at point M (see Fig. 4.11) ($\bar{J} = J/J_M; \bar{x} = x/2r_e$; *dashed lines* along MN, *solid lines* along PQ. **a** α = 0.05, **b** α = 0.125, **c** α = 0.2, **d** α = 0.25)

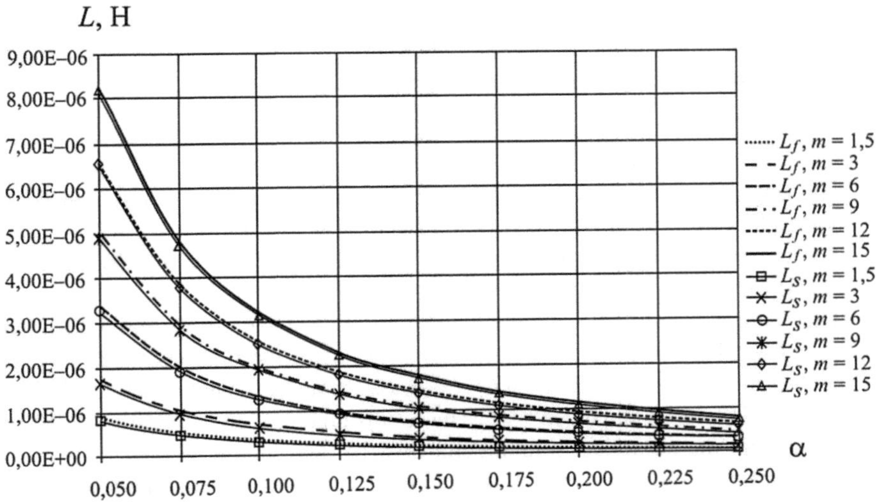

Fig. 4.14 Inductance of toroid conductors, calculated analytically (L_a) and numerically (L_n) as a function of parameters m and α

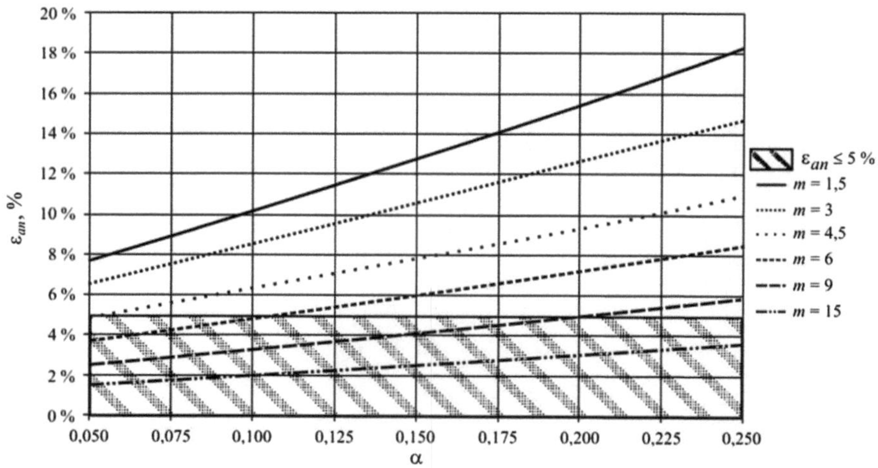

Fig. 4.15 Percent difference between inductance values given by Eq. (4.9) and calculations with FLUX, as a function of m and α

Equation (4.8) is recommended by Vologdin for $\alpha \leq 0.1$. For $\alpha = 0.1$, Eqs. (4.7) and (4.8) give a difference with numerical solution of 5.75 %, while for $\alpha = 0.05$ the difference is 2.23 %. Thus, for engineering calculations, within an error not exceeding 6 %, we can recommend the use of Eqs. (4.7) and (4.8) only for $\alpha \leq 0.1$.

The diagrams of Figs. 4.16 and 4.17 give the values of the resistance R and the reactance X of toroid conductors calculated with numerical simulations.

Figures 4.18 and 4.19 give the skin effect coefficients k_r and k_x for the resistance and the reactance, as a function of m and α. Skin effect coefficients are defined by the relations $k_r = R/R_{dc}$ and $k_x = X/R_{dc}$, where R_{dc} is the DC resistance of a straight cylindrical conductor of radius r_e and length $\ell = 2\pi r_0$.

Conclusions

- The ring effect increases with the increase of α, and it is more pronounced in conductors of large cross-section (i.e., large values of m)
- The current density distributions obtained with Eqs. (4.7) and (4.8) are coincident for values of $\alpha \leq 0.25$.
- The current density distributions given by Eqs. (4.6), (4.9) and (4.10) are coincident for values $\alpha \leq 0.25$. In this range of α, the difference between the analytical solutions solutions (4.6), (4.9), (4.10) and the numerical calculations (for $m = 15$) does not exceed 5 %. The results given by Eq. (4.8), for $\alpha \leq 0.125$, differ from the numerical solution (for $m = 15$) less than 8 %.
- The current density distribution in the cross-section of the conductor along the direction PQ is non-linear and symmetrical about the centre of the cross-section.

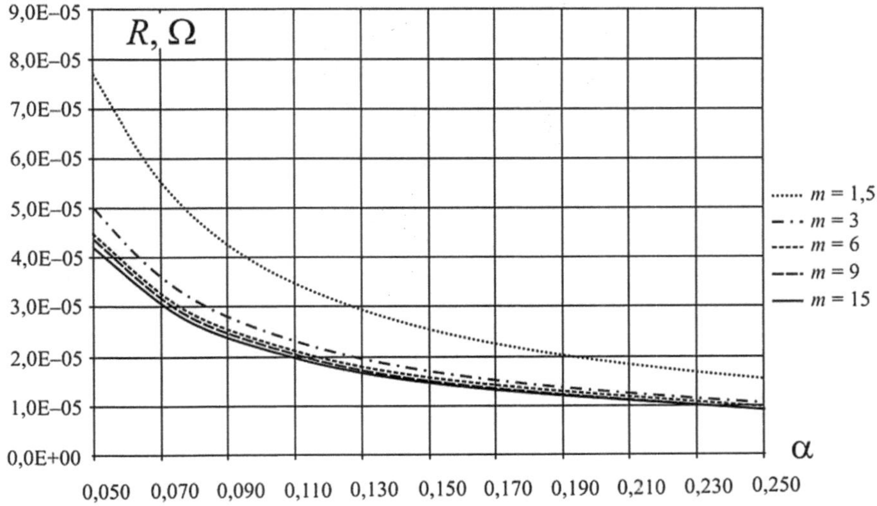

Fig. 4.16 Toroid conductor resistance as a function of m and α

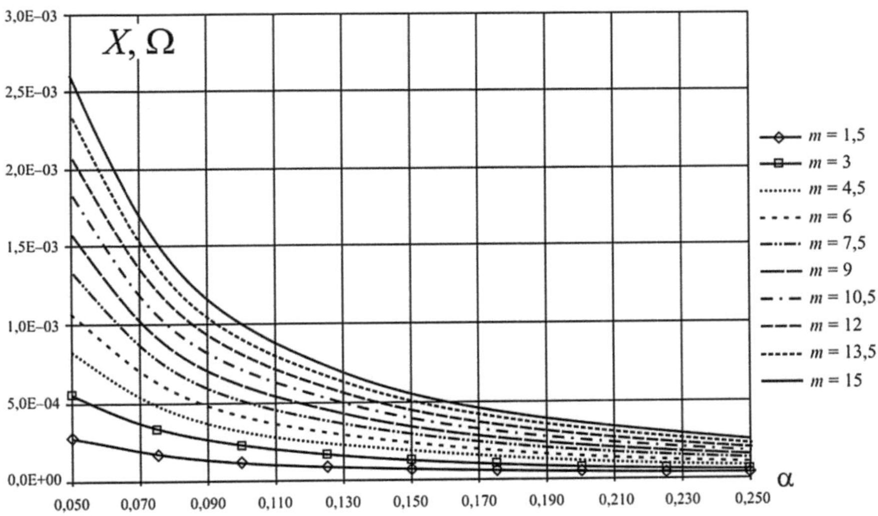

Fig. 4.17 Toroid conductor reactance as a function of m and α

- With the increase of α, the current density distribution along the line PQ becomes more uniform. This effect is more pronounced for relatively small values of m.

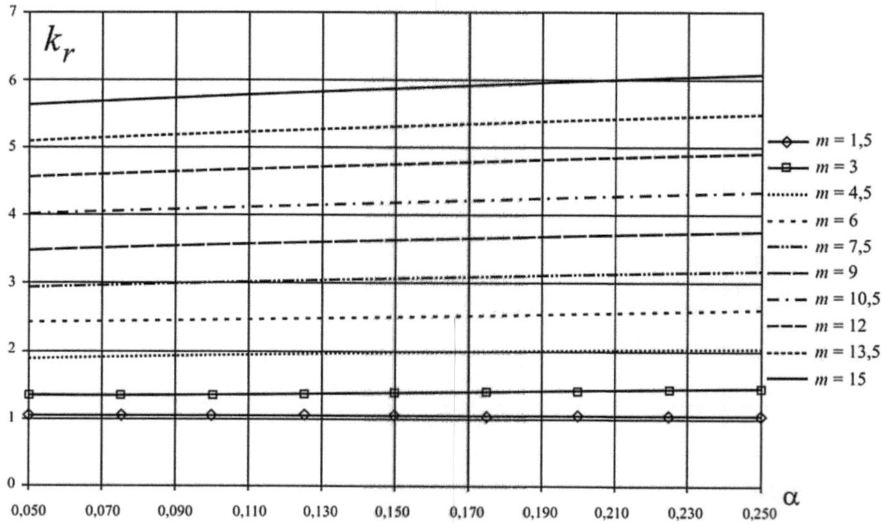

Fig. 4.18 Coefficient k_r as a function of m and α

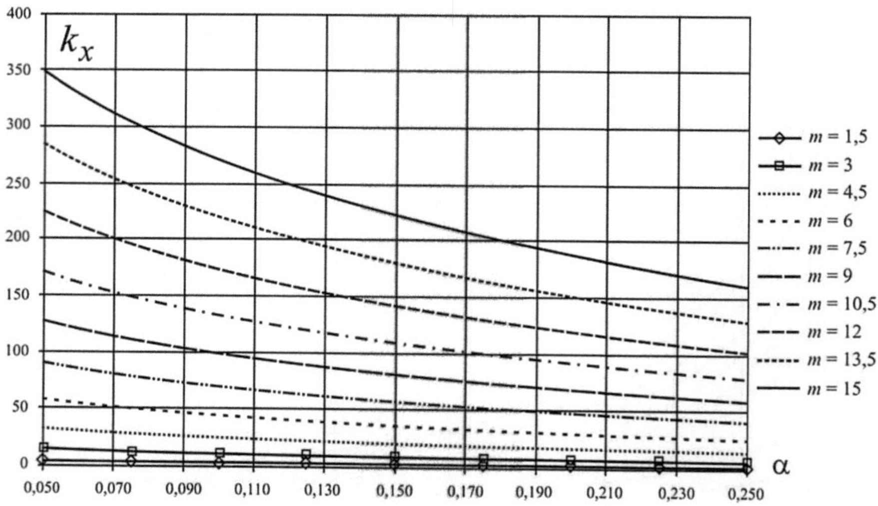

Fig. 4.19 Coefficient k_x as a function of m and α

- The current density distribution along MN is non-linear and non-symmetric about the center of the conductor cross-section, as a consequence of the ring effect.
- Discrepancies between inductance values calculated with Eq. (4.11) and numerical results do not exceed 5 %.

- The skin effect coefficient k_r of toroidal conductors, in the range $0.05 \le \alpha \le 0.25$, is weakly dependent on α. Increasing the value of m from 1.5 to 15, the coefficient k_r increases by 6 times.
- The coefficient k_x decreases non-linearly with α; this decrease is more pronounced for $m = 15$. With the increase of m from 1.5 to 15, the average value of the coefficient increases by 300 times.

4.2.2 Resistance Heating of Curved Cylindrical Work-Pieces

Current Density and Specific Power Distributions

Practical application of electromagnetic and thermal processes can be found in resistance heating of curved ferromagnetic pipes of large cross-section (e.g. 200 mm diameter) used in high-pressure steam lines of thermal power plants or in cogeneration plants.

The current density distributions along the perimeter of curved cylindrical workpieces with arbitrary angle of curvature have been presented by the authors in Refs. [1, 19–23]. These distributions have been obtained as a result of a statistical analysis of experimental data for ferromagnetic workpieces and can be described by the formula (similar to Eq. (4.8)):

$$J_{me} = J_{me,av}\left(\frac{1 - c_1\left(1 - \frac{1-\alpha^2}{1-\alpha\cos\beta}\right)}{1 - c_1\left(1 - \sqrt{1-\alpha^2}\right)}\right)^{\frac{\varepsilon}{\varepsilon_0}} \tag{4.12}$$

with: $J_{me,av} = 1.68 \cdot I/(2\pi r_e \delta_{e)}$—average current density along perimeter of a cylindrical ferromagnetic workpiece of radius r_e, taking into account the dependence $\mu_r = f(H)$; $\delta_e = \sqrt{2\rho/\omega\mu_e}$; $\alpha = r_e/r_0$; ε—angle of curvature of the workpiece, $\varepsilon_0 = 2\pi$—angle of curvature of toroid conductor; and $c_1 = \ln 4 - 0.5 + \ln\frac{1+\sqrt{1-a^2}}{a}$.

Figure 4.20 shows the azimuthal distributions of the current density at the surface of a bent conductor, obtained by Eq. (4.12), and experimental data for an angle of curvature $\varepsilon = \pi/2$. The comparison of the distributions has been done for different average temperature values in the conductor cross-section.

The analysis of calculated and experimental results shows that at the initial stage of the heating, when the temperature of the workpiece is in the range 20–100 °C, the difference between experimental and calculated values of the current density is less than ±8 %. When the workpiece is heated at 700 °C the difference does not exceed 15 %. Thus, Eq. (4.12) may be used for approximate calculations of the current density distribution on the surface of curved ferromagnetic workpieces of circular cross-section.

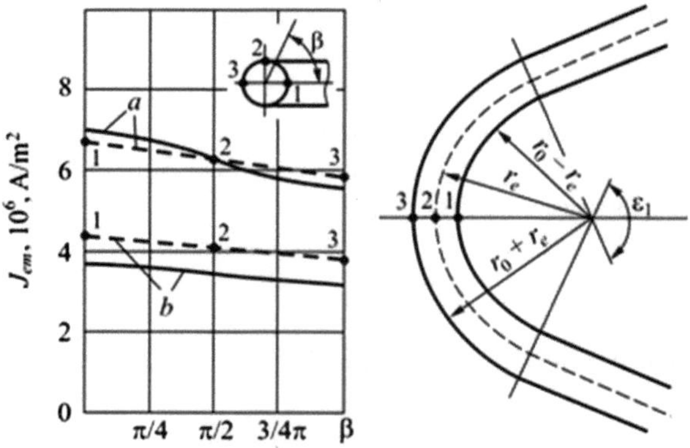

Fig. 4.20 Current density distribution on the surface of a curved workpiece, as a function of the azimuthal coordinate β, for $\varepsilon = \pi/2$: *a* $\vartheta_{av} = 20$ °C; *b* $\vartheta_{av} = 700$ °C; *dashed lines* experimental, *solid lines* calculated by Eq. (4.12)

Equation (4.12) is also useful since it covers two extreme cases. In fact, for $\varepsilon = 0$, it reduces to the relationship $J_{me} = J_{me,av}$, which is characteristic of the current density distribution on the surface of a straight cylindrical ferromagnetic conductor [24]. For $\varepsilon = 2\pi$ the equation becomes the same obtained for a toroid conductor of circular cross-section [15].

We analyse now the behaviour of the current density J_{me} as a function of the ratio $\alpha = r_e/r_0$ and the bending angle α. To this end, we rewrite Eq. (4.12) in relative units:

$$\overline{J}_{me} = \frac{J_{me}}{J_{me,av}} = \left(\frac{1 - c_1\left(1 - \frac{1-\alpha^2}{1-\alpha\,\cos\beta}\right)}{1 - c_1\left(1 - \sqrt{1-\alpha^2}\right)} \right)^{\frac{\varepsilon}{\varepsilon_0}}.$$

We have considered bending angles between $\varepsilon = 0$ and $\varepsilon = 2\pi$, and α in the range 0.025–0.225.

Figure 4.21 gives the values of $\overline{J}_{me}$ as a function the bending angle α and the parameter ε, at the surface points $\beta = 0$ and $\beta = \pi$, where the maximum difference of the current density on the surface of the workpiece occurs.

The shaded area of Fig. 4.21 defines the region of values of geometric parameters ε and α, where—within an accuracy of 10 %—the curvature of the workpiece can be neglected.

Figure 4.22 shows the azimuthal distributions of the current density on the surface of a curved workpiece as a function of its geometrical parameters ε and α. The polar coordinate system used in this case, gives with the angle β the azimuthal position of a point on the surface, while the amplitude of the radial vector gives the relative current density $\overline{J}_{me}$.

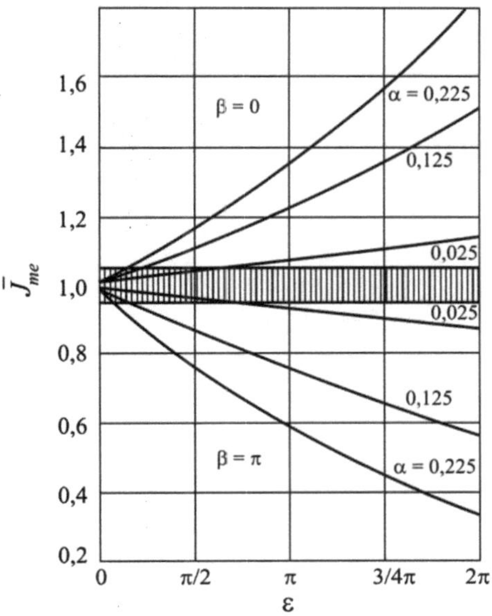

Fig. 4.21 Distributions of relative current density $\overline{J}_{me}$ at workpiece surface points $\beta = 0$ and $\beta = \pi$, as a function of angle of curvature ε and parameter $\alpha = r_e/r_0$

The distributions are shown for three values of the bending angle, ε = 0 – π – 2 π and three values of the parameter α = 0.025–0.125–0.2.

It is evident from the graphs that the unevenness of the distribution of the current density along the perimeter increases with the increase of α.

We can observe that for values of $\alpha \le 0.025$ it is possible to use Eq. (4.12) with ε = π. In this case the maximum error in the evaluation of J_{me} for angles of curvature ε = 0 and ε = 2π, is lower then 10 % in comparison with the values calculated with ε = π.

In the calculation of the temperature fields in DRH of solid cylindrical curved workpieces, it should be taken into account the radial and azimuthal distribution of the power density. We present in the following formulas giving such distributions in non-magnetic and magnetic workpieces.

Non-magnetic Workpieces

The current density distribution on the cross-section of the workpiece can be obtained by the relationship:

$$J_{em} = J_{em,av=} \cdot K(\beta) \cdot K(r)_{=},$$

where $K(\beta) = \overline{J}_{em}$—function which characterises the azimuthal distribution of the current density along the workpiece perimeter; $J_{em,av=} = I/(\sqrt{2}\pi r_e \delta_e)$—average value along perimeter of the current density at the workpiece surface, calculated with the average value of resistivity ρ in the heating cycle and relative permeability $\mu_r = 1$.

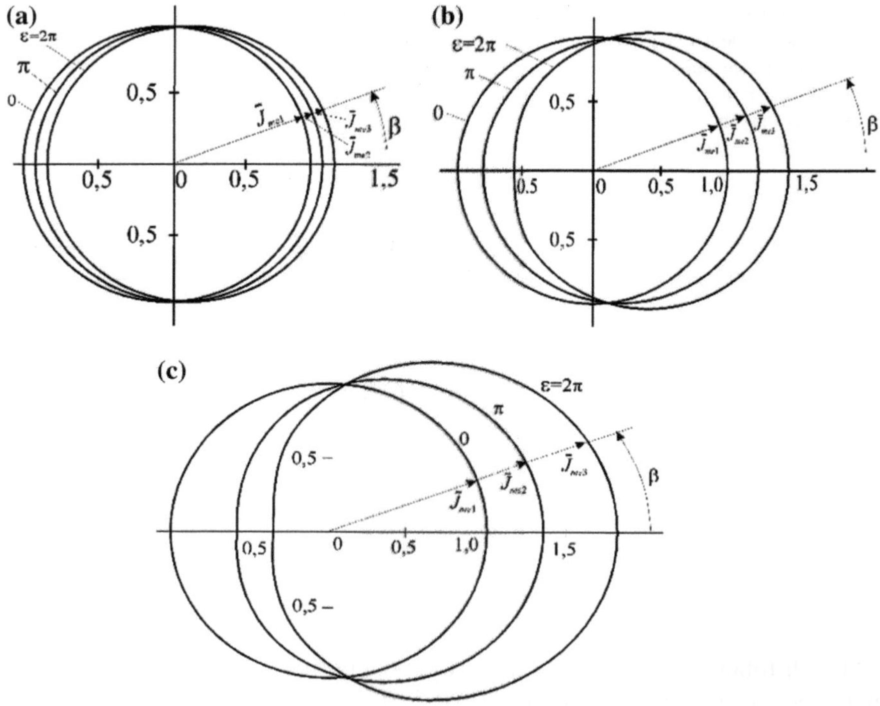

Fig. 4.22 Diagram of azimuthal distributions of the relative current density $\overline{J}_{me}$: **a** $\alpha = 0.025$; **b** $\alpha = 0.125$; **c** $\alpha = 0.2$

In this case the function $K(r)_{=}$, characteristic of the radial distribution of current density with relative permeability $\mu_r = 1$, can be obtained by the formulae:

- for $m > 10$ [18]

$$K(r)_{=} = \frac{e^{-\frac{(1-\xi)m}{\sqrt{2}}}}{\sqrt{\xi}},$$

with $\xi = r/r_e$;

- for $m < 10$ [3]

$$K(r)_{=} = \sqrt{\frac{ber^2(m\xi) + bei^2(m\xi)}{ber^2(m) + bei^2(m)}}$$

The power density of heating sources can be evaluated by the equation:

$$w(\beta, r)_{=} = \rho(\vartheta) \cdot J^2_{em,av=} \cdot K(\beta)^2 \cdot K(r)^2_{=}, \quad (4.13)$$

Ferromagnetic Workpieces

For ferromagnetic workpieces the current density distribution in the cross-section can be determined by [1]:

$$J_m = J_{em,av\approx} \cdot K(\beta) \cdot K(r)_{\sim},$$

where $J_{em,av\approx} = 1.68 \cdot I/(2\pi r_e \delta_e)$—average value along the perimeter of the surface current density, calculated with the average value of resistivity ρ during the heating cycle and relative permeability $\mu_e = f(H, \vartheta)$, evaluated with the magnetic field intensity H_e at the workpiece surface; $\delta_e = 503\sqrt{\rho/\mu_e f}$—penetration depth, calculated with constant value of resistivity ρ and the relative permeability μ_e; $K(r)_{\sim} = 1 - [(r_e - r)/(1.457 \cdot \delta_e)]$—function which characterises the radial distribution of the current density with variable relative permeability [1].

In this case the distribution of the heating sources in the cross-section of the bent workpiece can be determined by the formula [23]:

$$w(\beta, r)_{\approx} = \rho(\vartheta) \cdot J^2_{em,av\approx} \cdot K(\beta)^2 \cdot K(r)^2_{\approx}. \quad (4.14)$$

Equations (4.13) and (4.14) take into account the variations of resistivity and magnetic permeability with temperature and, as regards permeability, also with the magnetic field intensity. They allow to obtain the distribution of internal heating sources in the cross-section of the curved conductor as a function of the curvature angle ε, within the range 0–2π, and the ratio between the external radius of the workpiece r_e and the radius of curvature of its axis r_0.

These distributions of internal heating sources, functions of the coordinate β and the radius *r*, can be recommended for the solution of the thermal problem in the DRH of cylindrical workpieces with any angle of curvature.

4.2.3 Equalization of Temperature Distribution in the Cross-section of Bent Conductors

In resistance heating of curved cylindrical workpieces, temperature differences exist not only in radial direction but also along the perimeter of the workpiece cross-section. The maximal temperature occurs on the external surface of the workpiece, on the radius of curvature $r_{0i} = r_0 - r_e$, while the minimal one is experienced on the radius $r_{e0} = r_0 + r_e$ (Fig. 4.23).

This temperature difference is present regardless of the type of current (constant or variable) and power source (stabilized voltage or current) [20, 21].

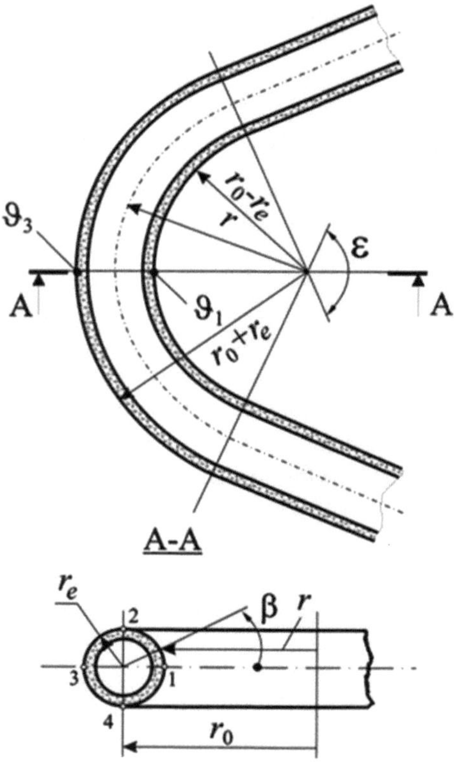

Fig. 4.23 Geometry of the heated workpiece

In the following, the temperatures ϑ_1 and ϑ_3 are evaluated on points 1 and 3 of Fig. 4.23. The heated workpiece is a curved ferromagnetic tube with external diameter $d = 0.159$ m, angle of curvature $\varepsilon = 90°$, radius of curvature of the central axis $r_0 = 0.64$ m and wall thickness $t = 0.02$ m.

The temperature difference $\Delta\vartheta_{13} = \vartheta_1 - \vartheta_3$ along the workpiece perimeter is determined mostly by the heating rate.

Figure 4.24 shows a diagram giving the temperature difference $\Delta\vartheta_{13}$ as a function of the r.m.s. value of the supply current *I*, at industrial frequency. The curve $\Delta\vartheta_{13} = f(I)$ refers to the average final temperature $\vartheta_{fin} = 600$ °C.

The experimental setup and test methods are described in [19]. As can be seen from Fig. 4.24, the increase of the supply current produces an increase of the temperature difference. Typically, DRH processes are very fast (heating rates of more than 10 °C/s), in most cases giving rise in the cross-section of bent cylindrical workpieces to temperature differentials which are above the maximum values admissible by the technological requirements.

But also in processes with heating rates lower than 1° C/s, unacceptable temperature differences may occur.

To heat up to 700 °C the workpiece used in the experimental tests without thermal insulation, it was necessary a current value exceeding 11 kA, and the final

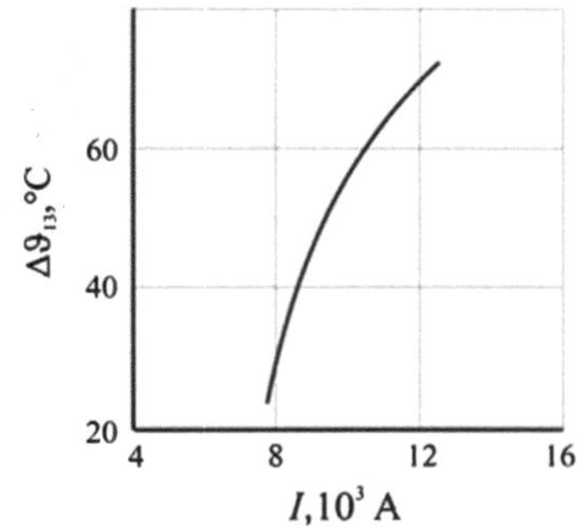

Fig. 4.24 Values of temperature difference $\Delta\vartheta_{13}$ as a function of current for $\vartheta_{av,\,fin} = 600$ °C

temperature difference $\Delta\vartheta_{13,fin}$ exceeded 70 °C, i.e. a value which is well above the admissible technological standard limit ($\Delta\vartheta_{stand} \leq 20$ °C).

For this reason in the DRH of curved workpieces it is necessary to adopt special methods for reducing the temperature differential in the cross-section.

The methods for equalization of the temperature field in the cross-section can be conventionally sub-divided into two groups (as shown in Fig. 4.25) [25].

The first group refers to processes which modify the heat transfer conditions from the surface of the heated body; the second is based on the redistribution of the internal heating sources in the cross-section of the workpiece.

The methods belonging to the first group can be implemented in two ways: either by modifying the convective or radiation heat flux or by changing the boundary conditions, using an external thermal insulation.

The modification of the convective heat flux from the surface can be achieved by a convenient *"spatial layout"* of the workpiece. In this case on the surface of the workpiece can be achieved an uneven heat flux, through a variable heat transfer coefficient along the perimeter. This method can be used for a narrow range of geometrical parameters of the workpiece (i.e. for $\varepsilon < \pi/8$ and $\alpha < 0.075$).

The heat flux by radiation from the surface of the workpiece can be modified by:

(1) use of a passive heat shield;
(2) a screen, heated by an independent power source;
(3) a movable screen located in the plane of the heated body [26].

In case of straight workpieces, temperature equalization is achieved by heat shields covering the whole external surface of the heated billets. Their use reduces the temperature differential and increases the thermal efficiency. The use of such screens reduces the temperature differential by 1.2–1.5 times [27], while the thermal efficiency—for heating rates in the range 1–5 °C/s—increases of 40–50 % [25].

In curved workpieces, the use of the heat shields helps to equalize the temperature differential along perimeter, but cannot significantly improve thermal efficiency.

The second method of the first group for temperature equalization, is to apply thermal insulation of uniform or non-uniform thickness. Uniform insulation is used

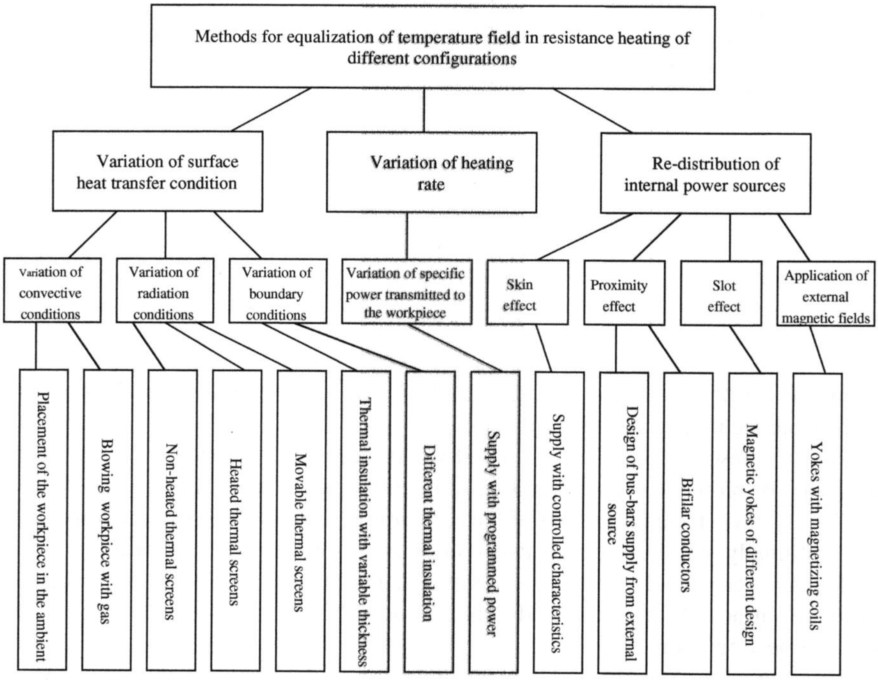

Fig. 4.25 Classification of methods for equalization of temperature field in resistance heating of workpieces of different configurations

to equalize temperature in straight workpieces, variable insulation thickness in the curved ones. The experience has shown the effectiveness of this solution at low rates of heating, i.e. in the range 0.1–1 °C/s.

For example, the thermal efficiency of a process at heating rate of 0.2 °C/s, can be increased of about 40 % with respect to heating processes without insulation, of 60 % with insulation. Correspondingly the temperature differential can be reduced from 70 to 20 °C [20, 28].

The second group of methods for temperature equalization is based on the control of the skin, proximity and slot effects or the application of an external electromagnetic field.

For rapid resistance heating of ferromagnetic cylindrical straight bars (with heating speed higher than 10 °C/s) an effective method of equalization of the temperature field is to use a DC power supply with a superimposed AC component.

In this case, the ratio between the AC and the DC parts of the current in the workpiece is chosen in such a way that the basic energy for heating is produced by the DC part of the current, while the energy due to the AC part, which is mainly concentrated in the surface layer, is used for equalizing the temperature over the cross-section.

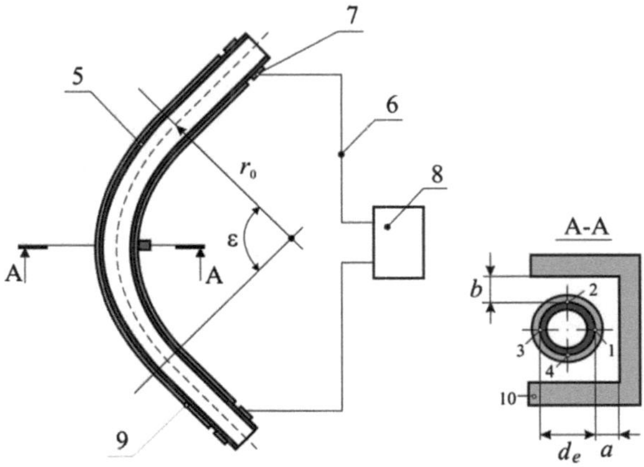

Fig. 4.26 Experimental setup: *1–4* temperature and voltage sensors; *5* heated tube; *6* current leads; *7* water-cooled contact, *8* stabilized current source; *9* thermal insulation; *10* magnetic core

The proximity effect can be used for reducing the temperature differential in curved cylindrical workpieces or in the width of flat sheet blanks. Several studies have shown the effectiveness of this solution [29, 30].

Slot effect is another very effective method for compensating the ring effect in applications at industrial or high frequency [26, 31].

The method which uses the application of an external electromagnetic field [32, 33] is also a promising option for temperature equalization. However, its application requires the development of magnetic cores of special design, placed around the heated workpiece.

It should be noted that the use of methods that modify the distribution of the internal heat sources does not preclude the use of uniform insulation on the outer surface of the product, for increasing thermal efficiency.

Temperature Equalisation Using Movable Magnetic Yokes (Slot Effect)

The application of magnetic cores is one of the most effective methods for equalization of the temperature field in curved workpieces.

Experimental and theoretical studies have analysed the effect of magnetic U-shaped cores on the temperature redistribution [25, 26, 34].

Experimental tests were carried out using the setup shown in Fig. 4.26, with a stabilized current $I = 6.5$ kA at industrial frequency. The workpiece was constituted by a curved ferromagnetic tube with external diameter $d_e = 0.159$ m, wall thickness $t = 0.02$ m, bending angle $\varepsilon = 90°$, bending radius $r_0 = 0.64$ m, and external thermal insulation.

The special design of the magnetic yokes (see Fig. 4.26) allows to change the gap between the surface of the tube and the load in the ranges $a/d_e = 0.15–\infty$ and $b/d_e = 0.2–\infty$. The experiments are described in Ref. [1].

Figures 4.27 and 4.28 shows, as a function of time, the variations of the current density *r.m.s.* values, the specific power and the temperature measured at points 1 and 3.

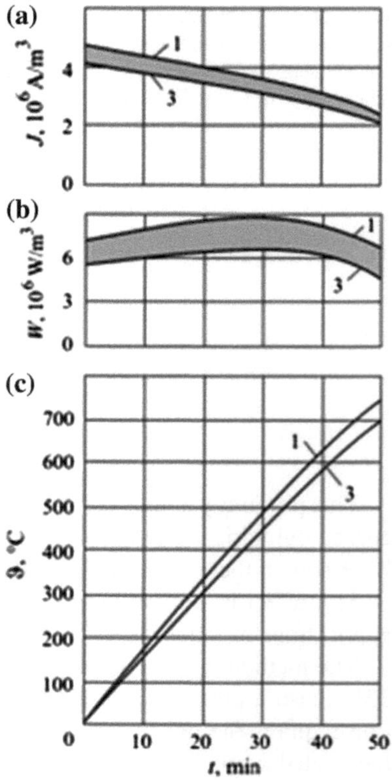

Fig. 4.27 Variations during heating of J, w, ϑ ($a/d_e = \infty$; $b/d_e = \infty$)

The reduction of the current density during the heating process is due to the increase of the penetration depth into the ferromagnetic material of the tube.

Comparing the curves of $J = f(t)$ and $\vartheta = f(t)$ at points 1 and 3, we can observe that the difference $\Delta J_{13} = J_1 - J_3$ of the current density values at the workpiece surface, for $a/d_e = \infty$ and $b/d_e = \infty$, is greater during the whole heating transient than the difference ΔJ_{13} occurring with $a/d_e = 0.15$ and $b/d_e = 0.785$.

The temperature difference between points 1 and 3 at the end of the heating in the first case is $\Delta\vartheta_{13} = 52$ °C, while in the second is $\Delta\vartheta_{13} = 10$ °C, which means that the application of U-shaped magnetic cores has significantly modified the current density and temperature distributions in the workpiece cross-section.

A series of experiments has been done for analysing the influence of the gap between magnetic core and surface of the workpiece on the temperature differential. The results are presented in Figs. 4.29 and 4.30.

This study refers to a workpiece constituted by a curved ferromagnetic tube, provided on external surface of uniform thermal insulation. U-shaped magnetic cores made of transformer steel sheets were installed in the middle section of the curvature (Fig. 4.26).

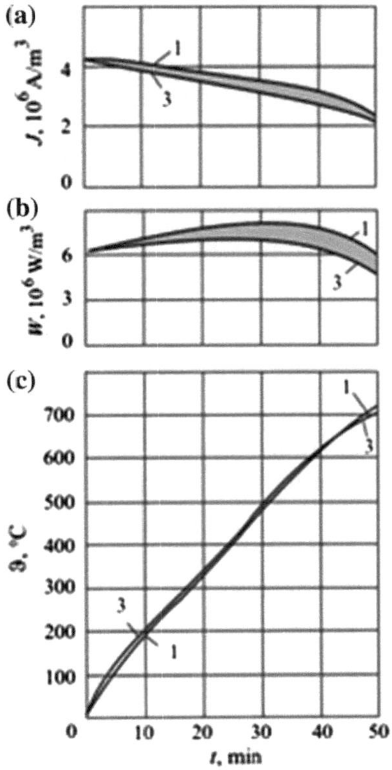

Fig. 4.28 Variations during heating of J, w, ϑ ($a/d_e = 0.15$; $b/d_e = 0.785$)

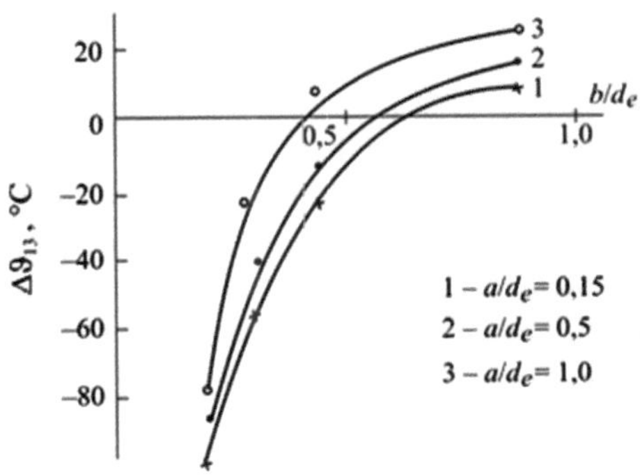

Fig. 4.29 Variation of maximum temperature difference $\Delta\vartheta_{13}$ with position of magnetic core

By their design, it was possible to vary the distances a and b within the ranges $a/d_e = 0.15 \cdots \infty$ and $b/d_e = 0.2 \cdots \infty$, with d_e—tube external diameter. The heating process was carried out up to a final average temperature of 750 °C, with heating current of 6.5 kA at 50 Hz.

For assessing the influence of the magnetic yokes in redistributing internal heating sources and temperature field, the ratios a/d_e and b/d_e were selected as

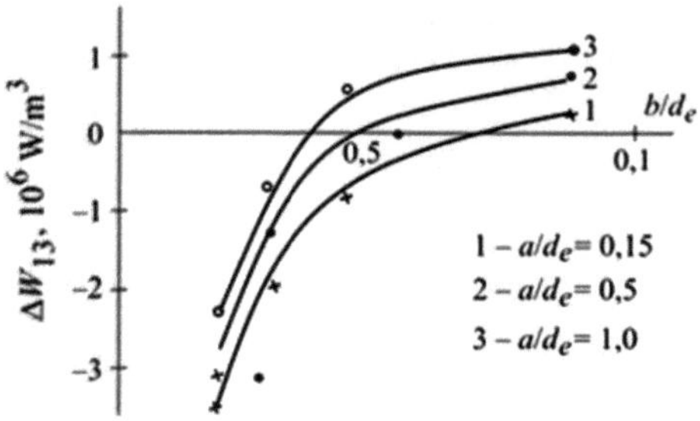

Fig. 4.30 Variation of maximum difference of the heating sources Δw_{13} with the position of the magnetic core

design variables and the difference $\Delta w_{13} = w_1 - w_3$ between power densities of points 1 and 3 and the temperature difference $\Delta\vartheta_{13} = \vartheta_1 - \vartheta_3$ as dependent parameters.

From the experimental data have been obtained the following empirical relationships:

$$\Delta\vartheta_{13} = 27.28 + 14.89\left(\frac{a}{d_e}\right)^{1.61} - 15.68\left(\frac{b}{d_e}\right)^{-1.33} + 0.718\left(\frac{a}{d_e}\right)^{1.61} \cdot \left(\frac{b}{d_e}\right)^{-1.33}, {}^\circ\mathrm{C} \tag{4.15}$$

$$\Delta w_{13} = \left[0.587 + \left(\frac{a}{d_e}\right)^{0.75} - 0.538\left(\frac{b}{d_e}\right)^{-1.3} + 0.0567\left(\frac{a}{d_e}\right)^{0.75} \cdot \left(\frac{b}{d_e}\right)^{-1.3}\right] \cdot 10^6, \mathrm{Wm}^3 \tag{4.16}$$

Figures 4.29 and 4.30 show the comparison of the experimental data (*points*) and the values given by Eqs. (4.15) and (4.16) (*solid lines*). Since the differences do not exceed 10 %, we recommend these equations for preliminary calculations.

Numerical simulations of electromagnetic and thermal processes in a system constituted by a curved cylindrical workpiece and U-shaped magnetic cores have been done with the software package FLUX [12]. The results of the experimental tests previously described, are used to check the reliability of the computer model [26, 34].

Calculations and tests refer to a workpiece constituted of a ferromagnetic tube, $d_e = 159$ mm diameter, $t = 20$ mm wall thickness, angle of curvature $\varepsilon = 90°$, radius $r_0 = 0.64$ m, with external thermal insulation made of kaolin wool reinforced with fiberglass. U-shaped laminated magnetic cores are located along the internal part of the tube curvature, according to the schematic of Fig. 4.31.

The supply of the system is done with constant current of 6.2 kA at mains frequency.

Due to the symmetry, the numerical model represents only one-fourth of the real system, as shown in Fig. 4.31.

Fig. 4.31 Schematic of the ferromagnetic workpiece (*A*), thermal insulation (*B*) and laminated magnetic core (*C*)

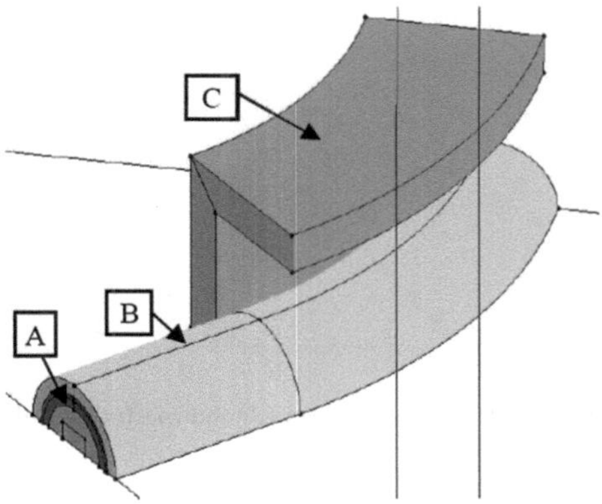

Figure 4.32a, represents the boundaries of the model, defined by the line SP_1, corresponding to the position of the contact system, and the line SP_2, corresponding to the plane of symmetry of the ferromagnetic tube.

Figure 4.32b shows the mesh used in the 3D FEM model in the region of the ferromagnetic workpiece. For the solution of the electromagnetic problem, tangential magnetic field conditions have been applied at the boundaries SP_1 and SP_2, and in the plane of the central axis of the tube conditions of normal magnetic field and tangential electric field.

The solution has been carried out with impressed constant current at industrial frequency. The coupled solution of the quasi-stationary electromagnetic and transient thermal problems is done iteratively, taking into account the temperature dependence of physical properties of the workpiece material (ρ, c, λ) and the thermal insulation (λ_t, c_t).

The experimental setup, where it is visible the location of thermocouples on the surface of the curved tube and the power supply contacts, and the position of temperature and voltage sensors in the plane of the magnetic core, is described in Ref. [1].

The comparison of experimental data and the results of numerical calculations, given in Fig. 4.33, shows a quite good agreement.

The results have demonstrated the effectiveness of the use of magnetic cores for equalization of the temperature field in the DRH heating of curved sections of cylindrical workpieces. Moreover, the combined use of magnetic cores and thermal insulation can positively affect the energetic parameters of the heating installation [20, 21].

Further heating tests have been done for improving the installation power factor, by variations of the supply current from 2.5 to 6.3 kA, the temperature from 20 to 700 °C, and the relative positions a/d_e and b/d_e of the tube and the magnetic core in the range 0.2 to ∞.

The diagrams of Fig. 4.34, give the values of power factor as a function of the average temperature in the workpiece, for different intensities of supply current and

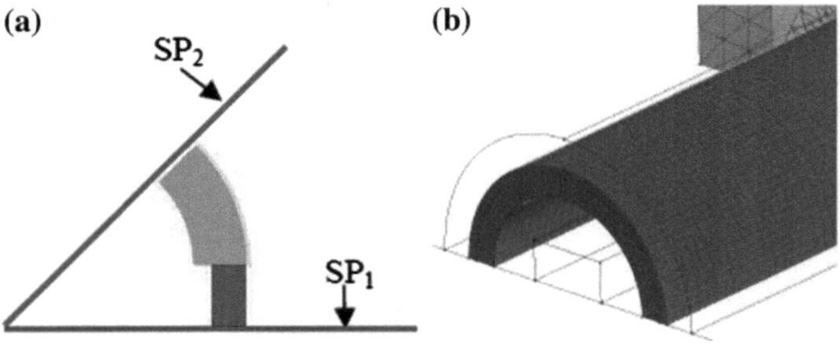

Fig. 4.32 Boundary lines (a) and mesh (b) of the 3D FEM model of curved ferromagnetic tube and magnetic core

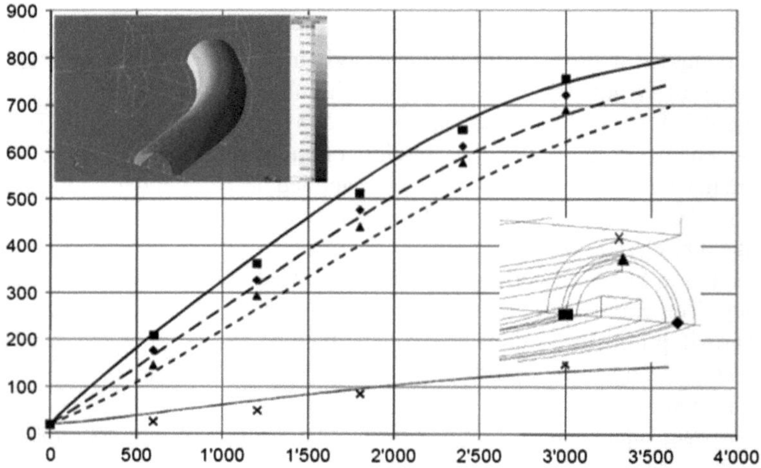

Fig. 4.33 Comparison of calculated and experimental data of heating transient experimental data (from thermocouples, placed on the tube as shown in the diagram): *points*-experiments; *lines*-calculations. The measuring point "*X*", is placed on the surface of thermal insulation

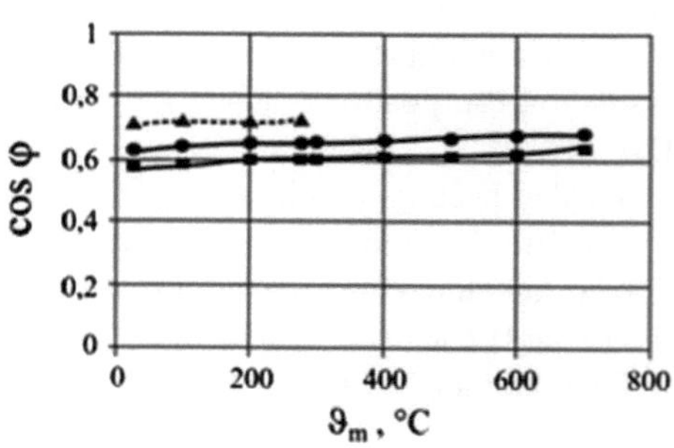

Fig. 4.34 Power factor as a function of average temperature of workpiece, for different r.m.s. values of supply current and positions of magnetic cores: *dotted lines* I = 2.5 kA, $a/d_e = \infty$, $b/d_e = \infty$; *solid lines* ● I = 6.3 kA, $a/d_e = \infty$, $b/d_e = \infty$; ■ I = 6.3 kA, a/d_e = 0.2, b/d$_e$ = 0.2

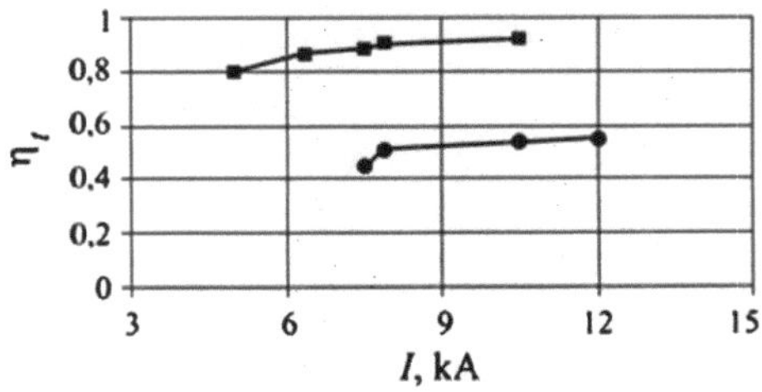

Fig. 4.35 Thermal efficiency as a function of (constant) heating current: ■ with thermal insulation; ● without thermal insulation

positions of the cores. It appears that heating with lower current values leads to an improvement of the power factor (for $I = 2.5$ kA, cos φ = 0.7–0.72; for $I = 6.3$ kA, cos φ = 0.62–0.68).

Such increase is explained in [3] for the heating of straight ferromagnetic steel tubes as follows. With the current of 2.5 kA the value of the magnetic field intensity is $H_{me} = 7$ kA/m and the relative magnetic permeability of the workpiece is $\mu_r = 180$, while with $I = 6.3$ kA the corresponding values are $H_{me} = 17$ kA/m and $\mu_r = 80$ respectively.

In [3] it is shown that with the increase of μ_r, the resistance of the ferromagnetic tube increases more rapidly than its reactance.

As shown by the figure, the value of cos φ is considerably influenced by the presence of the magnetic cores used for equalization of the temperature field. When the cores are placed close to the workpiece (i.e. reducing the ratios a/d_e and b/d_e) the power factor becomes lower.

In Fig. 4.35 the diagrams of the thermal efficiency η_T are given as a function of the heating current. The thermal efficiency is obviously higher when the workpiece is provided with an external thermal insulation, than without it. Varying the heating current from 5 to 10.5 kA, the thermal efficiency with thermal insulation increases from 0.8 to 0.92, while without it a variation of the heating current from 7.5 to 12 kA produces an improvement of thermal efficiency from 0.45 to 0.55.

It must be noted that during these tests it was not possible to reach in the workpiece the temperature of 600 °C with current of 4.5 kA with thermal insulation and 7.2 kA without insulation, since in these conditions the power transformed into heat in the workpiece is comparable with thermal losses.

For this reason when the heating current approximates the above values, the thermal efficiency diminishes and tends to 0. This trend is similar to the one observed in the heating of straight workpieces [20, 21].

4.3 Transverse Flux Induction Heating

Since the early applications of induction heating before hot working, it was recognized that the heating of non-ferrous strips, made of stainless steel, aluminum, brass, copper and magnesium, could be relevant in industrial processes.

It was also realized that "longitudinal flux heating" (LFH), well suited for heating bars and billets, was not satisfactory for strips which do not have magnetic properties.

Limitations of longitudinal flux heating are [35]:

- only a small part of the magnetic flux produced by the inductor passes through the cross-section of a non-magnetic strip. In fact, most of the flux is located in the gap between strip and inductor (since it is proportional to the ratio of cross-sections of strip and gap);
- radiofrequency currents are required to obtain relatively good coil efficiency; the theory shows that good efficiency is obtained when for ratio between strip thickness (t) and penetration depth (δ) is $t/\delta \geq 3$.

As a result, longitudinal flux heating of thin strips is characterized by low efficiency, low power factor, relatively low induced power density and the necessity to use more expensive frequency converters.

As an example, for strips of non-magnetic material (copper, aluminum) with thickness less than 1 mm, the optimal frequency is above 150 kHz, for stainless steel strips—above 2 MHz.

A special method of induction heating was proposed to overcome these drawbacks. The method is called "*transverse flux heating*" (TFH) because the magnetic flux is directed perpendicular to the surface of the strip (Fig. 4.36a) compared to longitudinal (axial) flux heating (Fig. 4.36b), where the magnetic flux is tangential to the strip surface.

A transverse flux inductor has usually a magnetic structure with two opposing poles (Fig. 4.37) produced by excitation coils located in the slots. The coils connection is realized in order to produce at any instant magnetic poles of opposite signs, which forces the magnetic flux to cross the strip.

The main advantages of transverse flux induction heating is the possibility to use frequencies much lower (50 kHz–10 Hz) than in a longitudinal magnetic field and to achieve high electrical efficiency.

On the other hand, this method has some disadvantages, in particular:

- uneven distribution of induced power sources on the strip width, which can result in overheating or under-heating of the strip edges;
- strict dependence of the temperature uniformity on the shape of the inductor coils and the material properties of the heated tape, the geometric dimensions of the tape and its location inside the inductor;
- complexity of inductor design;
- presence of vibrations.

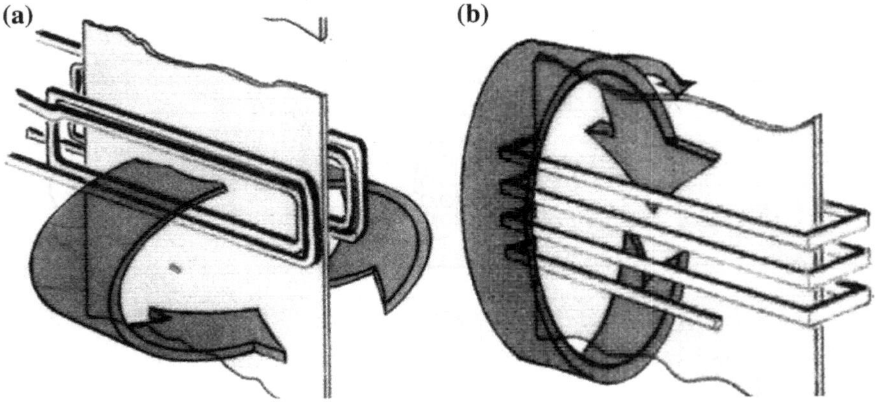

Fig. 4.36 Transverse (**a**) and longitudinal (**b**) flux heating of strips

The idea of TFH was already proposed in the fifties of the 20th century [36], but its acceptance in the industry has been very slow, because of the difficulty to achieve uniform heating of the strip.

In fact this technique can produce severe current concentration on the strip edges with excessive localized heating. This effect becomes more pronounced with frequency, strip thickness and pole pitch. In order to optimize the design of the inductor regarding temperature uniformity in the strip, coil efficiency and power factor, it is necessary to control the effects of many design variables, i.e. air-gap, pole width, slot dimensions, pole pitch, strip thickness, material resistivity and current frequency.

The problem becomes even more complex if it is required to adapt the same inductor to different strip widths.

Many papers have dealt with the analysis and design of this type of heaters [37–46], but the problem found solution only when 2D and 3D numerical calculation programmes became available and have been coupled with optimization techniques [47–71].

In fact, the design of a TFH inductor requires the use of coupled 3D electromagnetic and thermal calculations, which became possible only in the beginning of the 21st century, with the availability of powerful computers and suitable numerical FEM software packages. However, in practice, a real 3D system optimization is still today cumbersome, due to the large number of variables and the huge calculation time required.

For preliminary investigations even today it can be more convenient to use simple tools, like analytical and two-dimensional numerical codes. With these tools it is possible to find some basic dependencies, which can be used as starting values for more accurate 3D numerical calculations, and to identify convenient ranges of variables for the final optimization.

These tools are based on 2D programs which allow to analyze the influence of design parameters on the integral characteristics of the heaters. Short calculation

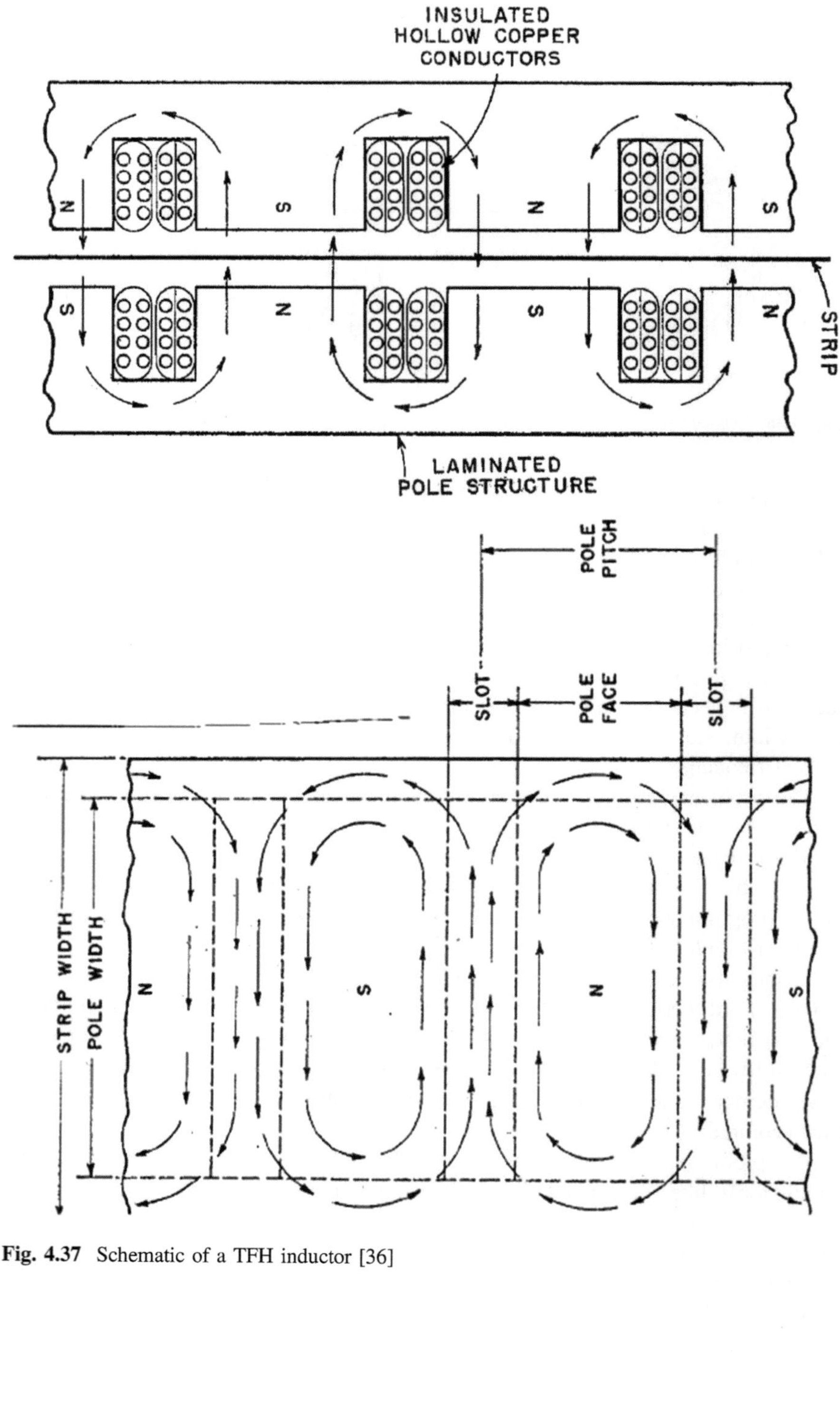

Fig. 4.37 Schematic of a TFH inductor [36]

time and little amount of input data make them very useful in the first steps of the inductor design.

Among these simple tools, the one based on an analytical solution given by Haubitzer has proved to be very efficient [38].

This solution makes reference to the schematic representation of the heater shown in the Fig. 4.38a–c. The heater is supposed to be constituted in the direction x of the strip movement by an infinite series of ideal coils, identical to the one sketched in the Fig. 4.38b, c, equally spaced along x of the quantity $2f$ (see Fig. 4.39).

The solution is obtained with the following simplifying assumptions:

- coil conductors of negligible thickness in z direction
- sinusoidal currents flowing in the inductor coils
- ideal magnetic yokes of infinite magnetic permeability and no losses, located in the regions $(z > h)$ and $(z < -h)$,
- constant resistivity ρ and relative magnetic permeability μ_r of the strip material
- inductor width $2(b + d)$ not exceeding the strip width L.

The last assumption is less restrictive as it may appear at first glance, since this condition is usually fulfilled in most practical cases for avoiding overheating of the strip edges.

With the previous hypotheses, the *m.m.f.* produced by one inductor coil can be represented by the function $g(x, y)$ shown in Fig. 4.39, which is assumed to be periodical along x and y, and can be therefore expressed by the following double Fourier-series expansion:

$$\begin{aligned} f.m.m.(x,y) &= NI \cdot g(x,y) \\ &= NI \sum_{i=1,3,\ldots}^{\infty} \sum_{k=1,3,\ldots}^{\infty} \alpha_{ik} \cos\left(\frac{2\pi}{\tau} ix\right) \cos\left(\frac{2\pi}{\sigma} ky\right) \end{aligned} \tag{4.17}$$

where: N, I—number of turns of each coil and r.m.s. value of the current flowing in it;

$$\tau = 2(a+c+f); \sigma = 2(b+c+\Delta); \Omega = \frac{2\pi}{\tau} i; \Psi = \frac{2\pi}{\sigma} k;$$

$$\left.\begin{aligned} \alpha_{ik} &= \frac{32}{\sigma\tau\Delta\Omega\Psi} \\ &\left\{\frac{1}{\Psi-\Omega}\cdot [\sin(\Psi(b+c)-\Omega(a+c)) - \sin(\Psi b - \Omega a)] \right. \\ &\left. + \frac{1}{\Psi+\Omega}[\sin(\Psi b + \Omega a) - \sin(\Psi(b+c)+\Omega(a+c))]\right\}, \\ &(for\ \Psi \neq \Omega) \end{aligned}\right\} \tag{4.17a}$$

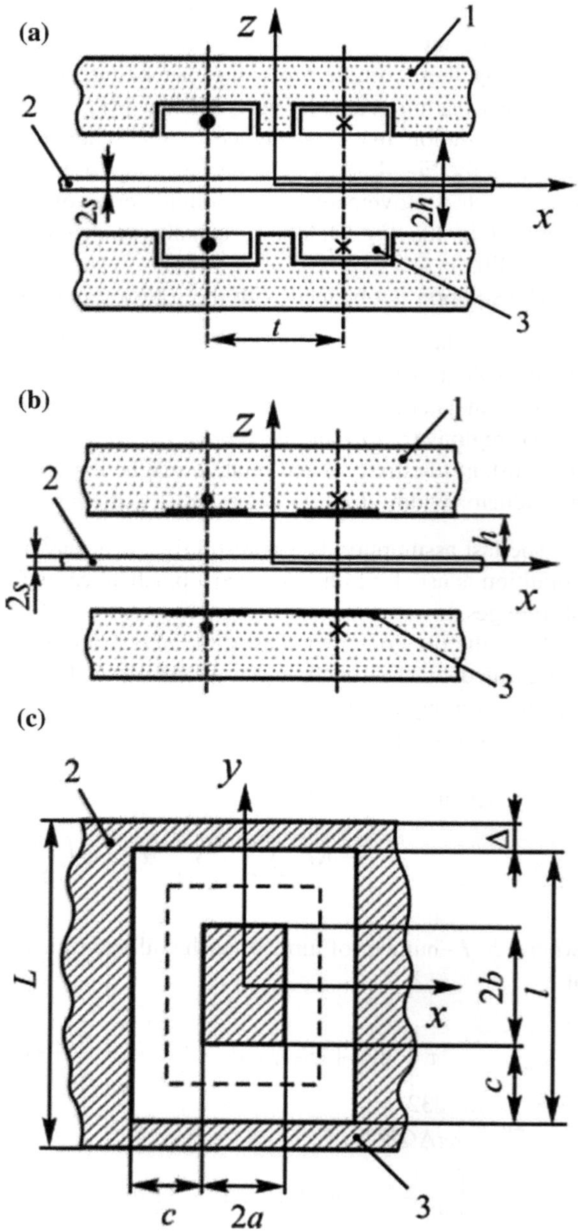

Fig. 4.38 Schematic of the heater used for the analytical solution (**a** real geometry (cross-section); **b** ideal geometry (cross-section); **c** ideal coil geometry (*upper view*); *1* magnetic yoke; *2* strip; *3* inductor coil; $t = (2a + d)$ pole pitch; *2h* air gap; *2s* strip thickness; **a**, **b**, **d** coil dimensions; **c** distance between edges of strip and inductor coil)

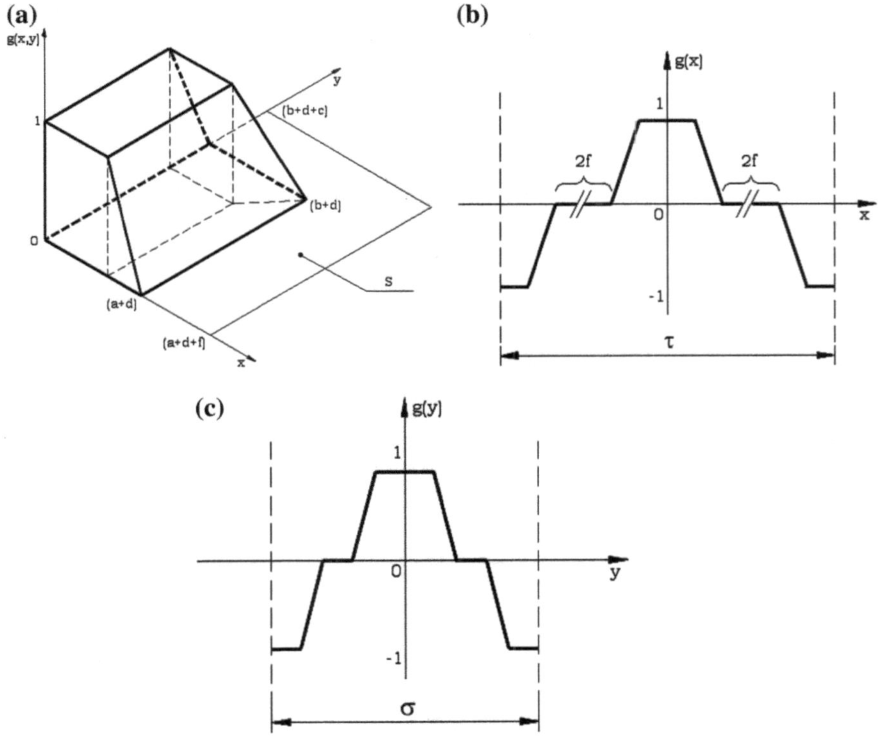

Fig. 4.39 Normalized *m.m.f.* distributions of the coil of Fig. 4.38

$$\left.\begin{aligned}\alpha_{ik} &= \frac{32}{\sigma\tau\Omega^2}\left\{\cos(\Omega(b-a))\right.\\ &\left.-\frac{1}{2\Omega d}\left[\sin\left(\Omega(a+b+2c)\right)-\sin\left(\Omega(b+a)\right)\right]\right\},\\ &(for\ \Psi=\Omega)\end{aligned}\right\} \tag{4.17b}$$

The problem requires the solution of the differential equations in terms of the magnetic vector potential A, taking into account the boundary conditions for the components of the magnetic field intensity H at $z = h$:

$$\left\{\begin{aligned}\nabla^2 A_x - j\frac{\omega\mu}{\rho}A_x &= 0\\ \nabla^2 A_y - j\frac{\omega\mu}{\rho}A_y &= 0\\ H_x(x,y,h) &= -K_y(x,y)\\ H_y(x,y,h) &= -K_x(x,y)\end{aligned}\right. \tag{4.18}$$

with: ρ, μ—resistivity and absolute magnetic permeability of the strip material; $\omega = 2\pi f$—angular frequency of the inductor current;

$$K_x = \frac{\partial f.m.m.(x,y)}{\partial y} = -\frac{2\pi NI}{\sigma} \times \sum_{i=1,3,\ldots}^{\infty} \sum_{k=1,3,\ldots}^{\infty} k\alpha_{ik} \cos\left(\frac{2\pi}{\tau} ix\right) \sin\left(\frac{2\pi}{\sigma} ky\right)$$

$$K_y = \frac{\partial f.m.m.(x,y)}{\partial x} = \frac{2\pi NI}{\tau} \times \sum_{i=1,3,\ldots}^{\infty} \sum_{k=1,3,\ldots}^{\infty} i\alpha_{ik} \sin\left(\frac{2\pi}{\tau} ix\right) \cos\left(\frac{2\pi}{\sigma} ky\right).$$

In this way the following equations can be derived for the components of the electric field intensity in the air-gap between the inductor coil and the strip (region denoted by suffix *1*) and in the strip (suffix *2*): [38]

$$E_{1x} = j\omega\mu \frac{\pi\tau}{\sigma} NI e^{j\omega t} \sum_{i=0}^{\infty} \sum_{k=0}^{\infty} k\left[A_{ik} e^{\left(-a_{ik}\frac{z}{\tau}\right)} + B_{ik} e^{\left(a_{ik}\frac{z}{\tau}\right)}\right] \cdot \cos\left(\frac{i\pi}{\tau} x\right) \sin\left(\frac{k\pi}{\sigma} y\right),$$

$$E_{1y} = -j\omega\mu\pi NI e^{j\omega t} \sum_{i=0}^{\infty} \sum_{k=0}^{\infty} i\left[A_{ik} e^{\left(-a_{ik}\frac{z}{\tau}\right)} + B_{ik} e^{\left(a_{ik}\frac{z}{\tau}\right)}\right] \cdot \sin\left(\frac{i\pi}{\tau} x\right) \cos\left(\frac{k\pi}{\sigma} y\right)$$

with:

$$a_{ik} = \pi\sqrt{i^2 + \left(\frac{k\tau}{\sigma}\right)^2};\ d = (h - s)$$

$$A_{ik} = \left[\frac{\alpha_{ik} e^{\left(a_{ik}\frac{h}{\tau}\right)}}{a_{ik}}\right] \times \frac{\mu a_{ik} ch\left(\gamma_{ik}\frac{s}{\tau}\right) - \gamma_{ik} sh\left(\gamma_{ik}\frac{s}{\tau}\right)}{\gamma_{ik}\left(e^{2a_{ik}\frac{d}{\tau}} + 1\right) sh\left(\gamma_{ik}\frac{s}{\tau}\right) + \mu a_{ik}\left(e^{2a_{ik}\frac{d}{\tau}} - 1\right) ch\left(\gamma_{ik}\frac{s}{\tau}\right)},$$

$$B_{ik} = \left[\frac{\alpha_{ik} e^{\left(a_{ik}\frac{h-2s}{\tau}\right)}}{a_{ik}}\right] \times \frac{\mu a_{ik} ch\left(\gamma_{ik}\frac{s}{\tau}\right) - \gamma_{ik} sh\left(\gamma_{ik}\frac{s}{\tau}\right)}{\gamma_{ik}\left(e^{2a_{ik}\frac{d}{\tau}} + 1\right) sh\left(\gamma_{ik}\frac{s}{\tau}\right) + \mu a_{ik}\left(e^{2a_{ik}\frac{d}{\tau}} - 1\right) ch\left(\gamma_{ik}\frac{s}{\tau}\right)}$$

and

$$E_{2x} = j\omega\mu\frac{\pi\tau}{\sigma}NIe^{j\omega t} \times \sum_{i=0}^{\infty}\sum_{k=0}^{\infty} kC_{ik}ch\left(\gamma_{ik}\frac{z}{\tau}\right)\cos\left(\frac{i\pi}{\tau}x\right)\sin\left(\frac{k\pi}{\sigma}y\right)$$

$$E_{2y} = -j\omega\mu\pi NIe^{j\omega t} \times \sum_{i=0}^{\infty}\sum_{k=0}^{\infty} iC_{ik}ch\left(\gamma_{ik}\frac{z}{\tau}\right)\sin\left(\frac{i\pi}{\tau}x\right)\cos\left(\frac{k\pi}{\sigma}y\right)$$

with:

$$\gamma_{ik} = \sqrt{\pi i^2 + \left(\frac{\pi k\tau}{\sigma}\right)^2 + j\frac{\omega\mu\tau^2}{\rho}},$$

$$C_{ik} = \frac{2\alpha_{ik}e^{\left(a_{ik}\frac{d}{\tau}\right)}}{\gamma_{ik}\left(e^{2a_{ik}\frac{d}{\tau}}+1\right)sh\left(\gamma_{ik}\frac{s}{\tau}\right) + \mu a_{ik}\left(e^{2a_{ik}\frac{d}{\tau}}-1\right)ch\left(\gamma_{ik}\frac{s}{\tau}\right)}$$

In the different regions of the model, similar equations can be derived for the components of the magnetic field intensity, the current density and, in particular, the distribution of the volumetric power density w in the strip, which constitutes the input data for the calculation of the heating transient:

$$w = \frac{1}{2\rho}\left(E_{2x}E_{2x}^{*} + E_{2y}E_{2y}^{*}\right) \tag{4.19}$$

The previous equations allow the calculation of the electromagnetic field distributions and integral parameters of the system for different strip characteristics, as a function of the design variables, i.e. air-gap, pole width, slot dimensions, pole pitch and frequency of the supply current.

The equivalent impedance $\dot{Z}_{e0}$ of a single coil—supposed to have negligible own resistance—can be obtained through the calculation of the active and reactive power $(P + jQ)$ in the region $(-h \leq z \leq h; -\tau/2 \leq x \leq +\tau/2.)$ Their values are derived as the flux of the Poynting vector through the surfaces at $z = \pm h$ facing one inductor:

$$P + jQ = \frac{\cos t}{\tau\sigma}\int_{-\tau/2}^{+\tau/2}\int_{-\sigma/2}^{+\sigma/2}\left[E_{1x}(x,y,h)K_x^{*}(x,y) - E_{1y}(x,y,h)K_y^{*}(x,y)\right]dxdy \tag{4.20}$$

$$\dot{Z}_{e0} = \frac{P + jQ}{I^2} \tag{4.21}$$

The active power induced in the strip by a single coil is therefore represented by the real part P of Eq. (4.20).

For the hypothesis of linear characteristics of the strip material (ρ = cost. and μ = cost.) the presence of a series of different inductor coils along the strip length can be accounted for with a superposition procedure.

In the following some typical dependencies of the heater characteristics from design parameters will be illustrated with reference to a design example developed using a software based on the above analytical solution.

The example refers to a pre-heater used for increasing the production rate of a gas-fired annealing installation for brass strips, 420 mm width, 0.8 mm thickness, 10 m/min velocity, which must be heated from 20 °C to about 300 °C before entering in the gas-fired annealing zone [53].

The first design step is the calculation of the power that must be transformed into heat in the strip in order to obtain the required production rate; considering the physical characteristics of the strip material[1] and assuming a tentative value of the inductor-load system efficiency of 90 %, this power can be estimated to be about 60 kW.

4.3.1 Preliminary Choice of Design Parameters

The subsequent step is the preliminary tentative choice of some design parameters according to the following rules suggested in the bibliography [43]:

a. the value of the air-gap $2h$ should be assumed as small as possible from the point of view of the electrical efficiency, but cannot be reduced below certain limits imposed by the space requirements for thermal insulation between inductor and strip and the necessity of assuring a free movement of the strip across the inductor, taking also into account the oscillations of the strip during the process. A suggested compromise is $2h/t \leq 0.5$.
b. choice of the exciting frequency according the relationship

$$f = 2.1 \times 10^6 \cdot \frac{\rho}{t^2} \cdot \frac{h}{s} \cdot k \tag{4.22}$$

where k is a coefficient, dependent on the ratio $(2h/t)$, ranging from 0.42 to 1.90.

[1] Data of brass CuZn37—density: 8,440 kg/m^3; specific heat capacity: 404 J/kgK; electrical conductivity: 13.5 × 10^6 S/m.

c. the optimum ratio between slot width c (which must contain the inductor coil and the insulation) and pole pitch t should be in the range $0.4 \leq c/t \leq 0.75$. Increasing this ratio towards the value 0.75 produces a more uniform power density distribution, but considerably reduces the total power transferred to the strip. An acceptable compromise can be $c = t/2$.

Choosing, according to point (a), an air-gap value of 0.04 m, the condition $t \geq 0.08$ m is obtained for the pole pitch.

As shown by Eq. (4.22), the increase of the pole pitch t allows to reduce the supply frequency.

Based on the previous considerations, the analysis of the heater characteristics has been developed in the following range of parameters:

- exciting frequency: 100–1,400 Hz [according to Eq. (4.22)]
- pole pitch: from 0.08 to 0.20 m
- ratio c/t: from 0.4 to 0.75.

The results of the preliminary analysis done with the above parameters are summarized in the diagrams of Fig. 4.40. They are in excellent agreement with those obtained by 2D numerical methods [53].

From the diagrams the following conclusions can be drawn:

- the maximum of electrical efficiency is not very pronounced and therefore high efficiency values can be reached in a wide range of frequencies;
- on the contrary, in the same frequency range, the power factor has a more distinct maximum;
- with the increase of the pole pitch, the maximum values of efficiency and power factor are shifted towards lower frequencies and at the same time become higher.

Choice of Frequency and Pole Pitch Values

According to the above results, for the case under consideration, the design frequency can be conveniently chosen in the range 300–500 Hz and the pole pitch between 160 and 200 mm.

When the frequency and the pole pitch have been defined, the data of Fig. 4.40c allow to calculate the ampere-turns NI required for inducing the required active power in the strip.

In particular, with $f = 300$ Hz and $t = 160$ mm, a value NI $\approx$ 4,900–5,000 A is necessary in order to obtain a total induced power of 60 kW. This value of ampere-turns can be realized with a 3 turns inductor coil carrying 1,650 A.

The calculation results made with these design parameters are summarized in Fig. 4.41.

Figure 4.41a gives the resulting power density distribution in half strip: in particular it shows that the power density distribution is always strongly uneven and that it can be distinguished a central "*regular zone*" with two peeks located in front of the coil conductors normal to the movement direction, and an "*edge area*" characterized by a pronounced single peek.

Fig. 4.40 **a** Power factor as a function of frequency, for different values of pole pitch t $c/t = 0.5$; ----t = 200 mm; ___ t = 160 mm; **...** t = 120 mm; · - · - · t = 80 mm. **b**, **c** Characteristics of the heater, as a function of frequency, for different values of pole pitch t **b** coil efficiency; **c** active power in the strip with NI = 1,000 A $c/t = 0.5$; ----t = 200 mm; ___ t = 160 mm; **...** t = 120 mm; · - · - · t = 80 mm

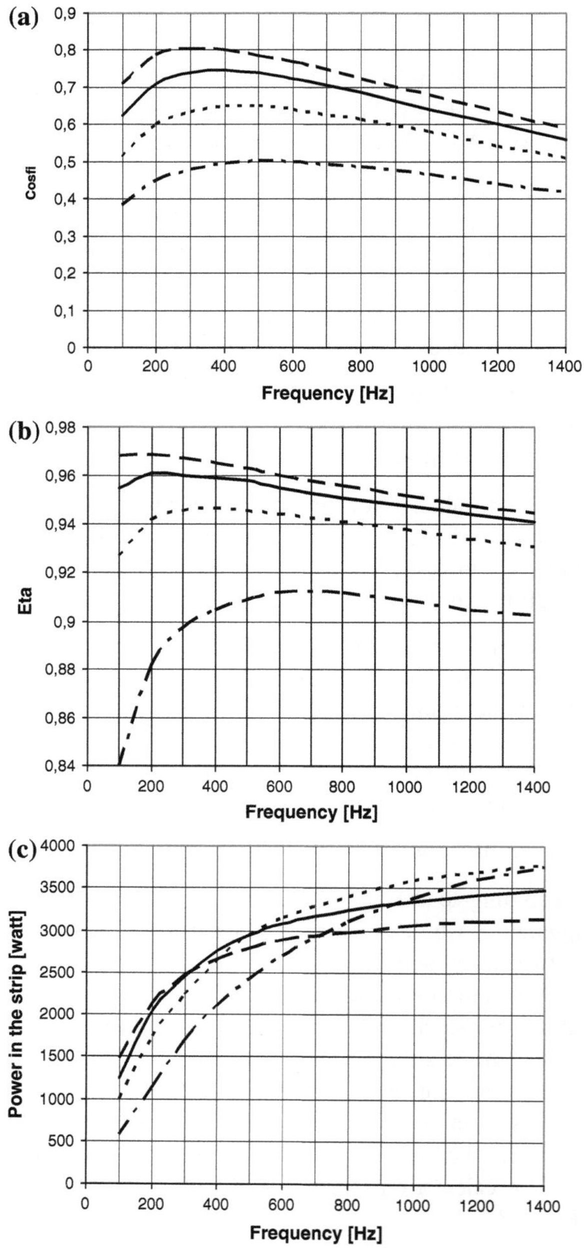

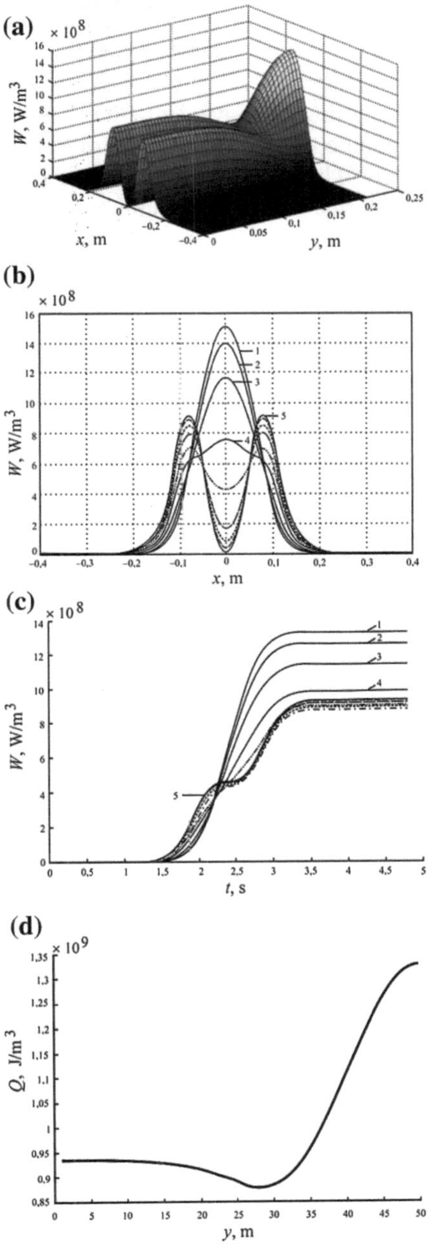

Fig. 4.41 Typical TFH system characteristics: **a** power density distribution in half strip; **b** power density along *equi-spaced lines* in the movement direction; **c** integral of power density distribution along the *lines* of *b* **d** energy distribution in the strip cross-section at the exit of the inductor (0—strip axis; 50—strip edge) $f = 300$ Hz; $c/t = 0.5$; $t = 160$ mm; $NI = 5{,}300$ A; Inductor power: 82.8 kVA; Power factor: 0.729; Induced power in the strip: 60.1 kW

Figure 4.41b shows the same distribution in the direction of the movement, along five lines equi-spaced between the strip axis and the edge.

In Fig. 4.41c the integral of the power density along the same lines is shown as a function of time; taking into account that the velocity of the strip is constant, it can be considered also representative of the energy transferred to the unit volume of the strip during its movement along one line.

Figure 4.41d gives the distribution in the half cross-section of the total energy received by the strip at the exit of the inductor coil. The curve shows that the energy transferred to the edge region is considerably higher in comparison with the one in the central regular zone, with consequent overheating of the edge.

Influence of the Ratio C/T

As previously mentioned, the suggested range for the choice of the ratio *c/t* is 0.4–0.75.

The influence of this choice is illustrated by the diagrams of Fig. 4.42, which give the values of power factor and power induced in the strip, at frequency

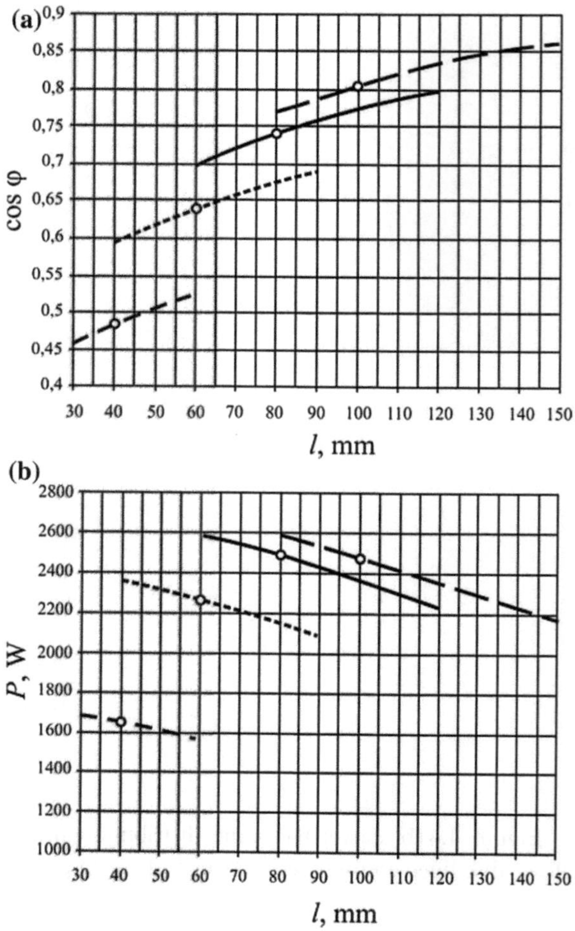

Fig. 4.42 Power factor **a** and induced power **b** as a function of the coil width *c*: $f = 300$ Hz; $c/t = 0.4$–0.75; $t = 80$–120–160–200 mm; $NI = 1{,}000$ A

$f = 300$ Hz, for values of $t = 80–120–160–200$ mm, as a function of the coil width c. The points on the curves show the values corresponding to the ratio c/t equal to 0.5.

The diagrams show the following:

- the pole pitches $t = 80$ and 120 mm give the lowest values of power factor and power induced in the strip;
- high power factor values, in the range 0.75–0.85 can be obtained only with pole pitch $t = 0.16$ and 0.2 m and ratios $c/t \geq 0.5$
- the same pitch values 0.16–0.2 m give also the highest values of the induced power
- for $t = 0.16–0.2$ m and $c/t \geq 0.5$ the power induced in the strip is practically the same.
- from the above, it may be concluded that the pole pitch value can be conveniently chosen between 0.16 and 0.2 m.

As regards the uniformity of the energy transfer in the strip, the diagrams of Fig. 4.43 give the relative energy distribution in half strip cross-section at the exit of the inductor. The curves show that for f = 300 Hz the peak of the energy transferred in the edge region is practically the same for t = 160 mm and t = 200 mm, but it is lower than the ones at other frequency values.

Influence of the Distance Between Strip and Coil Edges

The distance Δ between the strip and coil edges has a strong influence on the induced power density distribution and, as a consequence, on the relative distribution of the energy received by the strip during its movement inside the inductor.

This influence is illustrated by the curves of Fig. 4.44, which give the distribution in the cross-section of the total energy received by the strip at the exit of the inductor. Neglecting power losses from the surface and heat conduction inside the strip, this distribution is indicative of the temperature pattern at the end of heating.

In particular, the curves show the following: (1) with a convenient choice of the distance Δ it is possible to reach a sufficiently uniform energy distribution, (2) this distribution is particularly affected by the relative position between inductor and strip edges.

For the case under examination the best energy distribution is obtained with $\Delta = 10$ mm, with maximum deviations from the regular zone values of +4 % and −9 %. Moreover, as shown in Fig. 4.45, for $\Delta = 10$ mm the variation of the power factor is practically negligible and the reduction of the total induced power in comparison with the case $\Delta = 0$ is relatively small (i.e. about 7–8 %).

4.3.2 Final TFH System Characteristics

Taking into consideration the results of Figs. 4.44 and 4.45, the final characteristics of the system have been evaluated with the design data of Table 4.2.

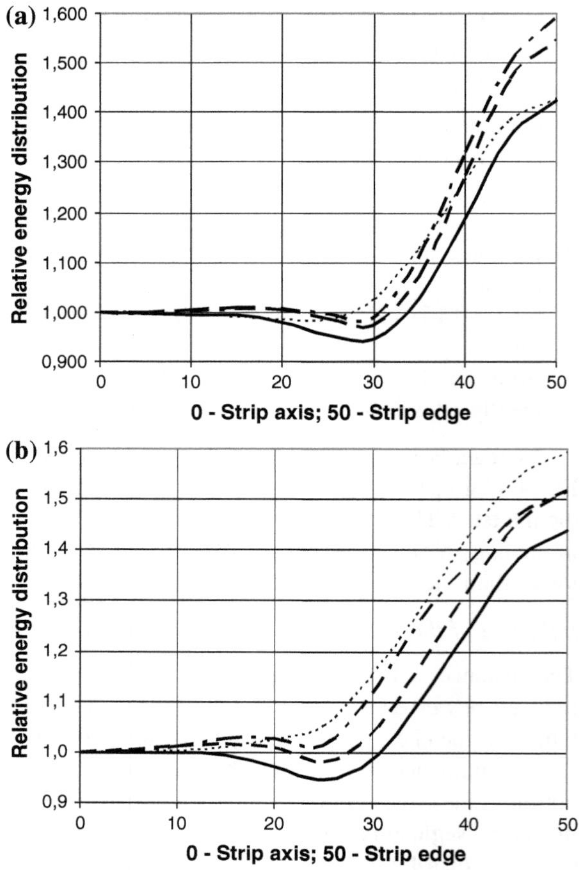

Fig. 4.43 Relative energy distribution in half strip cross-section at the exit of the inductor: ······ f = 100 Hz;— f = 300 Hz;—‐—f = 600 Hz; —·—f = 1,200 Hz; **a** t = 160 mm; **b** t = 200 mm; c/t = 0.5

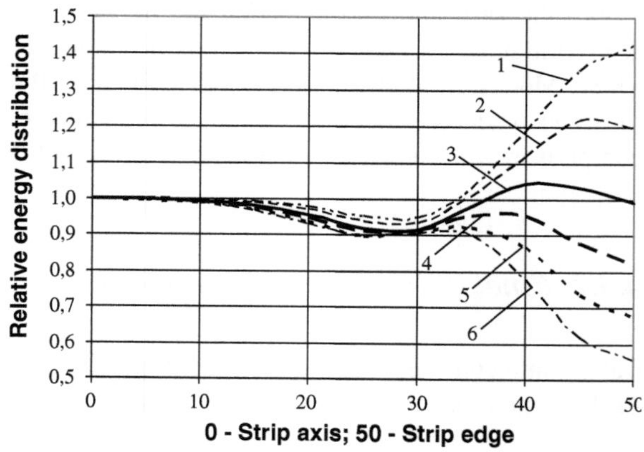

Fig. 4.44 Relative energy distribution in half strip cross-section at the exit of the inductor for various distances Δ between coil and strip edges t = 160 mm; c/t = 0.5; f = 300 Hz; curves from 1 to 6: Δ = 0–5–10–15–20–25 mm

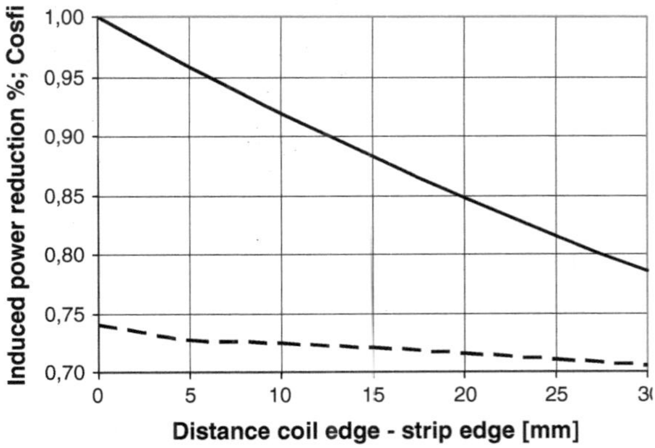

Fig. 4.45 Power factor (*dashed line*) and % variations of total power induced in the strip compared to the case $\Delta = 0$ (—*solid line*) as a function of the distance Δ between coil and strip edges

Table 4.2 Design parameters of the TFH heater

Frequency (Hz)	Pole pitch (mm)	Coil width (mm)	Distance Δ (mm)	Coil amp-turns NI (A)
300	160	80	10	5130

The corresponding electrical characteristics are summarized in the diagrams of Fig. 4.46.

The diagram of Fig. 4.46a shows the power density distribution over half strip; Fig. 4.46b gives the same distributions along equi-spaced lines in the direction of the strip movement.

In the regular region the lines are characterized by two peeks corresponding to the positions in front of the coil conductors in the y direction, while the lines lying in the edge region are characterized by a single central peek.

The curves of Fig. 4.46c shows the energy supplied to the strip along the same lines equi-spaced in the y direction, during the movement of the strip inside the inductor; they indicate that at the exit of the inductor the values of the total energy supplied are nearly the same.

This behavior is confirmed by the diagram of Fig. 4.46d, which gives the relative energy distribution in the y direction (from axis to edge) at the exit of the inductor, referred to the value on the axis of the strip. The curve shows that the maximum deviation of energy values from those of the regular region is between −8 % and +4 %.

4.3.3 Calculations of Heating Transients

Taking into account that in most industrial applications it is required uniform heating of the strip and that TFH produces intrinsically uneven power density

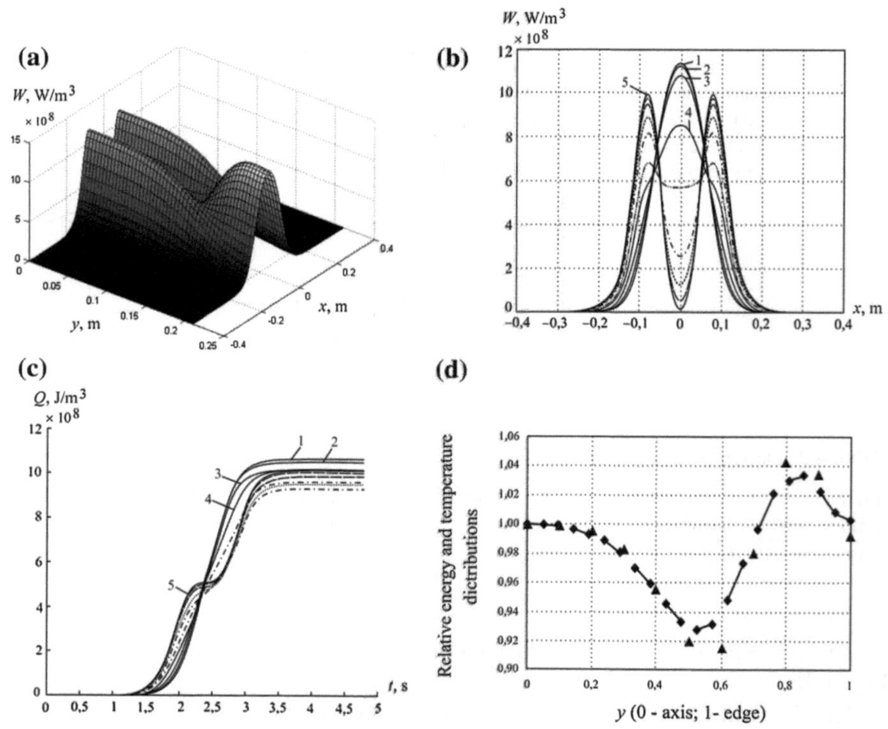

Fig. 4.46 Characteristics of the TFH heater corresponding to the design parameters of Table 4.2 (Induced power in the strip: 70.4 kW; power factor: 0.725) **a** Power density pattern over half strip **b** Power density along equi-spaced lines in the x direction **c** Energy supplied to the strip along the lines of figure **b**, **d** Relative energy and temperature distributions at the exit of the inductor (◆—temperature; ▲—energy)

distributions, it becomes of fundamental importance to check the resulting heating transient and the final temperature distribution in the strip.

To this end a FE model has been implemented which, starting from the previous calculated results, allows the solution of the transient heating problem and the analysis of the thermal results.

The diagrams of Fig. 4.47 illustrates the results obtained for the design example considered in this chapter:

- The curves of Fig. 4.47a shows the temperature increase during the movement across the inductor along five lines equi-spaced in the y direction ($y = 0$; $y = L/8$; $y = L/4$; $y = 3L/8$; $y = L/2$); they have the same shape of the energy curves of Fig. 4.46c. In the same figure it is also shown the temperature difference between strip edge and axis.
- Figure 4.46b gives the curves of the maximum and minimum temperature in the strip cross-section and their difference during the heating process.

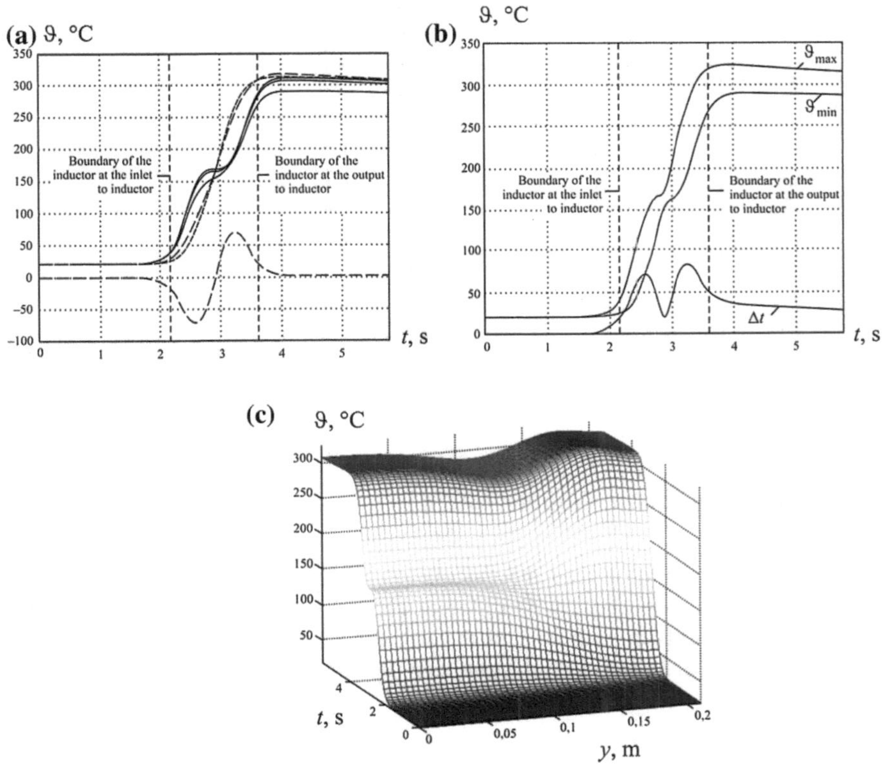

Fig. 4.47 Heating transient in the strip during TFH: **a** and **b** temperature distributions along lines in the x direction and maximum temperature difference along y; **c** 3D temperature pattern

- Figure 4.46c shows a 3D representation of the temperature distribution in half strip during the heating transient. In particular, it shows that the final temperature distribution in the y direction corresponds quite well to the energy curve given in Fig. 4.46d.
- Finally, the difference between maximum and minimum temperature confirm that an appropriate inductor design allows to keep this value within the limits usually required by the subsequent working operation.

The comparison between electromagnetic and thermal results allows us to conclude that the electromagnetic solution which makes use of analytical equations is a very convenient tool for the preliminary design of TFH heaters.

However, since this solution is based on some strong simplifying assumptions, the final design refinement should be always made by means of 3D numerical solutions which can take into account the variations of the material parameters with temperature and allow an accurate evaluation of thermal losses from the strip surface.

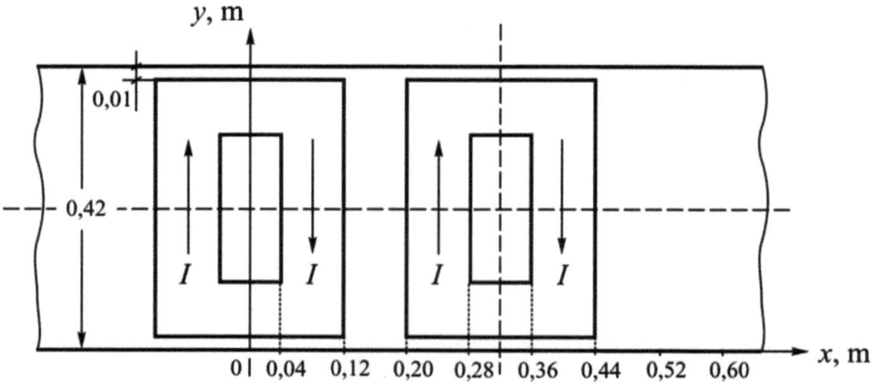

Fig. 4.48 Inductor system with multiple coils

4.3.4 Other Inductor Geometries

The analytical program allows the calculation of different inductor systems, e.g. inductors with multiple coils, with different geometries in the movement direction and different currents in each coil, or the so called "butterfly" geometry.

The following diagrams will illustrate two typical examples of such capability of the program.

(a) *Inductor system constituted by multiple coils* (e.g. *two coils as in* Fig. 4.48)

This system can be used, when enough space is available in the x direction, either for increasing the total induced power in the strip without increasing the inductor current, or for reducing the heating speed and the maximum temperature difference in the strip cross-section during the heating transient.

An example of this second possibility is illustrated by the results of Fig. 4.49 which refers to the geometry of Fig. 4.48 and the value of the ampere-turns NI = 3.720 A in each coil.

The comparison with the results of Fig. 4.46, which refers to the heating of the same brass strip with nearly the same induced power, shows that nearly the same temperature distribution has been obtained at the exit of the inductor (Fig. 4.46d, e), with a slower increase of the temperature and a reduced temperature differential during the heating (Fig. 4.46f).

(b) "*Butterfly*" *coil*

This type of inductor system is sketched in Fig. 4.50 and is characterized by two adjacent coils with currents flowing in the same direction in the central conductors.

Assuming for each coil the same geometrical dimensions of the previous example (see Fig. 4.48) and suitable values of Δ for the two coils, it is possible to optimize the energy distribution at the exit of the inductor, as show in the example of Fig. 4.51.

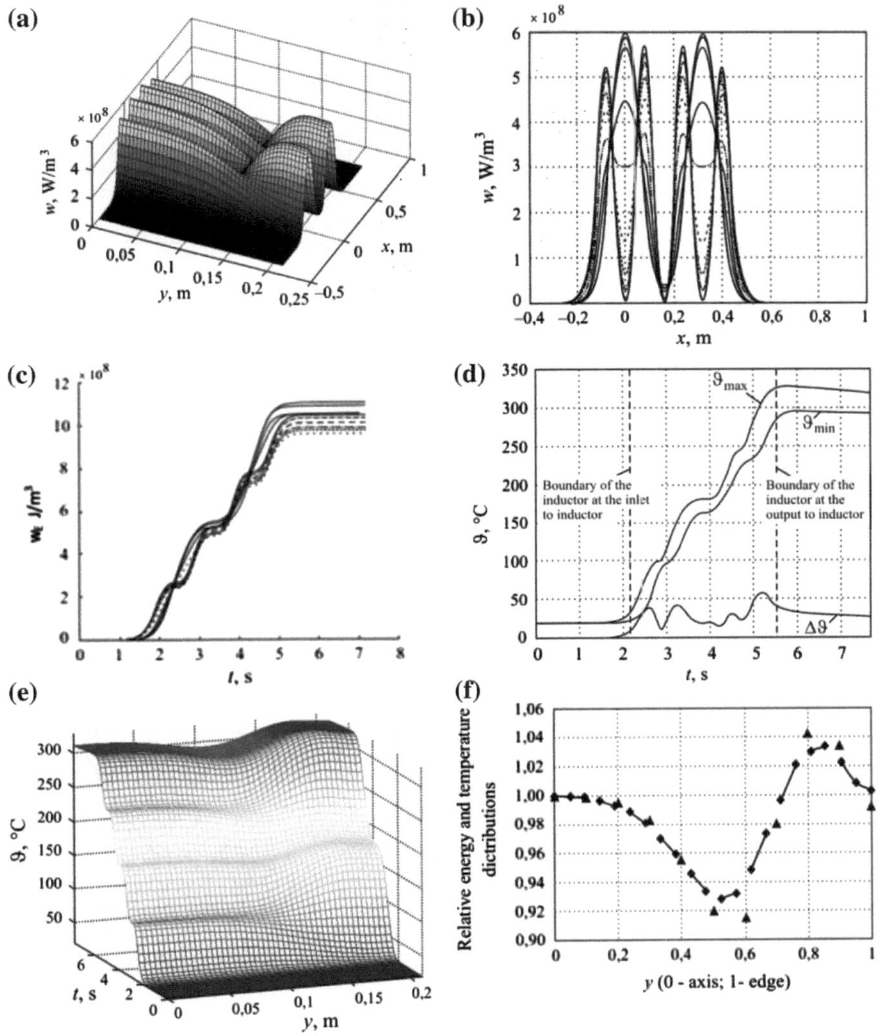

Fig. 4.49 Electrical and thermal results for the inductor of Fig. 4.48: **a** power density distribution; **b** power density along lines in the movement direction; **c** energy transferred to the strip along the lines of figure **b**; **d** max. and min. temperature in the strip and their difference during heating; **e** 3D temperature distribution during heating; **f** relative energy and temperature distributions in the strip cross-section at x = 520 mm (NI = 3.720 A in each coil; total induced power: 62.8 kW; power factor: 0.712)

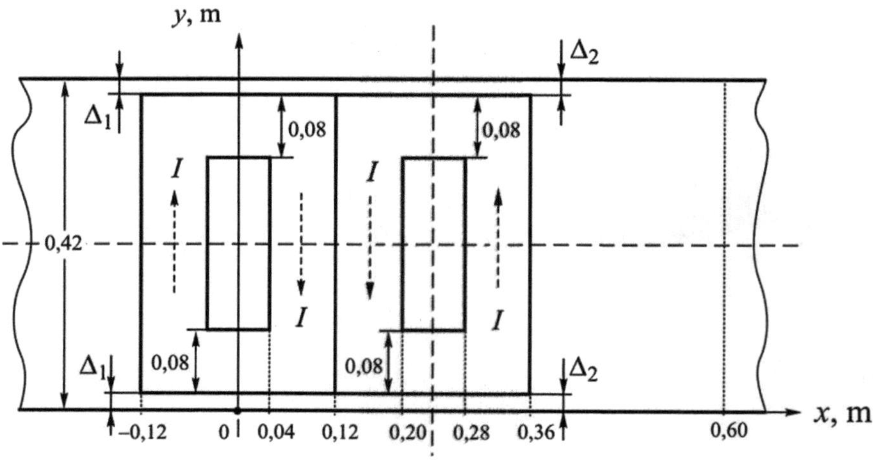

Fig. 4.50 Inductor system with "butterfly" coils and different values of Δ_1 and Δ_2 (mm)

Fig. 4.51 Relative energy distribution at the inductor's exit with different values of Δ_1 and Δ_2 (in mm) (*dotted lines* $\Delta_1 = 0$, $\Delta_2 = 0$; *solid lines* $\Delta_1 = 6$, $\Delta_2 = 0$; *dots and point lines* $\Delta_1 = 10$, $\Delta_2 = 0$)

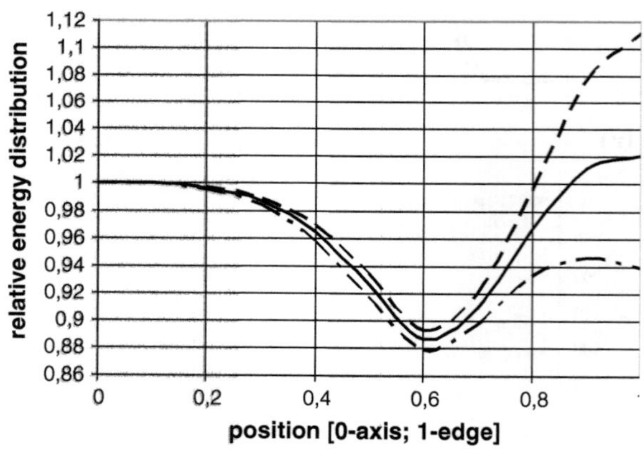

According to this figure, a good solution is obtained with $\Delta_1 \approx 6$ mm and $\Delta_2 = 0$; with these values the heating characteristics are those of Fig. 4.52.

The diagrams show the following:

- the power density distribution of Fig. 4.52a is characterized in the regular zone by three peaks; in particular, the central one—which lies in front of the coil central conductors—is twice more intensive than the two lateral ones. As regards the two peaks in the edge regions, different distances of coil edges to the strip edge have been assumed in order to obtain a sufficiently uniform energy distribution during heating (Fig. 4.52c) and in the *y* direction at the exit of the inductor, as shown in Fig. 4.52f).

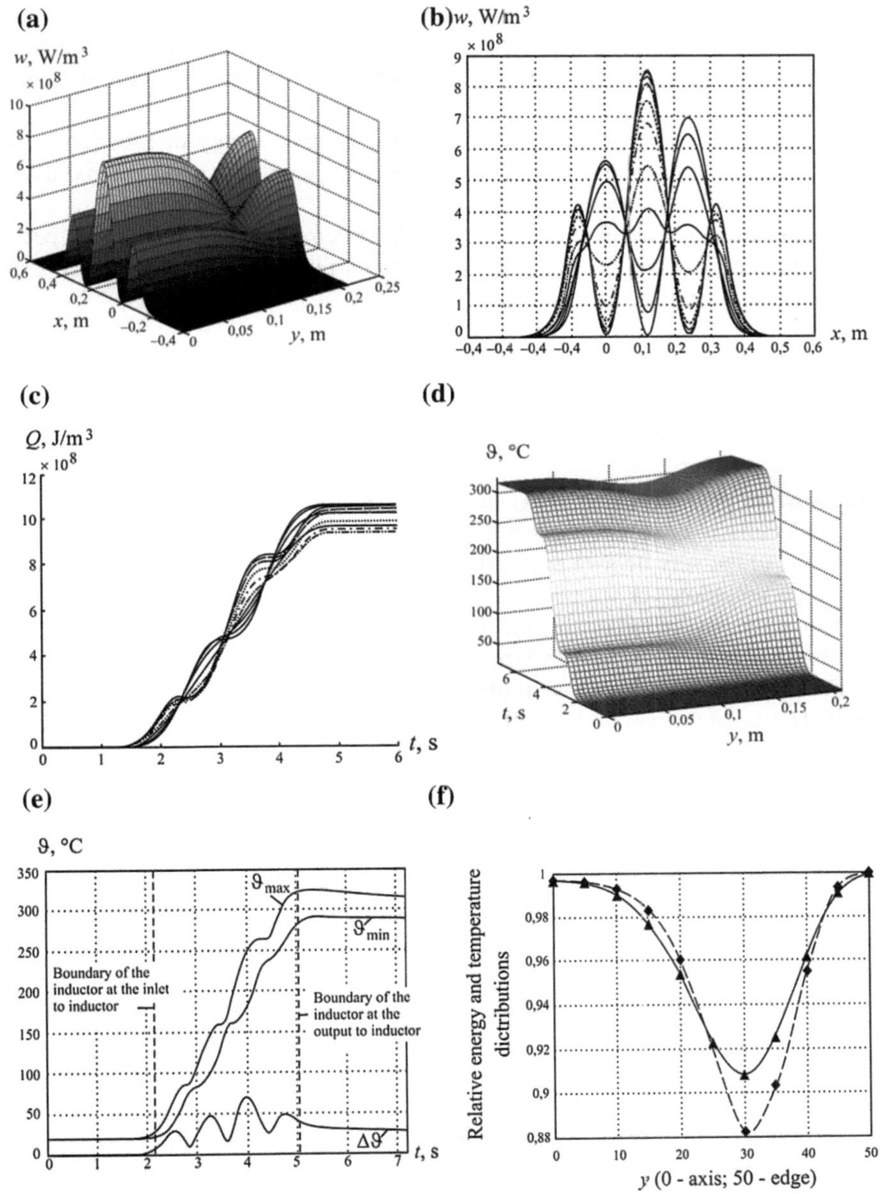

Fig. 4.52 Electrical and thermal characteristics of the inductor of Fig. 4.50 with Δ_1 = 6 mm and Δ_2 = 0: **a** power density distribution; **b** power density along lines in the movement direction; **c** energy transferred to the strip along the lines of figure **b**; **d** 3D temperature distribution during heating; **e** max. and min. temperature in the strip and their difference during heating; **f** relative energy and temperature distributions in the strip cross-section at x = 600 mm; (NI = 3.300 A in each coil; total induced power: 60.2 kW; power factor: 0.772)

- Figure 4.52d–f show that the maximum final temperature differential in the strip cross-section is about 9 % of maximum temperature. This difference reaches a peak value of about 70 °C when the strip passes in front of the coil central conductors.
- Figure 4.52f confirms that the curve of the relative energy distribution in the strip cross-section at the exit of the inductor represents with good approximation also the final temperature distribution.
- the integral parameters of the inductor-strip system show that the required power of about 60 kW can be induced with a lower value of coil ampere-turns NI and with a better power factor of the system.

4.3.5 Recent Developments and Conclusions

As shown in previous paragraphs, the most difficult problem in utilizing transverse flux inductors is to obtain a good temperature uniformity across the width of the strip, due to the very uneven power density distribution and the short heating times which do not allow for temperature equalization through thermal conduction.

Moreover, an important feature for the application of this technology in the industry is the ability of heating strips of different widths within the same inductor.

For the solution of the above problems, in the last 15 years, the research activity on TFH has been mainly focused on the numerical solution of the 3D coupled electromagnetic-thermal problem of a moving conducting sheet under different inductor geometries, symmetrical or non-symmetrical with the strip axis, with or without magnetic yokes.

This activity developed progressively from simplified models with constant material parameters and no movement, to the more realistic ones. Today these models can take into account the non linear variations of all material characteristics with local temperature and—for magnetic strips—with the magnetic field intensity during the strip movement.

Theoretical work and mathematical modelling have been conducted at the University of Hanover in Germany, University of Padua in Italy, Electrotechnical University of St. Petersburg in Russia, EDF in France and several industrial companies [53, 64, 67].

The improvement of design procedures and the development of new types of inductors, capable of controlling the edge effects in strips of different widths, have allowed the realization of some plants which constitute important steps towards a wider diffusion of this technology in the industry. In particular,

- A 10 MW installation for reheating steel strips above 1,000 °C, has been built by SMS Elotherm in cooperation with the University of Hanover (Fig. 4.53a). Characteristic data of this installation are: 4 inductors, 2,500 kW each, 1,000 Hz, 4 m total length, 1,100–1,450 mm strip width, 1.4–3.5 mm thickness, average speed 20–100 m/min. The adjustment of the edge effects, for different strip

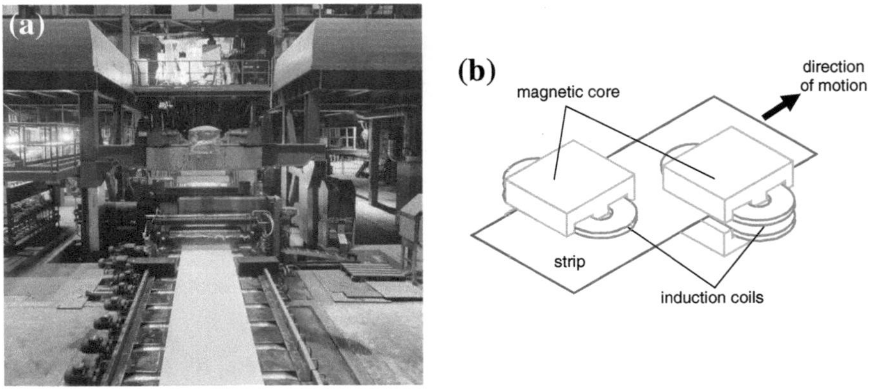

Fig. 4.53 10 MW transverse flux heater [67]

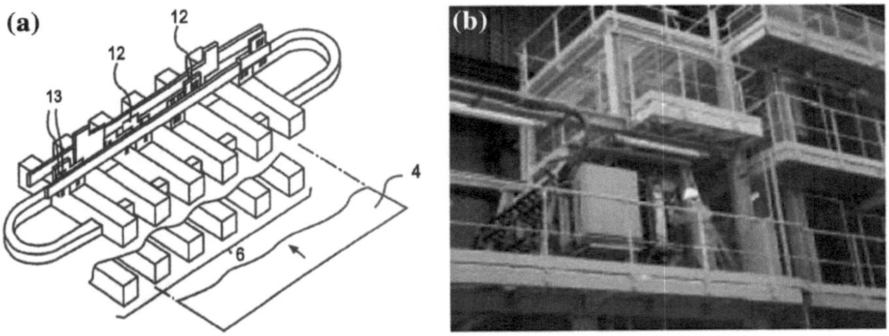

Fig. 4.54 3 MW Pilot Plant [47, 61]

widths, is obtained by moving the two coils of 4.53b) along the direction transversal to the strip movement. This solution is suitable for high-temperature high-power applications and allows to reach in thick strips temperature homogeneity of ±3 % [64].

- A 3 MW industrial pilot plant was designed and installed in France at the Arcelor Research centre of IRSID, by EDF (Electricité de France), IRSID, Arcelor (Maizières lès Metz) and CELES, Lautenbach (Fig. 4.54b). The plant is characterized by a very high specific power (up to 2.5 MW/m^2), very large mechanical air gap between the two parts of the inductor, capability of heating strips widths from 800 to 1,500 mm and thickness from 0.1 to 1.5 mm. In this case the inductor, shown schematically in Fig. 4.54a, adjusts the edge effects for different strip widths by moving the magnetic yokes, shown in the figure, along the direction transversal to the strip movement [47].
- A 600 kW, 3–4 kHz installation for the austempering of magnetic steel strips, has been built by the company ATE (VI, Italy) and designed in cooperation with

the laboratory for Electroheat LEP of the University of Padua. The data of this installation are: magnetic steel strip heated to 850–950 °C, widths 160–260 mm, thickness 0.5–1.2 mm, speed up to 27–30 m/min. To cope with different strip widths the inductor is constituted by a series of butterfly coils of different geometry as shown in Fig. 4.55a. Figure 4.55b, c shows the temperature distribution obtained in a strip, 200 mm. width, characterized by a temperature difference of 50 °C and maximum temperature of 965 °C (i.e. ±2.6 %).

Several other special inductors capable of controlling the edge effects when heating strips of different widths have been proposed and patented.

Among them we can mention as examples the following:

- the solution, patented with trademark VABID, refers to the drawing of Fig. 4.56a.
  In this solution each inductor coil, i.e. the upper and the lower one, is subdivided into two parts which can be moved in the direction normal to the movement. In this way one obtains an "effective" rectangular coil having a fixed pole pitch but transversal dimensions which can be adapted to the strip width. In the terminals of this rectangular coils, currents of equal amplitudes but opposite directions flow, and therefore their influence on the flux pattern is practically negligible. It allows to reach temperature homogeneity of ±2.5 % [67].
- the solution shown in Fig. 4.56b, is characterized by an inductor coil with constant position of its edges in relationship to the edge of the strip but has an

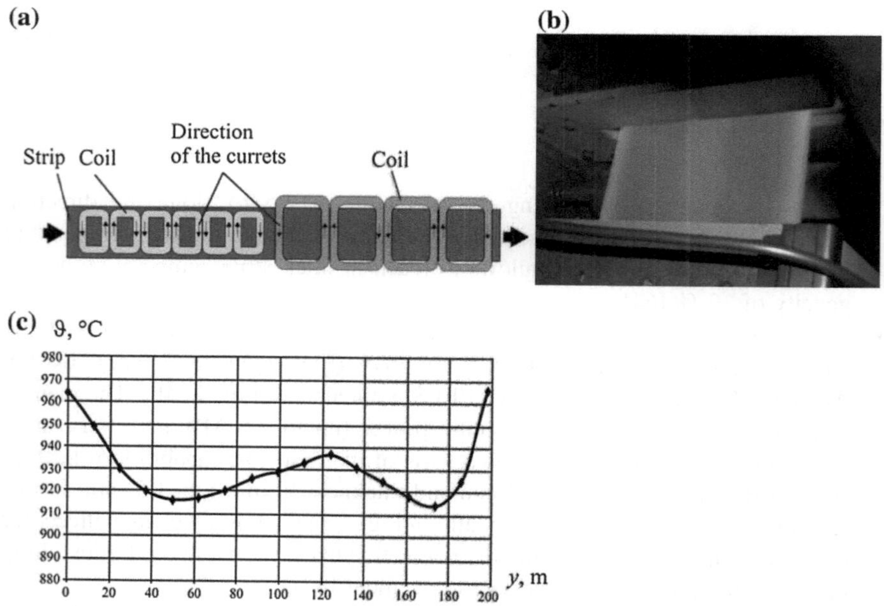

Fig. 4.55 600 kW installation for austempering magnetic steel strips: **a** inductor coils; **b**, **c** experimental temperature distribution

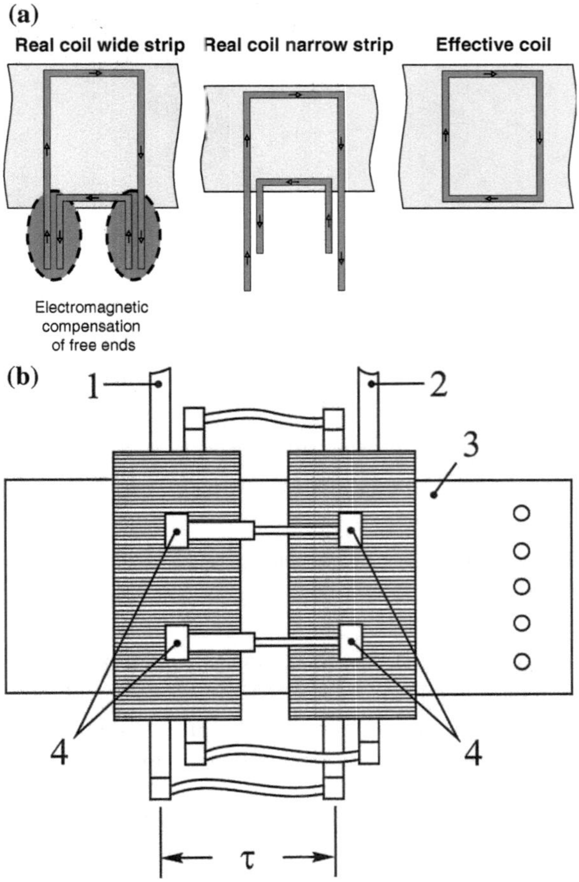

Fig. 4.56 Different arrangements of TFH inductors for heating strips of different widths in the same inductor

adjustable pole pitch useful for modifying the heating edge effect as the strip width changes [62].

- another promising solution based on the method named "*Zone Controlled Induction Heating*" (ZCIH) has been proposed in the last years for controlling the induced power distribution in induction heated loads [72]. The method is based on the use of inductors with multiple coils which are independently supplied with a set of currents of suitable amplitudes and phases, in order to obtain in the workpiece an optimized distribution of the induced power density, able to assure the required heating homogeneity [73].

The possibility of applying this technique to TF strip heating has been analysed by the optimization of a pilot installation characterized by an inductor with six independent coils suited for uniform heating of non magnetic steel strips with different width ranging from 500 to 1,000 mm (Fig. 4.57). The main feature of this

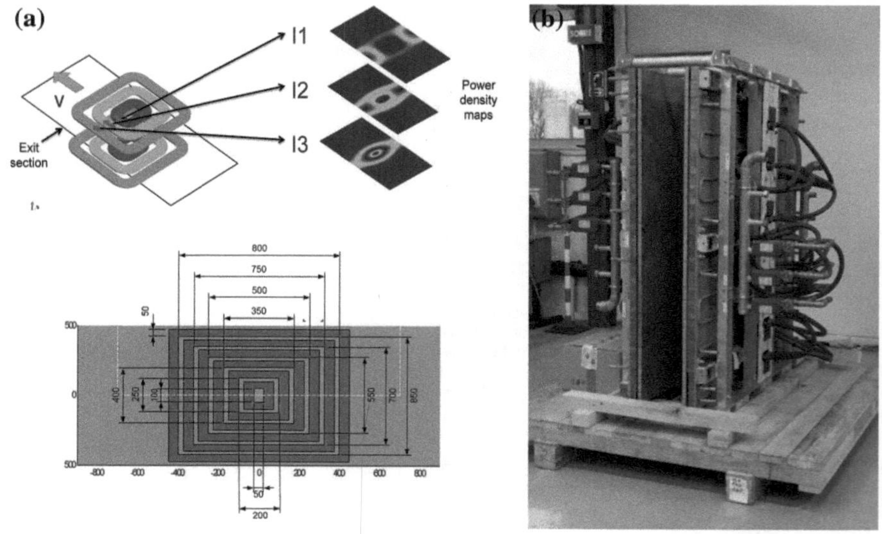

Fig. 4.57 **a** Schematic of inductor coils, load and yoke of pilot system; **b** test inductor (Courtesy of EDF)

inductor is that it can be adapted to the different strip widths without the need of moving parts [74].

The target performance of the pilot system was an average temperature rise in the strip of 300 °C and a maximum temperature gradient in the movement direction of 300 °C/m.

Preliminary analyses of this system have been developed with the programme TFH3D coupled with an optimization procedure which allows to determine the different sets of currents (amplitudes and phases) which will assure at the exit of the inductor uniform heating of strips with different widths.

The optimized patterns of induced power density and temperature shown in Fig. 4.58, have demonstrated the applicability of this technique to TFH since sufficiently uniform final temperature distributions in strip cross-sections (top region in Fig. 4.58b can be achieved.

Starting from the results of TFH3D computations, the final sets of currents for the excitation system are determined through a 3D FEM optimization, which takes into account the real inductor and yoke geometry.

As a conclusion to this chapter it can be stated that the TFH of strips is today a mature technology applicable to different heating processes in the metal industry.

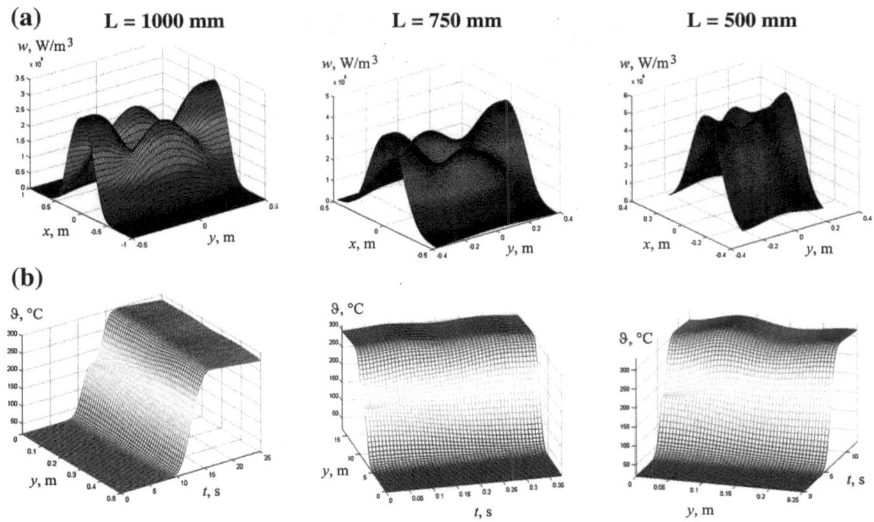

Fig. 4.58 Optimized characteristics of ZCIH for three strip widths (L) **a** induced power density distribution in the strip; **b** temperature during heating transient in half strip

4.4 Planar Circular Coils for Induction Heating

4.4.1 Analytical Solutions

The configuration under consideration, shown in Fig. 4.59, consists of an ideal planar circular coil of negligible thickness in axial direction, constituted by N series connected concentric turns, equally spaced in radial direction. This kind of inductor is commonly named pancake inductor. The coil, in which a sinusoidal current I flows, is placed facing the workpiece, a plate whose radius is much larger than that

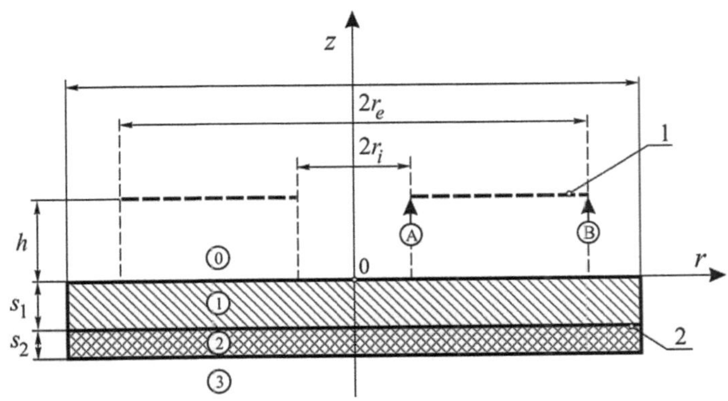

Fig. 4.59 Schematic of axis-symmetric planar system: **a** inductor; **b** bimetallic plate

of the coil and consists of two homogeneous layers with constant resistivity and permeability.

With reference to a cylindrical coordinate system (r, φ, z), by symmetry the magnetic field lies on r, z plane depending only on r and z coordinates, and not on azimuth φ. Current density and magnetic vector potential A have only the azimthal component, being nil the components on the r, z plane.

In the hypothesis of an ideal inductor consisting of a uniform layer of negligible axial thickness, the phasor of the magnetic vector potential $\dot{A}$ satisfies the following scalar equation [75–82]:

$$\frac{\partial^2 \dot{A}}{\partial r^2} + \frac{1}{r}\frac{\partial \dot{A}}{\partial r} - \frac{\dot{A}}{r^2} + \frac{\partial^2 \dot{A}}{2} = \begin{cases} 0 & \text{in air} \\ \dfrac{j\omega\mu_0\mu_{ri}}{\rho_i}\dot{A} & \text{in the } i\text{–th layer (i} = 1, 2) \end{cases} \tag{4.23}$$

where: $\dot{A} = \dot{A}_{m\varphi}/\sqrt{2}$; $\dot{A}_{m\varphi}$—complex amplitude of the azimuthal component of vector potential.

The vector potential in the various regions of the system geometry is given by the following equations [78]:

- Region 0: *[for* $0 \le z \le h$*]*

$$\begin{aligned} \dot{A}_0 &= \dot{A}_{0e} + \dot{A}_{0r} \\ &= \frac{\mu_0 N\dot{I}}{2(r_e - r_i)} \int_{k=0}^{\infty} \dot{C}_0(k) \cdot J_1(kr) \cdot e^{-k(h-z)} \cdot dk \\ &+ \int_{k=0}^{\infty} \dot{D}_0(k) \cdot J_1(kr) \cdot e^{-kz} \cdot dk \end{aligned} \tag{4.24}$$

with

$$\dot{C}_0(k) = \int_{r=r_i}^{r_e} r \cdot J_1(kr) \cdot dr \tag{4.24a}$$

The term A_{0e} of Eq. (4.24) represents the vector potential of the exciting field, i.e. the vector potential of the field produced by the inductor coil in absence of conducting materials in the regions 1 and 2. The second term A_{0r} represents the potential of the so-called reaction field, due to the current induced in the load.

- Regions 1 and 2: *[for* $0 \geq z \geq -s_1$ *and* $-s_1 \geq z \geq -(s_1+s_2)$*]*

$$\dot{A}_i = \int_{k=0}^{\infty} \left[C_i(k) \cdot e^{\dot{\beta}_i z} + D_i(k) \cdot e^{-\dot{\beta}_i z}\right] \cdot J_1(kr) dk \tag{4.25}$$

with: $\dot{\beta}_i^2 = k^2 + j\frac{\omega \mu_{ri} \mu_0}{\rho_i}$

- Region 3: *[for* $z \leq -(s_1+s_2)$*]*

$$\dot{A}_3 = \int_{k=0}^{\infty} \dot{C}_3(k) \cdot J_1(kr) \cdot e^{kz} dk \tag{4.26}$$

The unknown complex coefficients C_1, C_2, C_3, D_0, D_1, D_2 can be obtained by applying the continuity conditions of the vector potential and the radial component of the field intensity at the boundaries of the various regions, by means of the following system written in matrix form:

$$\begin{vmatrix} 1 & -1 & -1 & 0 & 0 & 0 \\ k & \frac{\dot{\beta}_1}{\mu_{r1}} & -\frac{\dot{\beta}_1}{\mu_{r1}} & 0 & 0 & 0 \\ 0 & e^{-\dot{\beta}_1 s_1} & e^{\dot{\beta}_1 s_1} & -e^{-\dot{\beta}_2 s_1} & -e^{\dot{\beta}_2 s_1} & 0 \\ 0 & \frac{\dot{\beta}_1}{\mu_{r1}} e^{-\dot{\beta}_1 s_1} & -\frac{\dot{\beta}_1}{\mu_{r1}} e^{\dot{\beta}_1 s_1} & -\frac{\dot{\beta}_2}{\mu_{r2}} e^{\dot{\beta}_2 s_1} & \frac{\dot{\beta}_2}{\mu_{r2}} e^{\dot{\beta}_2 s_1} & 0 \\ 0 & 0 & 0 & e^{-\dot{\beta}_2 (s_1+s_2)} & e^{\dot{\beta}_2 (s_1+s_2)} & -e^{-k(s_1+s_2)} \\ 0 & 0 & 0 & \frac{\dot{\beta}_2}{\mu_{r2}} e^{-\dot{\beta}_2 (s_1+s_2)} & -\frac{\dot{\beta}_2}{\mu_{r2}} e^{\dot{\beta}_2 (s_1+s_2)} & -ke^{-k(s_1+s_2)} \end{vmatrix} \cdot \begin{vmatrix} \dot{D}_0 \\ \dot{C}_1 \\ \dot{D}_1 \\ \dot{C}_2 \\ \dot{D}_2 \\ \dot{C}_3 \end{vmatrix} = \begin{vmatrix} -\frac{\mu_0 N\dot{I}}{2(r_e-r_i)} \dot{C}_0(k) e^{-kh} \\ \frac{\mu_0 N\dot{I}}{2(r_e-r_i)} k\dot{C}_0(k) e^{-kh} \\ 0 \\ 0 \\ 0 \\ 0 \end{vmatrix}$$

It should be noted that in order to evaluate the integral electrical parameters of the coil, the electric and magnetic field intensities in the region *0* and the induced power density distribution on the surface of the load, it is sufficient to know the complex coefficient D_0, while the values of C_1, C_2, and D_1, D_2 are necessary for the calculation of the current density and the power distributions inside the load.

In particular, from the previous system the following explicit relationship can be obtained for $D_0(k)$:

$$\dot{D}_0(k) = \frac{\mu_0 N \dot{I}}{2(r_e - r_i)} \cdot \dot{C}_0(k) \cdot \dot{F}(k) \cdot e^{-kh} \tag{4.27}$$

where

$$\dot{F}(k) = \frac{e^{\dot{\beta}_1 s_1}\left(k - \frac{\dot{\beta}_1}{\mu_{r1}}\right)\left[e^{\dot{\beta}_2 s_2}\left(k + \frac{\dot{\beta}_2}{\mu_{r2}}\right)\left(\frac{\dot{\beta}_1}{\mu_{r1}} + \frac{\dot{\beta}_2}{\mu_{r2}}\right) - e^{-\dot{\beta}_2 s_2}\left(k - \frac{\dot{\beta}_2}{\mu_{r2}}\right)\left(\frac{\dot{\beta}_1}{\mu_{r1}} - \frac{\dot{\beta}_2}{\mu_{r2}}\right)\right]}{e^{\dot{\beta}_1 s_1}\left(k + \frac{\dot{\beta}_1}{\mu_{r1}}\right)\left[e^{\dot{\beta}_2 s_2}\left(k + \frac{\dot{\beta}_2}{\mu_{r2}}\right)\left(\frac{\dot{\beta}_1}{\mu_{r1}} + \frac{\dot{\beta}_2}{\mu_{r2}}\right) - e^{-\dot{\beta}_2 s_2}\left(k - \frac{\dot{\beta}_2}{\mu_{r2}}\right)\left(\frac{\dot{\beta}_1}{\mu_{r1}} - \frac{\dot{\beta}_2}{\mu_{r2}}\right)\right]}$$
$$\frac{+e^{-\dot{\beta}_1 s_1}\left(k + \frac{\dot{\beta}_1}{\mu_{r1}}\right)\left[e^{\dot{\beta}_2 s_2}\left(k + \frac{\dot{\beta}_2}{\mu_{r2}}\right)\left(\frac{\dot{\beta}_1}{\mu_{r1}} - \frac{\dot{\beta}_2}{\mu_{r2}}\right) - e^{-\dot{\beta}_2 s_2}\left(k - \frac{\dot{\beta}_2}{\mu_{r2}}\right)\left(\frac{\dot{\beta}_1}{\mu_{r1}} + \frac{\dot{\beta}_2}{\mu_{r2}}\right)\right]}{+e^{-\dot{\beta}_1 s_1}\left(k - \frac{\dot{\beta}_1}{\mu_{r1}}\right)\left[e^{\dot{\beta}_2 s_2}\left(k + \frac{\dot{\beta}_2}{\mu_{r2}}\right)\left(\frac{\dot{\beta}_1}{\mu_{r1}} - \frac{\dot{\beta}_2}{\mu_{r2}}\right) - e^{-\dot{\beta}_2 s_2}\left(k - \frac{\dot{\beta}_2}{\mu_{r2}}\right)\left(\frac{\dot{\beta}_1}{\mu_{r1}} + \frac{\dot{\beta}_2}{\mu_{r2}}\right)\right]}$$

These equations can also be used for the calculation of the magnetic field intensity components H_{ir} and H_{iz}, the electric field intensity E_i and the current density J_i at various points of the system by the relationships:

$$\begin{aligned}\dot{H}_{ir} &= -\frac{1}{\mu_0 \mu_{ri}} \frac{\partial \dot{A}_i}{dz} \\ \dot{H}_{iz} &= \frac{1}{\mu_0 \mu_{ri}} \frac{1}{r} \frac{\partial}{dr}\left[r \dot{A}_i(r, z)\right]\end{aligned} \tag{4.28}$$

$$\dot{E}_i(r, z) = \rho_i \dot{J}_i(r, z) = -j\omega \dot{A}_i(r, z) \tag{4.29}$$

From Eqs. (4.28) and (4.29) the main system parameters can also be obtained:

- e.m.f. E'_N induced in an elementary turn of the coil at radius r:

$$\dot{E}'_N(r, h) = -j2\pi\omega r \cdot \dot{A}_0(r, h), \text{ for } (\mathrm{r_i} \le \mathrm{r} \le \mathrm{r_e}) \tag{4.30}$$

- total e.m.f. E_N induced across N turns of the coil:

$$\dot{E}_N = \int_{r=r_i}^{r_e} \dot{E}'_N \frac{N}{(r_e - r_i)} dr = -j2\pi\omega \frac{N}{(r_e - r_i)} \int_{r=r_i}^{r_e} r \dot{A}_0(r, h) dr \tag{4.31}$$

The value of the equivalent impedance $\dot{Z}_e$ of the inductor-load system can be evaluated taking into account that, assuming negligible the own resistance of the coil, the voltage applied must balance only the e.m.f. E_N:

$$\dot{Z}_e = -\frac{\dot{E}_N}{\dot{I}} = j\pi\omega\mu_0 \frac{N^2}{(r_e - r_i)^2} \cdot \left\{ \int_{k=0}^{\infty} \left[\dot{C}_0(k)\right]^2 dk + \int_{k=0}^{\infty} \left[\dot{C}_0(k)\right]^2 \dot{F}(k) e^{-2kh} dk \right\} \tag{4.32}$$

with $C_0\ (k)$ and $F(k)$ given by Eqs. (4.24a) and (4.27).

Despite the simplifying assumptions, the method has proved to be suitable for the analysis of the influence of electrical and geometrical parameters and the calculation of configurations with layers of different materials with constant resistivity and magnetic permeability.

The method has been also further developed and applied to the calculation of the electro-dynamic forces acting between inductor and load (see Sect. 4.10).

Numerical Evaluation of Analytical Solutions

The most difficult problem in the use of the previous analytical solutions is the numerical evaluation of the integrals of Eqs. (4.23)–(4.32). This evaluation can be conveniently done applying the Gauss-Laguerre quadrature formulae, which give the following relationships for the calculation of the main parameters:

- Current density in regions 1 and 2:

$$\dot{J}_i(r,z) = -j\frac{\omega}{\rho_i} \times \sum_{v=1}^{n} W_v e^{x_v} \left[\dot{C}_i(x_v) e^{\dot{\beta}_{vi} z} + \dot{D}_i(x_v) e^{-\dot{\beta}_{vi} z}\right] J_1(x_v r) \tag{4.33}$$

with: $i = 1, 2$; $\dot{\beta}_{vi} = \sqrt{x_v^2 + \frac{2j}{\delta_i^2}}$; n—integration order; x_v, W_v—abscissas and integration weights factors of Gauss-Laguerre integration; $J_1(x_v r)$—Bessel function.

- Induced e.m.f. in an elementary turn of the inductor at radius r:

$$\dot{E}'_N(r,h) = -j2\pi\omega r \left[\frac{\mu_0 N\dot{I}}{2(r_e - r_i)} \sum_{v=1}^{n} W_v e^{x_v} \dot{C}_0(x_v) J_1(x_v r) + \sum_{v=1}^{n} W_v e^{x_v} \dot{D}_0(x_v) J_1(x_v r) e^{-x_v h}\right] \tag{4.34}$$

- Impedance of the inductor-load system:

$$\dot{Z}_e = j\pi\omega\mu_0 \frac{N^2}{(r_e - r_i)^2} \cdot \left\{ \sum_{v=1}^{n} W_v e^{x_v} \left[\dot{C}_0(x_v)\right]^2 + \frac{1}{2h} \sum_{v=1}^{n} W_v \left[\dot{C}_0\left(\frac{x_v}{2h}\right)\right]^2 \dot{F}\left(\frac{x_v}{2h}\right) \right\} \quad (4.35)$$

It is therefore necessary, after the choice of the integration order n, to evaluate the corresponding values of C_0, C_1, C_2, D_0, D_1, D_2 and F for each value of x_v using Eqs. (4.24a), (4.27), (4.28). Depending on input values, it is sometimes difficult, from the point of view of the calculation, to evaluate the value of the integral (4.24a) to the required degree of accuracy; in any case, bearing in mind that the convergence is relatively slow as n increases, very high integration orders are necessary.

For this reason, the evaluation of $C_0(x_v)$ can be more conveniently made with the following equation which can be derived from Eq. (4.24a):

$$\dot{C}_0(x_v) = \frac{\sqrt{\pi}}{x_v} \Gamma\left(\frac{3}{2}\right) \cdot \{ r_e[J_1(x_v r_e)\dot{H}_0(x_v r_e) - J_0(x_v r_e)\dot{H}_1(x_v r_e)] - r_i[J_1(x_v r_i)\dot{H}_0(x_v r_i) - J_0(x_v r_i)\dot{H}_1(x_v r_i)] \} \quad (4.36)$$

with: Γ—Gamma funtion.

The accuracy that can be obtained with a given integration order n, can be checked by verifying the fulfilment of the following condition, valid for $r_i \leq r \leq r_e$:

$$\int_{k=0}^{\infty} k\dot{C}_0(k)J_1(kr)dk = \sum_{v=1}^{n} W_v e^{x_v} x_v \dot{C}_0(x_v) J_1(x_v r) = 1 \quad (4.37)$$

4.4.2 Examples of Results Obtained with the Analytical Solution

Among the large number of industrial applications of planar inductors, we present some results obtained with the previous analytical solution for the brazing or dishing of copper or aluminum bottoms on stainless steel saucepans by induction heating at medium frequency.

The results illustrate the influence of the electrical and geometrical parameters of the inductor-load system on the values of the equivalent impedance seen at the inductor terminals and the radial distributions of the induced current density in the layers of the load.

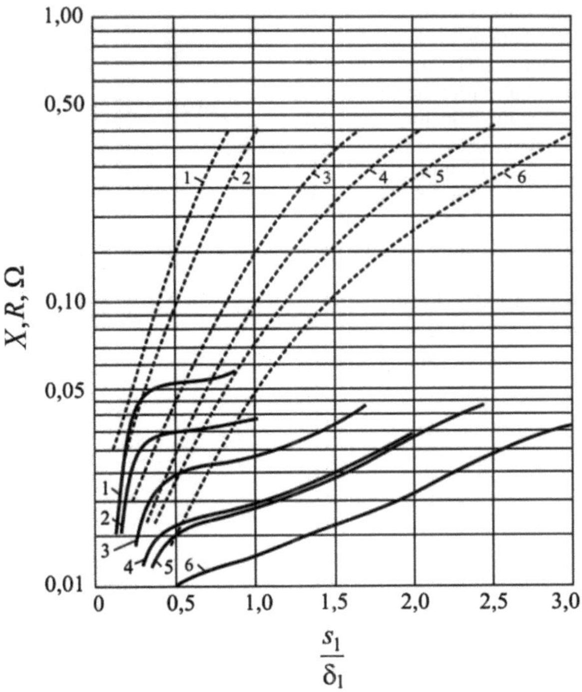

Fig. 4.60 Values of equivalent resistance R and reactance X as a function of the ratio s_1/δ_1 for various values of thickness and resistivity of layer1 (*solid lines*: R resistance; *dashed lines*: X reactance) (Inductor: $r_e = 100$ mm; $r_i = 0$; $N = 20$; Load: $h = 4$ mm; $s_2 = 1$ mm; $\rho_2 = 120$ μΩcm; $\mu_{r2} = \mu_{r1} = 1$; curves: (*1*) $s_1 = 1$ mm, $\rho_1 = 6$ μΩcm; (2) $s_1 = 1$ mm, $\rho_1 = 4$ μΩcm; (*3*) $s_1 = 2$ mm, $\rho_1 = 6$ μΩcm; (*4*) $s_1 = 2$ mm, $\rho_1 = 4$ μΩcm; (5) $s_1 = 3$ mm, $\rho_1 = 6$ μΩcm; (*6*) $s_1 = 3$ mm, $\rho_1 = 4$ μΩcm)

The curves of Fig. 4.60 give, for the system defined in the caption, the values of the equivalent resistance R and reactance X as a function of the ratio s_1/δ_1 for a gap h of 4 mm between the ideal inductor and the load surface.

They highlight the importance of selecting the frequency according to the dimensions and characteristics of the plate.

The curves of Fig. 4.61a, b shows the radial distributions of the induced current density, at various depths in the plate, for the configuration with $h = 4$ mm at frequencies of 2.5 and 10 kHz respectively.

The curves show the typical annulment of the current density on the axis of the configuration and in the region of the workpiece outside the edge of the coil (end effect of the inductor).

Other useful results for this application are given in Refs. [78, 79].

4.4.3 Examples of Numerical Results

For the real inductor-load configuration schematized in Fig. 4.62, the analytical solution cannot give any information on the current distribution in the turns of the inductor, on its resulting AC resistance and, finally, on the system efficiency and power factor.

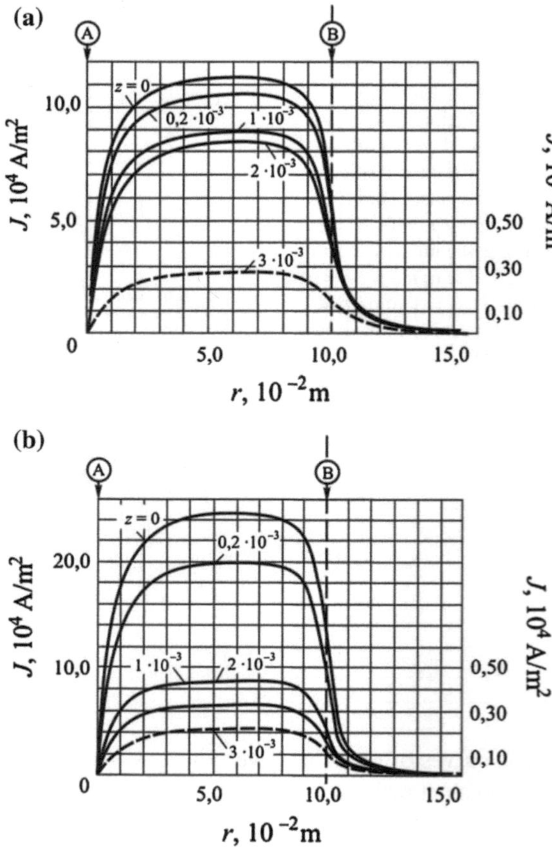

Fig. 4.61 Radial distributions of induced current density in the plate at a depth z from the surface facing the inductor $h = 4$ mm; Inductor: (A) $r_i = 0$; (B) $r_e = 100$ mm; $N = 20$; $I = 1$ A; **a** $f = 2.5$ kHz; Load: $\mu_{r2} = \mu_{r1} = 1$; $s_1 = 2$ mm, $\rho_1 = 4$ μΩ·cm; $s_2 = 1$ mm; $\rho_2 = 120$ μΩ·cm; **b** $f = 10$ kHz; Load: $\mu_{r2} = \mu_{r1} = 1$; $s_1 = 2$ mm, $\rho_1 = 4$ μΩ·cm; $s_2 = 1$ mm; $\rho_2 = 120$ μΩ·cm

Moreover, the analytical solution is applicable only to systems with coils with uniformly distributed ampere-turns and plates with radial dimension greater than that of the inductor.

When dealing with system geometries like the one shown in Fig. 4.62, in particular when the inductor has unevenly distributed ampere-turns in the radial direction, the use of numerical methods is mandatory for the analysis of the integral system parameters and the load end-effect.

Several numerical procedures for the calculation of cylindrically symmetric induction heating systems with non-magnetic or ferromagnetic loads are currently available.

An example of analysis of the load end-effect obtained with such procedures is shown in Fig. 4.63; it refers to an inductor constituted by 10 square copper tube water cooled turns (7.5 × 1 mm) supplied with current at various frequencies, placed in front of a two-layers load with different radial dimensions [79].

The solid curves show that when the plate has the same radial dimensions of the inductor, the resulting current density distributions are quite different from those

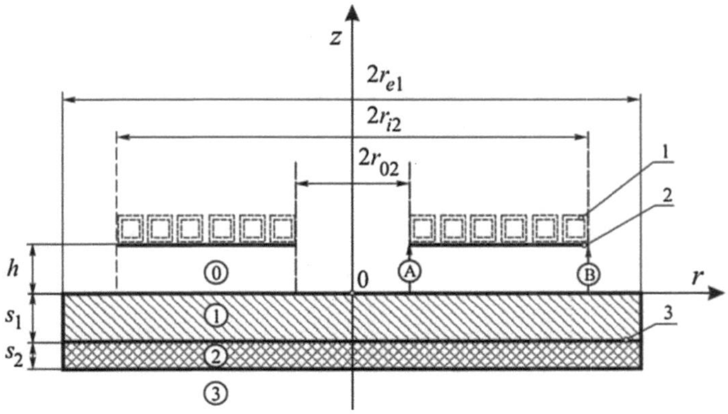

Fig. 4.62 Schematic diagram of the inductor-load system *1*. real inductor; *2*. idealized inductor; *3*. load constituted by two-layers of finite radial dimensions

Fig. 4.63 Radial distribution of induced current density at depth 0.2 mm from the surface of plates of different radial dimensions, placed in front of an inductor with uniform spaced turns, excited at 2–4–8 kHz. Inductor: $N = 10$ turns, square copper tube 7.5 × 1 mm; $\rho_i = 2$ μΩcm; I = 1,000 A; Load: $s_1 = 2$ mm; $\rho_1 = 8$ μΩcm; $s_2 = 1$ mm; $\rho_2 = 120$ μΩcm; $h = 4$ mm; **a** *Solid lines* $r_{e1} = r_{e2}$; **b** *Dashed lines* $r_{e1} >> r_{e2}$; *A*, *B*—as in Fig. 4.62

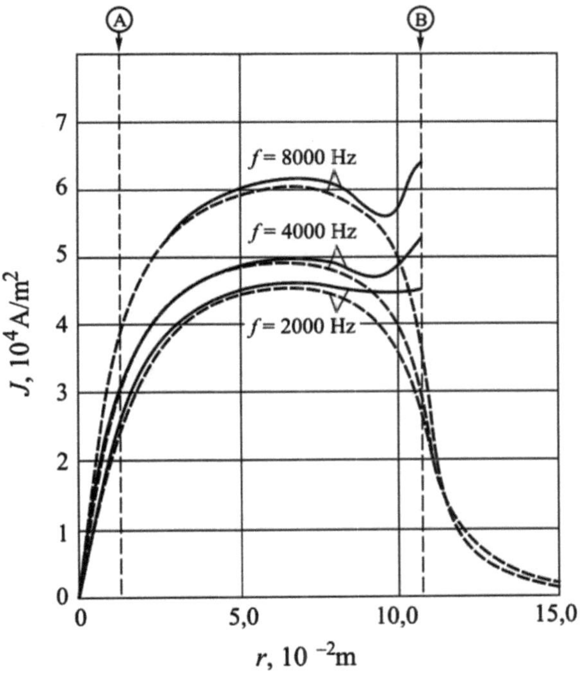

previously examined, notably in the region near the edge of the plate, with shapes and intensities which strongly depend on the frequency. In the same figure the dashed lines show the corresponding distributions for the same inductor in a plate whose radial dimensions are much larger than those of the inductor.

The values of integral parameters characterizing the systems at the various frequencies are given in Table 4.3.

Table 4.3 Values of integral parameters for systems of Fig. 4.63

Frequency f (kHz)	Inductor resistance R_i (mΩ)		Equivalent load resistance R_{12} (mΩ)		Reactance X (mΩ)		Power factor cos φ	
	(a)	(b)	(a)	(b)	(a)	(b)	(a)	(b)
2	5.80	5.69	10.57	9.48	29.14	28.34	0.490	0.472
4	7.21	7.07	11.26	10.20	52.83	51.48	0.330	0.318
8	8.14	7.97	12.74	11.73	101.09	98.53	0.202	0.196

(a) $r_{e1} = r_{e2}$
(b) $r_{e1} \gg r_{e2}$

Conclusions

Despite the simplifying hypotheses on which it is based, the analytical method has proved to be suitable for a general understanding of the influence of the electrical and geometrical parameters and the calculation of systems with layered loads having different constant values of resistivity and magnetic permeability.

On the other hand, it cannot give information on the current distribution in the inductor turns, its resulting AC resistance, the load end-effects and the system power factor and efficiency.

Moreover, it cannot take into account the "radial spacing" of turns and the axial current distribution in the wall thickness of the real inductor.

Finally, it cannot be applied to systems with inductors having non-uniform ampere-turn distribution or when the variations of the material properties with temperature and local magnetic field intensity must be taken into account [83–85].

4.5 Induction Heating of "Long" Cylindrical Workpieces with Inductors of Finite Axial Length

Analytical solutions of the electromagnetic field distributions in infinitely long systems for induction heating of solid cylindrical workpieces (Fig. 4.64a) or tubes (Fig. 4.64b) with external inductors have been considered in many papers [86–93].

This approach can be conveniently used also for systems with single- or multi-layer solid or hollow non-magnetic loads, with or without magnetic yokes, or for internal inductors with thin-walled tube loads [86, 91, 94, 95].

However, for complex systems with multi-layer cylindrical workpieces, provided or without magnetic yokes, the solution become cumbersome and make difficult the analysis of results for the high number of independent variables.

At the same time, the analytical methods, providing the solution in the form of convergent series or integral expressions, are suitable for calculations with commonly available software packages (like Math-Cad, MathLab).

We make reference in the following to the system of Fig. 4.64b, constituted by a cylindrical double-layer bi-metallic workpiece (in the following "*load*") and an

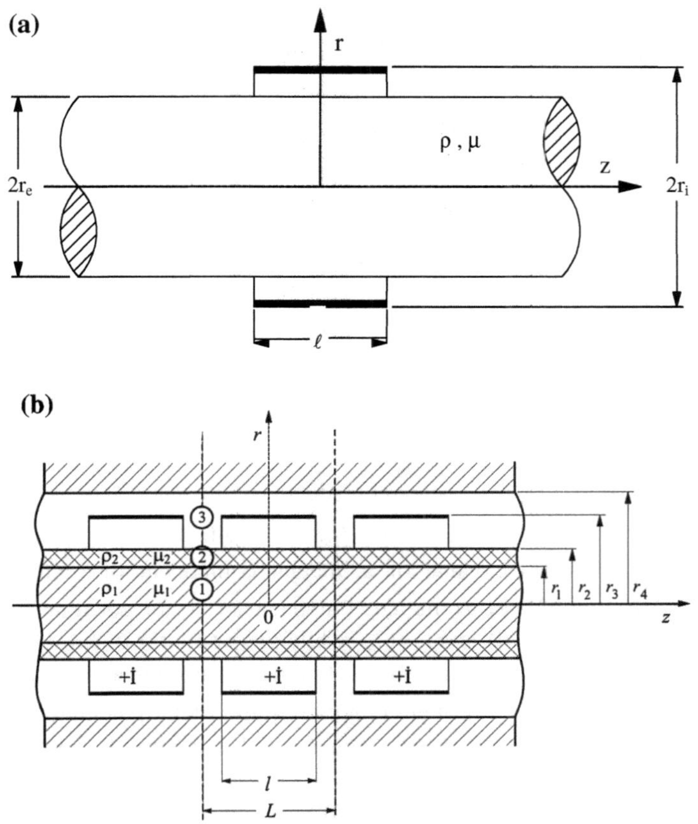

Fig. 4.64 **a** Schematic of cylindrical system with a single "short" inductor and infinitely long solid workpiece. **b** System with "long" cylindrical bi-metallic workpiece and inductor with multiple coils

internal or external inductor, provided with external or internal magnetic yoke. The inductor consists of a series of identical equidistant coaxial coils with the same number of turns *N*, in which a current *I* flows.

The distribution of the electromagnetic quantities in such system is obtained by the solution of Maxwell's Equations (2.1), with the following assumptions:

1. conduction current in the workpiece much larger than the displacement one, i.e. $\overline{J} > > \frac{\partial \overline{D}}{\partial t}$ or $\frac{\partial \overline{D}}{\partial t} = 0$;
2. within each workpiece layer the material is homogeneous, i.e. in the first layer is ρ_1 = const and μ_{r1} = const, in the second layer ρ_2 = const and μ_{r2} = const.;
3. the inductor current is sinusoidal, and hence all electromagnetic field quantities are also sinusoidal in time.
4. the sinusoidal variation of vectors $\overline{E}, \overline{H}, \overline{B}, \overline{A}$ in time, allows to represent them in complex form.

5. heated workpiece of infinite axial length.
6. the magnetic yokes of the system have infinite permeability, i.e. $\mu_r \to \infty$.

Taking as origin of a cylindrical coordinate system (r, φ, z) the centre of one coil (see Fig. 4.64b) and considering the periodicity of the field along z, the solution for the modulus of the vector potential of the system of Maxwell's equations can be written as follows:

$$\dot{A}_i = \left[C_0' I_1(\alpha_{0i}r) + C_0'' K_1(\alpha_{0i}r)\right] + \sum_{n=1}^{\infty}\left[C_n' I_1(\alpha_{ni}r) + C_n'' K_1(\alpha_{ni}r)\right]\cos(\lambda_n z) \tag{4.38}$$

where: C_n', C_n''—unknown coefficients (integration constants); $n = 1, 2, 3, \ldots$; $i = 1, 2, 3$—indexes of the different regions; $\lambda_n = \frac{\pi n}{L}$—constant of variables separation; $\alpha_{ni} = \sqrt{\lambda_n^2 + \frac{2j}{\delta_i^2}}$; $\delta_i = \sqrt{\frac{2\rho_i}{\omega\mu_0\mu_{ri}}}$—penetration depths in the material of the i-th layer of the load (i = 1.2 in case of Fig. 4.64b; I_1,K_1—modified Bessel functions of first order; $\dot{A} = \dot{A}_m/\sqrt{2}$—r.m.s. value of azimuthal component of vector potential; $\dot{A}_{m\varphi}$—complex amplitude of azimuthal component of vector potential.

For the air region (i = 3) is $\alpha_{n3} = \lambda_n$, while the first term of Eq. (4.38), which corresponds to the spatial harmonic for $n = 0$, must be substituted by $(C_0' r + C_0''/r)$.

In the third region it is convenient to represent the vector potential $\dot{A}$ as the sum of the excitation field $\dot{A}_e$ of an empty inductor coil without yoke and the reaction field $\dot{A}_r$ of load and yoke, i.e. $\dot{A}_3 = \dot{A}_{3e} + \dot{A}_{3r}$.

The vector potential of the exciting field is [86, 89]:

- for $r \leq r_3$

$$\dot{A}_{3e} = \frac{p_0}{2}\cdot\frac{r}{r_3} + \sum_{n=1}^{\infty} p_n I_1(\lambda_n r) K_1(\lambda_n r_3)\cos(\lambda_n z) \tag{4.39a}$$

- for $r \geq r_3$

$$\dot{A}_{3e} = \frac{p_0}{2}\cdot\frac{r_3}{r} + \sum_{n=1}^{\infty} p_n I_1(\lambda_n r_3) K_1(\lambda_n r)\cos(\lambda_n z) \tag{4.39b}$$

with: $p_0 = \mu_0 r_3 \frac{N\dot{I}}{2L}$; $p_n = \frac{2}{\pi}\cdot\mu_0 r_3 \frac{N\dot{I}}{2\ell}\cdot\frac{\sin(\lambda_n \ell)}{n}$.

On the basis of Eqs. (4.38) and (4.39) it is possible to obtain the solution for the vector potential in the various regions.

For the case of a system with external inductors the solution is:

$$\left.\begin{aligned} \dot{A}_1 &= \sum_{n=0}^{\infty} p_n F_n I_1(\gamma_n r)\cos(\lambda_n z) \\ \dot{A}_2 &= \sum_{n=0}^{\infty} p_n [L_n I_1(\beta_n r) + M_n K_1(\beta_n r)]\cos(\lambda_n z) \\ \dot{A}_3 &= \dot{A}_{3r0} + \sum_{n=0}^{\infty} p_n [T_n I_1(\lambda_n r) + S_n K_1(\lambda_n r)]\cos(\lambda_n z) \end{aligned}\right\} \quad (4.40)$$

with: $\gamma_n = \alpha_{n1}$; $\beta_n = \alpha_{n2}$; $\dot{A}_{3r0}$—vector potential of the reaction field for the spatial harmonic of order zero.

The value of $\dot{A}_{3r0}$ can be determined as follows:

- for external inductors with or without magnetic yokes

$$\dot{A}_{3r0} = \frac{p_0}{2} \cdot \frac{r_3}{r} \cdot T_0;$$

- for internal inductors without magnetic yokes

$$\dot{A}_{3r0} = \frac{p_0}{2} \cdot \frac{r}{r_3} \cdot T_0;$$

- for internal inductors with magnetic yokes

$$\dot{A}_{3r0} = \frac{p_0}{2} \cdot \frac{r_3}{r} \cdot \left[T_0 - \left(\frac{r}{r_3}\right)^2\right].$$

with F_n, L_n, M_n, S_n, T_n—coefficients (integration constants), which can be determined by the conditions of continuity of the vector potential and the tangential component of the magnetic field intensity at the boundaries between different regions ($r = r_1$, $r = r_2$), and the condition $H = 0$ at the surface of the magnetic yoke ($r = r_4$).

When the values of the modulus of the vector potentials $\dot{A}_i$ have been determined by Eq. (4.40), it's easy to evaluate the electric field intensity $\dot{E}_i$ and the current density $\dot{J}_i$ in any layer i-th:

$\dot{E}_i = -j\omega\dot{A}_i, \dot{J}_i = \frac{\dot{E}_i}{\rho_i};$

and the component of the vector of the magnetic field intensity $\dot{H}_i$

$\dot{\overline{H}}_r + \dot{\overline{H}}_z + \frac{rot\dot{\overline{A}}}{\mu_0}.$

The value of the equivalent impedance $\dot{Z}_e$ of one inductor coil can be obtained from the condition that the voltage $\dot{U}$ applied to the coil, must balance the voltage drop on the own resistance R_i of the windings and the e.m.f. $\dot{E}_N$, induced in the turns

$$\dot{E}_N = -j\omega\pi r_3 \frac{N}{\ell} \int_{-\ell}^{\ell} \dot{A}_3(r_3, z)dz.$$

For external inductors with magnetic yoke and resistance $R_i = 0$, we obtain:

$$\dot{Z}_e = -\frac{\dot{E}_N}{\dot{I}} = jX_{i0}\left\{\frac{\ell}{L}(1+T_0) + \left(\frac{2}{\pi}\right)^2 \frac{L}{\ell} \sum_{n=1}^{\infty} [I_1(\lambda_n r_3)K_1(\lambda_n r_3) + S_n K_1(\lambda_n r_3) + T_n I_1(\lambda_n r_3)] \frac{\sin^2(\lambda_n \ell)}{n^2}\right\} \tag{4.41}$$

where: $X_{i0} = \omega\mu_0\pi r_3^2 \frac{N^2}{2\ell}$—reactance of one empty inductor coil without end effects.

For internal inductors with magnetic yokes, in Eq. (4.41) the term (1 + T_0) must be replaced by T_0.

In case of external inductors without magnetic yokes in Eq. (4.41) is $T_n = 0$, while for internal inductors is $S_n = 0$.

To analyse the influence on the electrical and energetic characteristics of the inductor of system's geometry and frequency, it is convenient to represent the inductor impedance in the form:

$$\dot{Z}_e = R_i + R'_w + j(X_{i0} - \Delta X) = R_i + jX_{i0} + X_{i0}\left(\frac{r_2}{r_3}\right)^2 (F - jG) \tag{4.42}$$

with: X_{i0}—reactance of the empty inductor coil; R_i—own resistance of inductor coil windings; R'_w—resistance of the workpiece referred to the inductor; ΔX—variation of the inductor coil reactance from empty to loaded conditions; F, G—coefficients of active and reactive power.

Equations (4.41) and (4.42) are quite universal, since they allow to analyze the variation of the impedance due the introduction of solid or double-layer metallic cylindrical workpieces, single-layer or bimetallic tubes, in inductors with or without magnetic cores, as a function of the ratio between the axial length ℓ of the inductor coil and its inner radius r_3 and the frequency of the exciting current.

Since the coefficient F is characterized by very sharp variations, for convenience of graphical representation it is usual to represent it in the diagrams by the product mF, where $m = \sqrt{2}r_e/\delta$; r_e—radius of the surface of the workpiece facing the inductor (i.e. $r_e = r_2$—when heating solid or hollow cylinders with external inductors; $r_e = r_1$—when heating hollow cylinders with internal inductors); δ—penetration depth in the material of the workpiece (in the heating of a two-layered load, it is evaluated for the layer facing the inductor).

Figures 4.65 and 4.66 give diagrams of mF_0 and G_0 for the calculation of the resistance and reactance of external inductors with inside solid non-magnetic cylindrical load.

The diagrams cover a wide range of relative lengths of the inductor coil ($\ell/2r_3 = 0.1 - \infty$), parameters m (m = 0–20) and ratios of radii of the inductor and the heated workpiece ($r_3/r_2 = 1.05 - 1.65$).

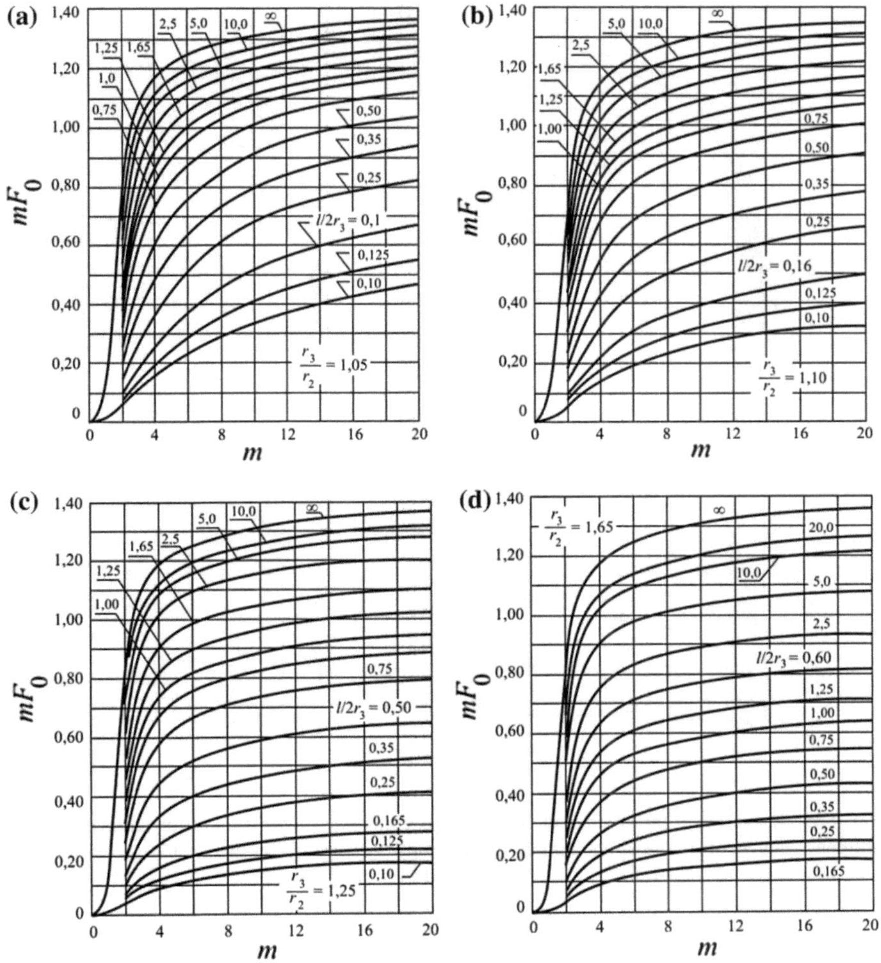

Fig. 4.65 Diagrams of coefficient mF_0 for solid cylinders, as a function of parameter m, for different values of the relative length $\ell/2r_3$ of the inductor coil, and various ratios r_3/r_2. **a** $r_3/r_2 = 1.05$; **b** $r_3/r_2 = 1.10$; **c** $r_3/r_2 = 1.25$; **d** $r_3/r_2 = 1.65$

Both coefficients increase with the increase of $\ell/2r_3$ and m, and the reduction of the ratio r_3/r_2.

With the increase of $\ell/2r_3$, the curves of mF_0 and G_0 are approaching the limit behavior for $\ell/2r_3 = \infty$, which corresponds to a system with inductor and load of infinite axial length. The functions mF_0 and G_0 for $\ell/2r_3 = \infty$ do not depend on the ratio r_3/r_2 and are related to the functions F_{cyl} and G_{cyl}, given in Chap. 2, characterizing active and reactive power in a solid homogeneous conducting cylinder of infinite axial length, by the relationships:

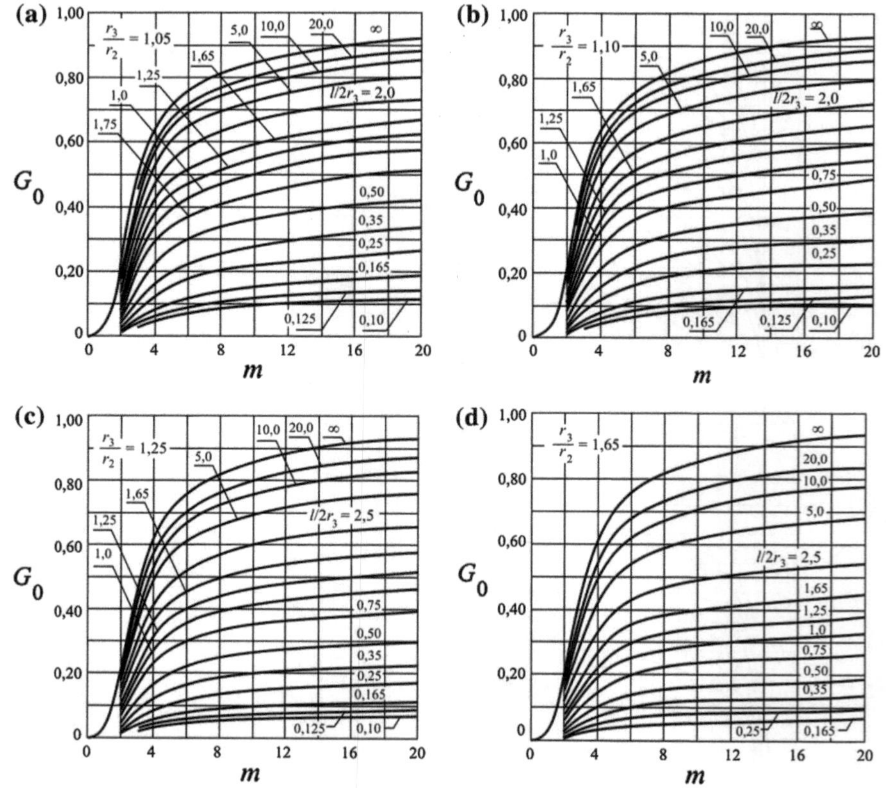

Fig. 4.66 Diagrams of the coefficient G_0 for solid cylinders, as a function of the parameter m, for different values of the relative length of the inductor coil $\ell/2r_3$ and various ratios r_3/r_2. **a** $r_3/r_2 = 1.05$; **b** $r_3/r_2 = 1.10$; **c** $r_3/r_2 = 1.25$; **d** $r_3/r_2 = 1.65$

$$F_{0\infty} = \frac{\sqrt{2}}{m} F_{cyl}, \; G_{0\infty} = \sqrt{2} G_{cyl}$$

This analytical method has been used in [95] to evaluate the resistance and the reactance of systems with solid bimetallic or hollow cylindrical single layer loads, as a function of the system geometrical dimensions and the frequency.

Figure 4.67 gives the coefficients m_2F and G for a solid two-layer cylinder with external inductor, as a function of the relative length of the inductor coil ℓ/r_3, the parameter m_2 (of the external layer), and two values of the ratios r_3/r_2 and r_1/r_2.

The calculations refer to the following properties of load materials: internal layer—$\rho_1 = 60 \times 10^{-8}\ \Omega{\cdot}\text{m}$; $\mu_{r1} = 10$ and $\mu_{r1} = 50$; external layer—$\rho_2 = 100 \times 10^{-8}\ \Omega{\cdot}\text{m}$; $\mu_{r2} = 1$.

This schematization can be representative of the intermediate stage of the heating of a ferromagnetic workpiece, like in induction hardening.

In fact, at the transition of Curie point, a surface layer is above Curie temperature and the material has lost its magnetic properties, while the core is still at a lower

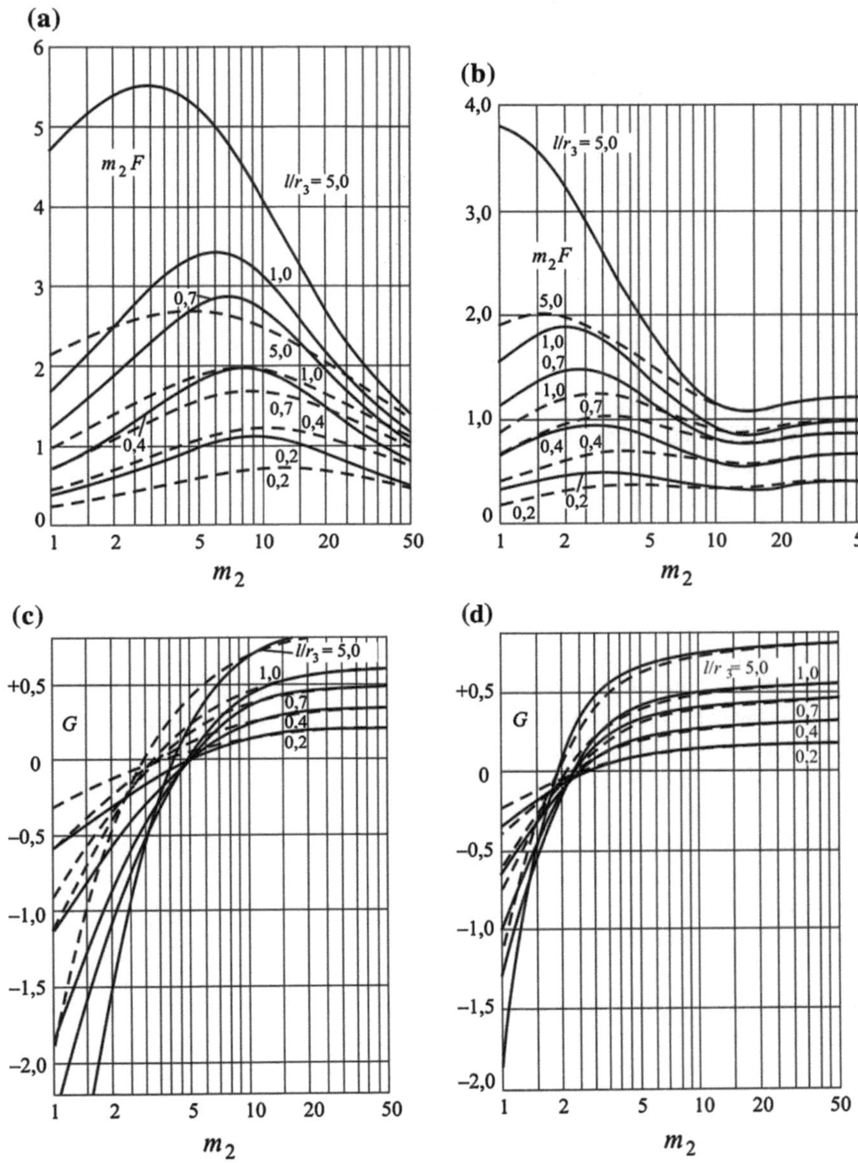

Fig. 4.67 Diagrams of coefficients F and G for a two-layer cylindrical workpiece as a function of relative length of the inductor ℓ/r_3 and parameter $m_2 = \sqrt{2}r_2/\delta_2$: $\rho_1 = 60 \times 10^{-8}$ Ω·m; $\mu_{r1} = 10$ (*dashed lines*); $\mu_{r1} = 50$ (*solid lines*); $\rho_2 = 100 \times 10^{-8}$ Ω·m; $\mu_{r2} = 1$; $r_3/r_2 = 1.25$; **a**, **c** $r_1/r_2 = 0.975$; **b**, **d** $r_1/r_2 = 0.85$

temperature, and the relative magnetic permeability is greater than unity, and dependent on magnetic field intensity.

The parameter $m_2 = \sqrt{2}r_2/\delta_2$ is calculated with the physical properties (ρ_2, μ_2) of the external layer.

As the diagrams show, with the increase of m_2 (i.e. with the increase either of the frequency of the exciting current, or the external radius of the workpiece) the reactance increases while the resistance of the load decreases.

In particular, with the increase of m_2 the reactance goes from negative to positive values. This transition occurs at different values of m_2, depending on the ratio ℓ/r_3 and the values of μ_{r1}. For the same value of r_3/r_2, with the reduction of r_1/r_2 (i.e. increasing the thickness of the external layer), the transition point shifts towards lower values of m_2.

In the negative region the curves of G corresponding to lower values of μ_{r1}, are higher than those calculated with higher values of μ_{r1}.

At low frequency ($m_2 < 5$ for $r_1/r_2 = 0.975$ and $m_2 < 3$ for $r_1/r_2 = 0.85$) the values of G are below zero, and the reactance of the inductor-load system is higher than that of the empty inductor due to the magnetizing effect of the internal magnetic region. On the contrary, with the increase of m_2 dominates the demagnetizing effect of the currents induced in the external layer, and the reactance of the loaded inductor is lower than that of the empty coil.

With the increase of the permeability of the internal layer, increases the resistance of the load in the whole range of variation of parameters m_2, ℓ/r_3, r_3/r_2, r_1/r_2. The law of variation of the reactance in this case is more complex.

Of practical use are also the diagrams of Fig. 4.68.

For the conditions described above, i.e. heating of double-layer cylindrical loads, representative of the intermediate regime in ferromagnetic steel cylinders, are given of the values of the ratio P_2/P between active power dissipated in the external layer and the total power induced in the workpiece, as a function of parameters ℓ/r_3 and m_2, and various values of magnetic permeability μ_{r1} [88, 95].

In the range of variation of ℓ/r_3 (from 0.2 to 5.0), when m_2 increases, the ratio P_2/P first decreases (for low values of m_2), reaches a minimum and then increases again towards unity. With the increase of the ratio ℓ/r_3, the minimum of P_2/P shifts toward lower values of m_2. At the lower values of μ_{r1} (dotted lines), the fraction of active power induced in the external layer (i.e. the value of P_2/P) decreases. With the increase of m_2, the curves calculated for different values of ℓ/r_3 and μ_r, converge in one line, which asymptotically approaches the value $P_2/P = 1$.

Thus, in the range $\ell/r_3 = 0.2 - 5$ the curves practically will merge:

- for $r_3/r_2 = 1.25$ and $r_1/r_2 = 0.975$:
  with $\mu_{r1} = 50$ when $m_2 \geq 8$; with $\mu_{r1} = 10$ when $m_2 \geq 10$;
- for $r_3/r_2 = 1.25$ and $r_1/r_2 = 0.85$
  with $\mu_{r1} = 50$ at $m_2 \geq 5.0$; with $\mu_{r1} = 10$ at $m_2 \geq 5.0$–6.0.

As the ratio r_1/r_2 decreases the region of coincidence of the curves of P_2/P, obtained with different values of μ_{r1}, will occur at lower values of m_2.

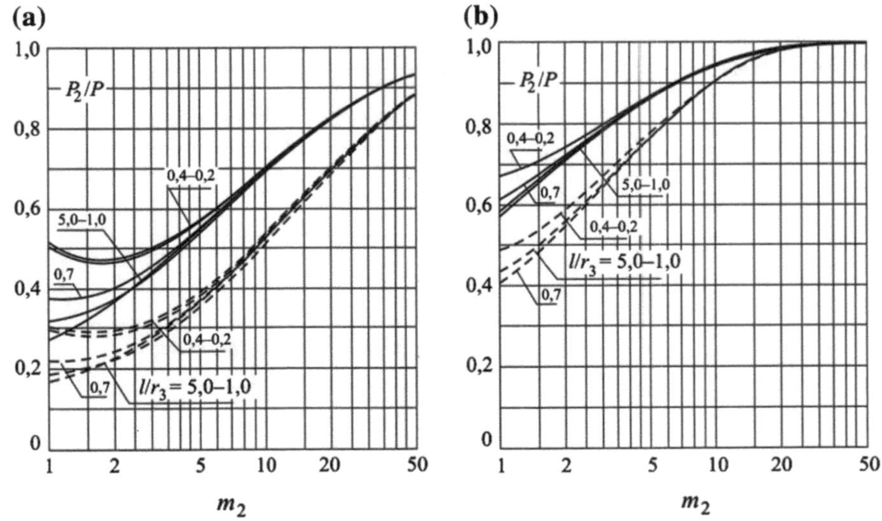

Fig. 4.68 Ratio P_2/P of active power in the external layer to total induced power, as a function of m_2 and ℓ/r_3: workpiece as in Fig. 4.67; $r_3/r_2 = 1.25$; **a** $r_1/r_2 = 0.975$; **b** $r_1/r_2 = 0.85$

The increase of magnetic permeability μ_{r1} leads to an increase of the fraction of power induced in the external layer (i.e. to the growth of P_2/P).

In paper [74] is given the coefficient k_P, defined as the ratio of the active power P_{1z} induced per unit length in a solid cylinder at the cross-section of coordinate z, and the active power P_{10}, induced in the cross-section at the centre of the inductor (at $z = 0$ in Fig. 4.69):

$$k_P = \frac{P_{1z}}{P_{10}} \approx \left| \frac{\dot{A}(r_2, z)}{\dot{A}(r_2, 0)} \right|^2 .$$

Figure 4.70 shows the variations of k_P with the relative distance h/ℓ between the turns of the inductor (or between two one-turn inductors) and the relative coordinate $2z/\ell$.

The increase of the distance between turns (or between inductors) reduces the uniformity of the induced power along z-axis.

Using the same approach described for the analytical calculations (Eqs. 4.41 and 4.42) in [95] has been developed the analysis of the equivalent resistance and reactance of metallic hollow workpieces (tubes) heated by internal inductors. Some results of this analysis are given in the diagrams of Fig. 4.71.

They show that increasing the distance between the internal surface of the workpiece and inductor (i.e. increasing the ratio r_1/r_3) the resistance of the charge decreases while the reactance increases.

With the increase of parameter m, the curves of G have three clearly defined regions: in the first one the variation of G are practically negligible; in the second

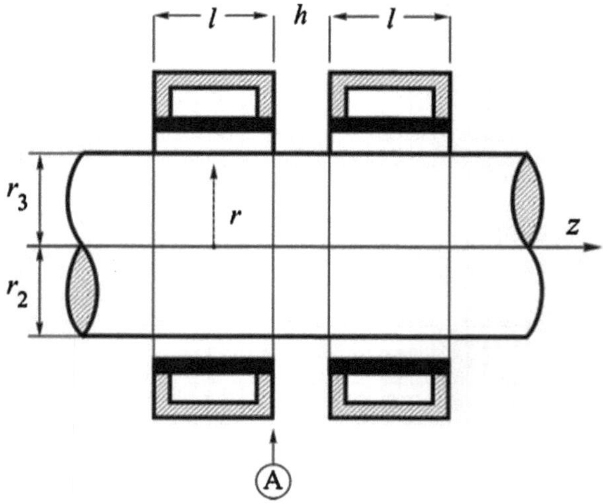

Fig. 4.69 Schematic of two inductor turns and a long solid cylindrical workpiece

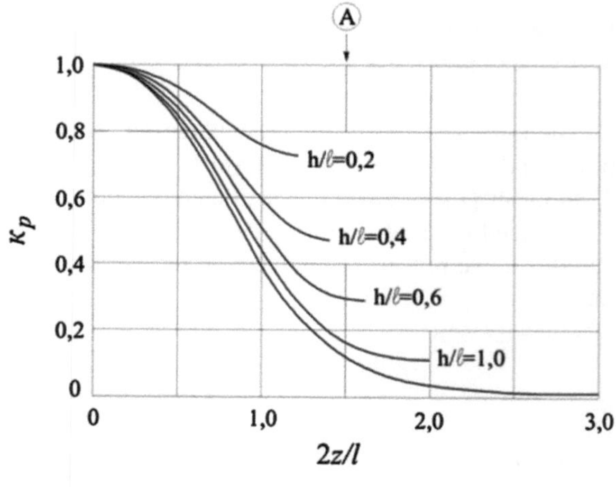

Fig. 4.70 Distributions of coefficient k_P as a function of the relative coordinate $2z/\ell$, and different values of h/ℓ

one occurs a sharp increase of G; in the third one the variation of G is low, approaching asymptotically a limit value. The transition from first to the second and from the second to the third region, with the increase of the ratios ℓ/r_3 and r_1/r_3, is shifted toward lower values of m.

Of particular interest is the analysis of the influence of an external magnetic yoke on the inductance of a coil without load; the increase of the inductance can be characterized by the coefficient k_y defined by the ratio:

$$k_y = \frac{X_y}{X_0 k_N},$$

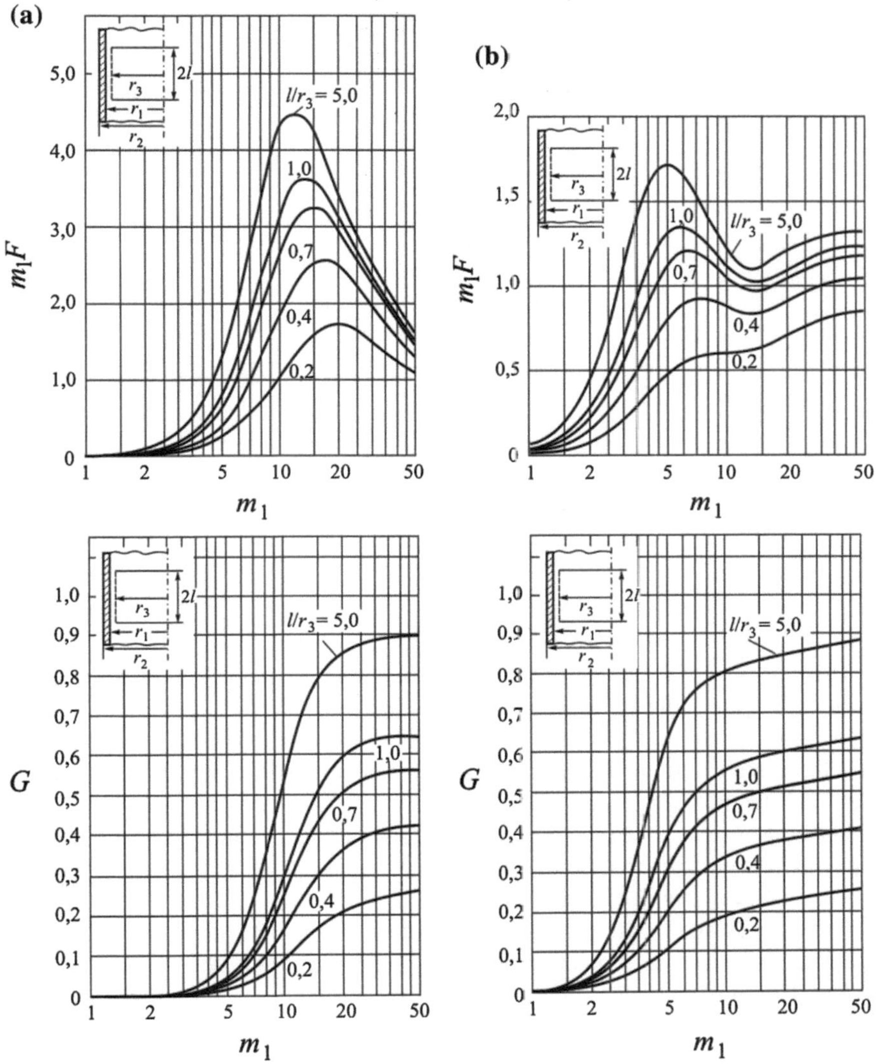

Fig. 4.71 Coefficients mF of the resistance and G of the reactance, in the heating of non-magnetic hollow metallic cylinders with internal inductors, as a function of parameter $m_1 = \sqrt{2}r_1/\delta$ and different relative inductor lengths ℓ/r_3: **a** $r_1/r_3 = 1.05$, $r_2/r_1 = 1.025$; **b** $r_1/r_3 = 1.05$, $r_2/r_1 = 1.15$

where X_y—reactance of the coil with yoke; X_0—reactance of the coil without load, considering it as a section of an infinitely long coil; k_N—coefficient of Nagaoka, which takes into account the finite dimensions of the empty coil.

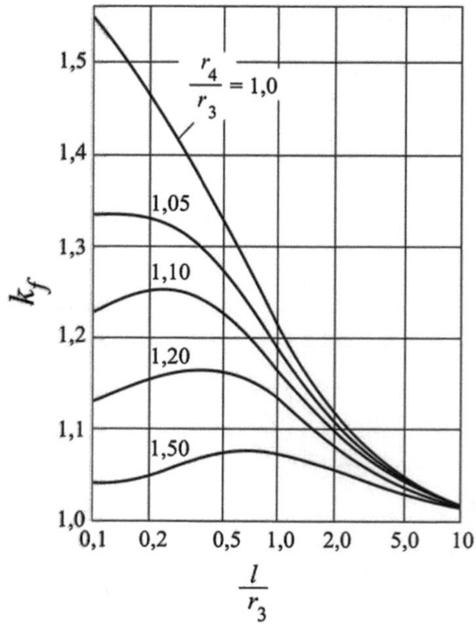

Fig. 4.72 Coefficient k_y of the influence of a magnetic yoke on the inductance of a coil, as a function of its relative length

The diagrams of k_y (given in Fig. 4.72) show that for "short" coils the magnetic yoke significantly affects the value of the inductance.

In particular, for $r_4 \neq r_3$ (with r_4—radius of the magnetic yoke; r_3—radius of the coil) there is a certain coil length at which this effect is maximum. For $r_4 \approx r_3$, this maximum tends to 2.0 for $l \to 0$ [89].

In [89] is also analysed the effect of the distance between inductor coils series or opposite connected, for different types of workpieces, as a function of the gap between inductor and magnetic shield.

Figure 4.73 shows an example of the variations of the resistance R and the reactance X of coils as a function of the distance h between them.

The diagrams show that the presence of the magnetic shield (at the radius r_4) significantly modifies the parameters of a single coil (for $h \gg r_3$), and the mutual influence between coils.

For coils very close to each other ($h = 0$), the external magnetic shield has no effect when they are series connected, while it greatly increases the values of R and X when they are opposite connected.

In the above example, the mutual influence in case of series connection is significant up to $h = 2r_3$ without magnetic shield ($r_4 \to \infty$) and almost disappears when $h > r_3$ for $r_4 = 1.05r_3$.

The analysis of the power distribution along the length of the workpiece has shown that with the use of a magnetic yoke the power outside the inductor's ends decreases faster than without it.

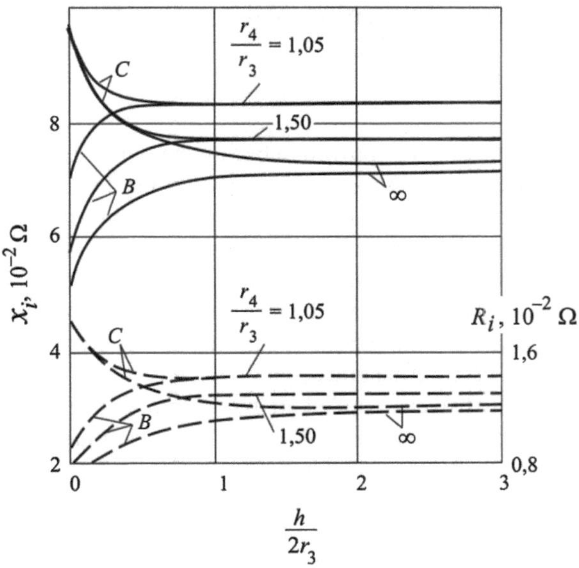

Fig. 4.73 Resistance and reactance of inductor coils as a function of the distance h between them in the heating of a solid non-magnetic cylinder, with different radii r_4 of the magnetic yoke at frequency $f = 2.075$ Hz; inductor coils: $r_3 = 5$ cm, $l = 5$ cm, $N = 10$, $r_0 = 3.14 \times 10^{-3}$ Ω; workpiece: $r_2 = 3.125$ cm, $\rho = 10^{-4}$ Ω·cm, $\mu = 1$: *C*—inductor coils series connected; *B*—opposite connected

It should be noted that, at constant applied voltage, the active power of the inductor rapidly increases (of about 15–30 %) when the distance between coils increases and then decreases, remaining above the value for $h = 0$.

These variations can be used to control the power absorbed by the inductor by the adjusting the distance between sections [88].

The analysis and the comparison with experimental results, allows us to draw the following general conclusions:

- increasing the distance h the resistance R decreases of 35 % for non-magnetic and 54 % for magnetic cylindrical workpieces; the reactance X of 22–26 %;
- the influence of magnetic yoke on the above parameters is practically nil for $h < 1.5 \cdot (r_3 - r_2)$, whereas it produces a considerable increase of their values in case of more spaced coils;
- in shield-less systems the mutual coupling between the coils is considerable up to about an inductor diameter for $\mu = 1$ and about two diameters for $\mu \gg 1$;
- the presence of a magnetic shield reduces the mutual coupling effect between coils; this effect becomes already negligible for values of h equal to twice the gap between load and shield (for $\mu = 1$), or three times when $\mu \gg 1$.

4.6 Pulse Induction Hardening of Complex Workpieces

4.6.1 Introduction

As known, induction heating has proved to be, among other methods of heating, the most suited for hardening processes, since it provides fast heating rates, precise and reproducible control of heating operation and allows to concentrate heating sources to selected areas of the work piece with consequent energy savings.

These characteristics are of particular interest in the case of surface hardening of steel and cast iron components, where the heating must be limited to a surface layer, in order to create in it a wear resistant martensitic structure while maintaining thermally unaffected the core material.

This process is extensively applied in the automotive industry where many different high stressed components (e.g. crankshafts, camshafts, gear wheels and others) require surface hardening at high production rates. Typical examples of such components are shown in Fig. 4.74 [96, 97].

All these components are characterized by complex geometries, types and sizes, variety of materials, which make difficult to achieve by induction heating the best service properties corresponding to a hardness surface layer more or less uniform along its contour.

Moreover, in this application induction heating must compete with other well established heat thermo-chemical treatments, like carburizing and nitriding. These processes fulfil the requirement of assuring a uniform hardened layer along the contour, but have the disadvantages of introducing considerable distortion of workpieces due to the relative high treatment temperatures and heating times. Moreover, thermo-chemical hardening does not enable process integration in the production lines and is not environmentally friendly because it makes use of polluting technical gases.

In the following, we make reference to gear-wheels as a typical case of induction hardening of complex geometries.

In spin induction hardening, only the part of gear that requires to be treated is brought to the transformation temperature by means of an inductor surrounding it. The metallurgical transformation is obtained by the subsequent quenching process. The gears are rotated during heating to ensure an even distribution of energy across their perimeter. In doing this it is possible to through harden the gear down to the

Fig. 4.74 Typical hardened components with complex geometrical shapes

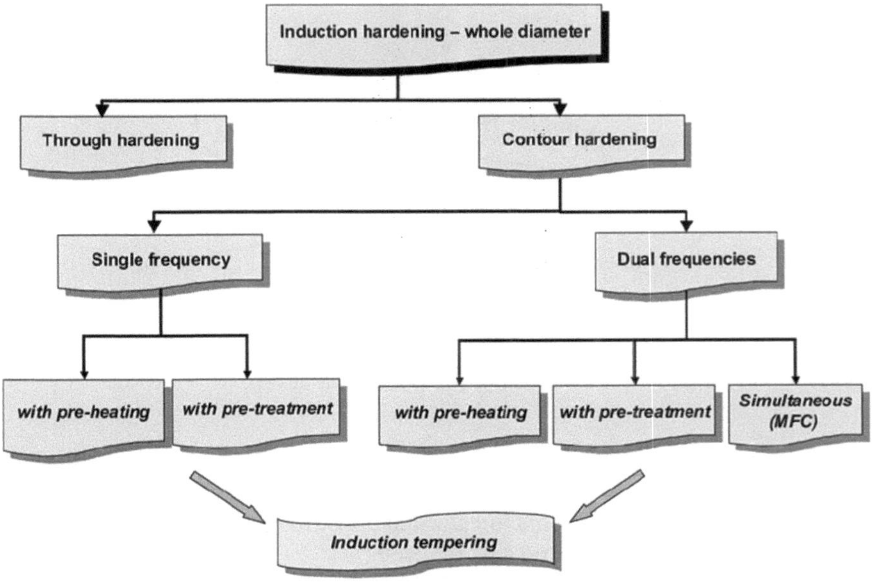

Fig. 4.75 Classification of different spin hardening methods

tooth root (producing almost the same result as in the case hardening) or to harden the outer surface at either uniform or irregular depth from the surface.

Although the best pattern is usually an uniform thickness of the hardened layer along the contour, different patterns may be acceptable or sometimes desirable depending on the application. In particular, non-uniform hardened layer may be accepted on the tips of the teeth where mechanical loading is very low, while on the flanks and roots it must be continuous in order to withstand the high contact pressure and tensile stress from loading.

The main parameters which play a dominant role in achieving the required hardness pattern are frequency, power density, heating time, quenching conditions and coil geometry.

As regards gear spin hardening by induction, different heating techniques can be used, as shown in Fig. 4.75 [98].

In particular, contour hardening can be obtained by single-frequency or dual-frequency processes.

4.6.2 Single-Frequency Processes

The "*single-frequency single-shot*" hardening process is a process in which all the teeth are simultaneously contour hardened by means of a single frequency exciting field. It requires high induced power densities and very short heating times in order

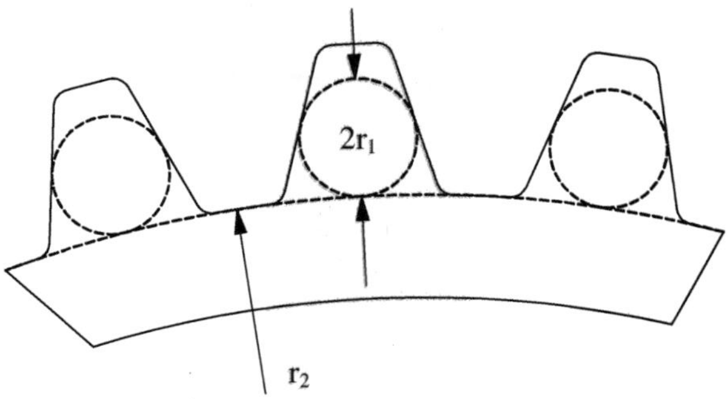

Fig. 4.76 Schematic of gear with ideal substitution of teeth by "equivalent" fictitious cylinders highlighting the presence of two different geometrical dimensions

to avoid heat diffusion from the surface to the core and the consequent through hardening of the teeth. Moreover, a convenient choice of the operating frequency is necessary in order to obtain suitable values and distribution of the induced power density at the flanks, tips and roots of the teeth.

In fact, the choice of the frequency must be done taking into account not only the "electrical" requirement of inducing high power values per unit surface. The designer must also consider that the areas exposed to the exciting field are quite different for flanks, tips and roots, and that the teeth have very different thermal capacities in comparison with the part of gear below the roots.

From the electrical point of view, the frequency value should be chosen with the purpose of concentrating most of the induced power within a surface layer of thickness comparable with the required hardened depth. This is obtained when the penetration depth is of the same order of magnitude of the hardened depth.

However, taking into account that the geometrical dimensions of the teeth are much smaller than the part below the gear root circle (see schematic of Fig. 4.76), and that the induced power density depends not only on frequency but also on the ratio between geometrical dimensions and penetration depth, it appears evident that the heating with a single frequency is a compromise solution for obtaining a reasonably good hardened profile at roots, flanks and tips of the teeth.

In fact, making reference to the schematic representation of the ring-gear of Fig. 4.76, in which each tooth is ideally substituted by an "equivalent" fictitious cylinder as suggested in [93], it can be easily seen that the geometry is characterized by two basic dimensions r_1 and r_2. As a consequence, the use of two different frequencies appears to be the natural solution for inducing the same power density per unit surface in the teeth (flanks and tips) and the roots of the gear.

Figure 4.77 shows schematically the equi-value lines of the induced current distributions in the cross-section of a gear placed in a cylindrical coil [99, 100].

It shows that, at high frequency, the point O in the tooth is affected by stronger heat sources than point O_1 in the root. In this case, an overheating of the teeth with respect to the root can occur. Conversely, at low frequencies, when the penetration

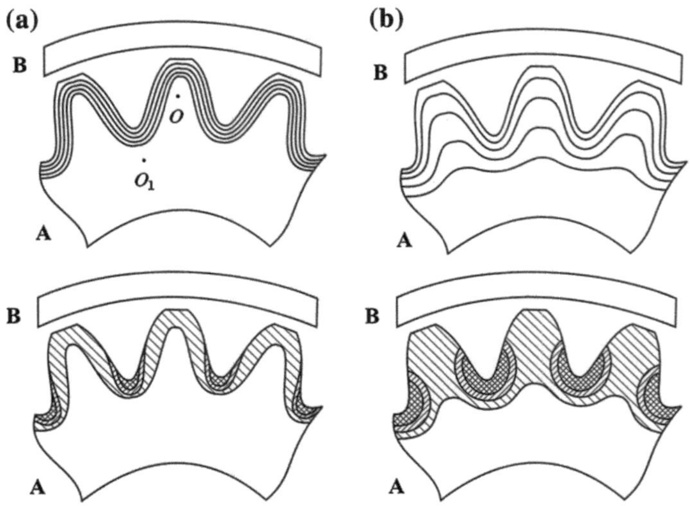

Fig. 4.77 Equi-value lines of induced current and patterns of induced power density in a ring-gear **a** above optimum frequency; **b** below optimum frequency; *A* ring-gear; *B* inductor)

depth is comparable with the module,[2] the power density at the top and flanks of a tooth diminishes, and the root reaches a temperature higher than the teeth. Therefore, there is an optimal intermediate frequency at which the power density distribution in teeth and roots produces optimal heating conditions. However, taking into account the different heating and cooling conditions of teeth and roots, at the optimal frequency the specific power in the roots must be approximately two times higher than in the teeth [101–103].

As stated before, if the frequency is significantly higher than the optimal one, only the tooth tips will be hardened. Vice versa, with frequency lower than the optimal one, only the roots are properly treated.

Figure 4.78 shows an example of the induced power density patterns in half cross-section of a tooth at different frequencies: it refers to the beginning and end of heating with the same exciting magnetic field intensity, at 10 and 500 kHz. The figure highlights the strong influence of the variations of physical characteristics of the steel with temperature.

Experiments and analytical studies have demonstrated that the contour hardening of the whole gear in a cylindrical inductor may be achieved with a single frequency only with a convenient choice of process parameters—frequency, specific power and heating time [93, 103, 104].

[2] The module of a wheel is defined as $M = \frac{p}{\pi}$ where p is the circular pitch, i.e. M is the distance along the pitch circle between the flanks of two adjacent teeth, in mm.

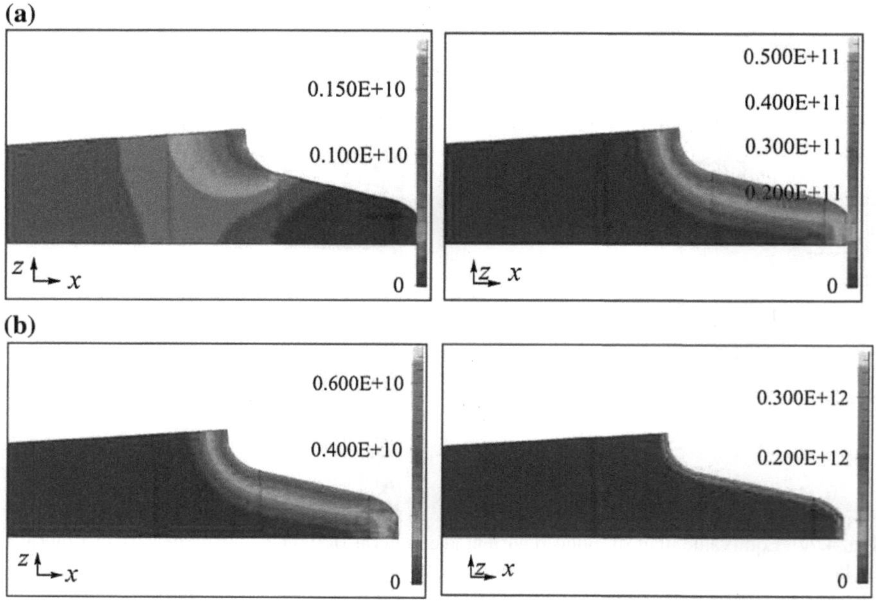

Fig. 4.78 Typical power density distributions in a ring-gear at medium- and high-frequency **a** beginning of heating (ρ = 20 μΩ·cm; μ_r = 10); **b** end of heating (ρ = 100 μΩ·cm; μ_r = 1)

The following formula has been proposed for a rough estimation of optimal frequency [93]:

$$f_{opt} = \frac{K}{M^2} \approx \frac{3 \cdot 10^5}{M^2} \text{[kHz]} \tag{4.43}$$

with M—module of the gear in mm.

Since the optimal frequency depends upon gear material, required hardened depth and pattern, gear geometry, tooth profile and other factors, the choice of the frequency is not crucial and several authors propose values of coefficient K in a relative wide range (i.e. 3–6 × 10^5) [93, 105].

In any case, the optimal frequency does not always guarantee to achieve the prescribed contour hardening (in particular for low modules) and appropriate heating times and surface power densities must be adopted. In particular, the smaller is the gear module, the higher must be the specific power and shorter the heating time as shown in Fig. 4.79, which refers to a uniform hardened depth approximately equal to one fourth of the tooth width [93].

Considering heat losses at the external surface and thermal conduction inside the gear during heating, above Curie point the specific power on the root of the tooth should be about twice the value at the top [103].

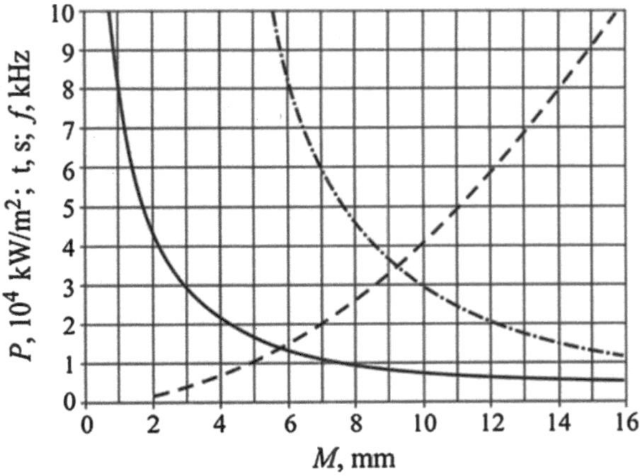

Fig. 4.79 Optimum heating parameters suggested for single-frequency single-shot hardening (*solid line* average surface specific power; *dotted line* heating time; *line with dots and points* frequency)

From the diagrams the following remarks can be made:

- for modules below 5 mm, frequencies above 10 kHz and heating times below 1 s must be used;
- hardening of gears with module lower than 3–4 mm require power density peaks above 5–6 kW/cm^2 and heating times of few hundreds of milliseconds.
- shallower hardened layers may be produced at frequency higher than optimal when very high power density is used. For automotive and machinery gears with module 5–10 mm, the frequency can be increased in the range 5–30 kHz.
- the above dependencies set strong constraints in designing hardening machines because they must operate at high power and frequency close to the optimal one. For gears of large diameter or low module, the requirement of high power becomes more critical and many attempts have been made to overcome this restriction.

In order to overcame these problems, single frequency contour hardening with pre-heating (also called *"Pulsing Single Frequency"* process) has been developed. The main process stages are illustrated in the cycle diagram of Fig. 4.80. It comprises a preheating stage at reduced power up to approximately 550–750 °C (temperature dependent on the material), a soaking stage and a short final heating period with higher specific power up to the hardening temperature, followed by quenching and a low-power heating stage for tempering [98, 106–108].

Pre-heating allows to attain a convenient heated depth at the roots of the gear without overheating the tooth tips. This heating and quenching cycle allows in many cases to reach the desired metallurgical results and to reduce distortion in the processed gear.

The preheating time, dependent on size and shape of the gear, is followed by a soaking period which allows to achieve a nearly uniform temperature distribution across the teeth. The final heating stage, which consists of a fast heating followed by a fast cooling, allows to realize a contour-like hardened profile.

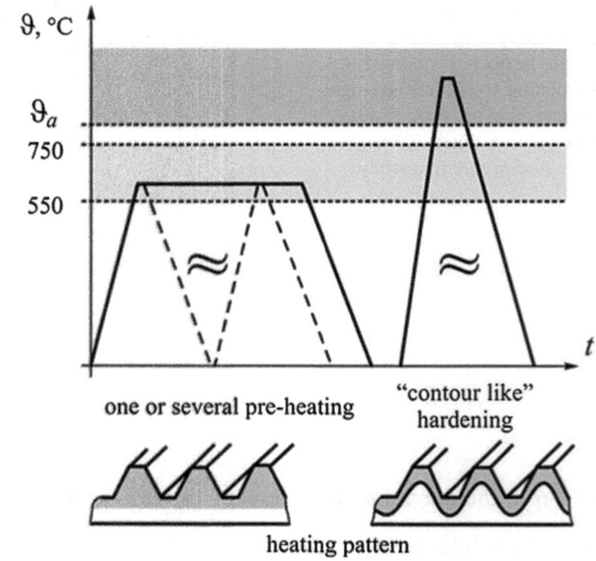

Fig. 4.80 Typical cycle of pulsing single frequency process [98]

It is well known that the microstructure conditions of the metal prior to hardening are of fundamental importance for the repeatability of mechanical and metallurgical results of the heat treatment. In particular, initial conditions characterised by homogeneous, fine-grained, quenched and tempered martensitic microstructure (with hardness of 30–34 HRC), are particularly favourable for assuring a fast and consistent response to the induction heat treatment, resulting in higher hardness and deeper case depth in comparison with a ferritic/perlitic initial microstructure.

If the initial microstructure is not the best suited for induction hardening, further improvements in the process result can be achieved by a "*pre-quenching and tempering*" operation before the actual hardening stage, according to the schematic time- temperature cycle with single frequency shown in Fig. 4.81.

During the pre-hardening phase, the material of the whole gear is firstly heated up to the austenitizing temperature. Then the piece is cooled by self-quenching, and the entire wheel undergoes a short "tempering". The actual hardening profile is finally produced by a fast heating stage.

Some examples of hardened profiles obtained with single frequency processes are shown in Fig. 4.82.

4.6.3 Dual-Frequency Processes

As already mentioned, the use of two different frequencies represents a solution for achieving an uniform hardened depth along the contour of the gear. However, the range of the gear diameters suitable for the dual frequency process is limited for economical and technical reasons to wheels with diameter $d \leq 250$ mm. In fact, the

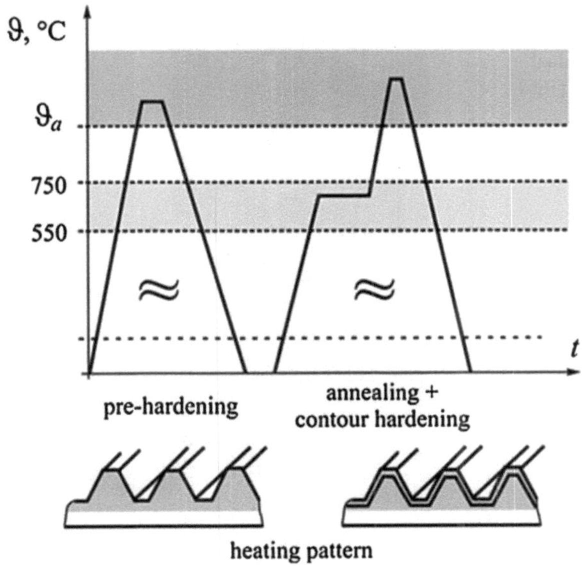

Fig. 4.81 Typical process cycle of single frequency induction contour hardening, with pre-quenching and tempering [98]

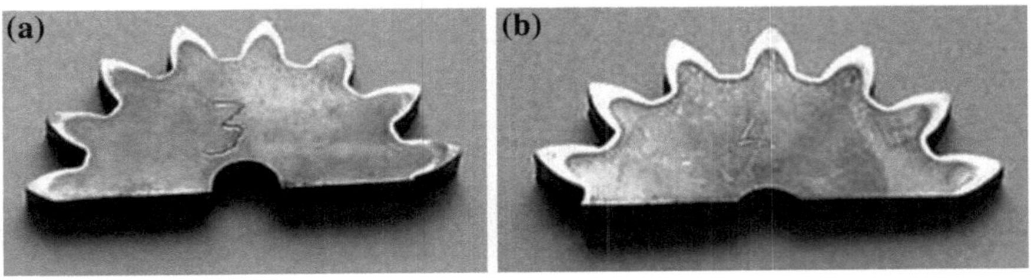

Fig. 4.82 Examples of hardened profiles obtained with single frequency processes **a** with pre-heating; **b** with pre-quenching and tempering (OD: 40 mm; thickness: 10 mm; module: 2.25; material: Ck53/1050)

process is effective only with very short heating times and corresponding high power densities, which require the use of frequency converters at medium and high frequency with output power ranging from several hundreds kW to some MW.

4.6.3.1 Processes in Two Steps with Separate Frequencies

In these processes, two different exciting frequencies are used in sequence: pre-heating is accomplished in the first step of the heating cycle by applying low power density at MF (usually in the range 3–10 kHz), while during the final heating stage

Fig. 4.83 Two inductors connected to separate converters for the dual frequency process [106]

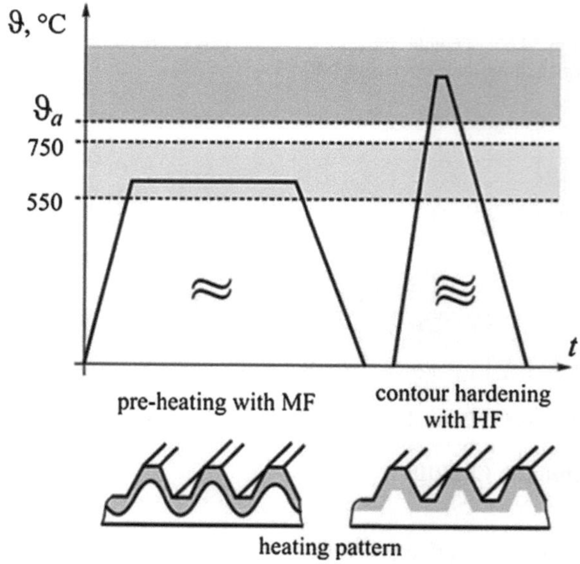

Fig. 4.84 Dual frequency contour hardening cycle with preheating stage [98]

high power density at HF is used for contour hardening. The selection of suitable values of frequency (in the range 30–450 kHz—depending on the type of gear, its size and material) and heating time for the final pulse allows to obtain the desired hardened depth.

In the great majority of cases two MF and HF frequency converters are connected to different coils, which are spatially separated, as shown in Fig. 4.83.

Figure 4.84 shows a typical cycle of the dual frequency contour hardening with a preheating stage.

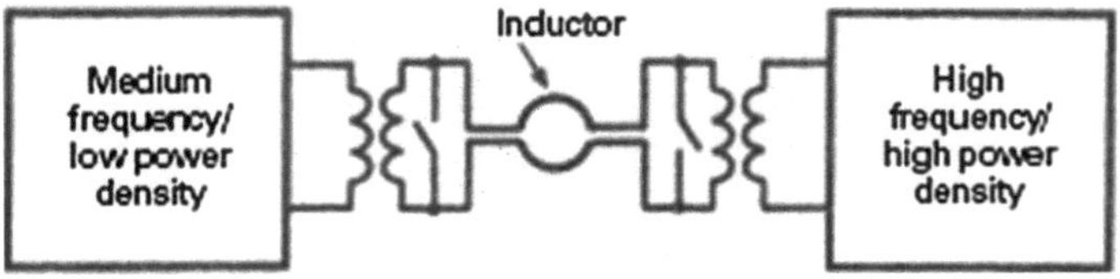

Fig. 4.85 Dual frequency scheme using one coil and two inverters [109]

The gear is preheated to a temperature below the critical temperature A_{C1}, usually 100–350 °C depending upon type and size of the gear, tooth shape, prior microstructure, required hardness pattern and distortion (which increases with preheating temperature) and available power source.

Obviously, the higher is the pre-heating temperature, the lower is the power for the final pulse. Increasing the pre-heating temperature can increase workpiece distortion.

After pre-heating, in the final stage the contour hardening at HF is carried out. Heating times are in the range from tenths of seconds to seconds, depending on the gear module; in this stage a very fine control of time and power is required. To avoid through heating of the tooth, the final austenitizing operation must be very short (often less than 1 s), which requires very fast transfer of the workpiece from one to the other inductor with high power switching speeds.

A refinement of the dual-frequency method uses one single inductor, which is first connected to the MF power supply circuit and then switched into the HF circuit, as shown in Fig. 4.85. A limit of this technique is the time required for power switching, typically of about 0.5 s.

As in the single frequency process, the heating cycle can comprise the pre-quenching and tempering stage of the tooth area before the hardening step; this stage is used for improving the final metallurgical results also when the prior microstructure of the steel is not particularly suitable for hardening.

The heating diagram for this case is the one of Fig. 4.86.

Examples of hardened profiles obtained with the heating cycles of Figs. 4.84 and 4.86 are given in Fig. 4.87.

4.6.3.2 Simultaneous Dual-Frequency Process

The development of solid state generators in the range 20–300 kHz allowed to improve the dual-frequency hardening process and, in particular, to supply power to a single inductor simultaneously at medium and high frequency. The waveform of the voltage (or the current) supplied to the inductor consists of a mixture of a high-frequency oscillation superimposed on a fundamental medium frequency, as shown as an example in Fig. 4.88.

Usually, the amplitude of the two oscillations can be controlled separately from 1 to 99 %. The power component of both frequencies, and therefore the hardening depth in the root and in the tooth flanks and tips, can be adjusted and set separately depending on the specific requirement.

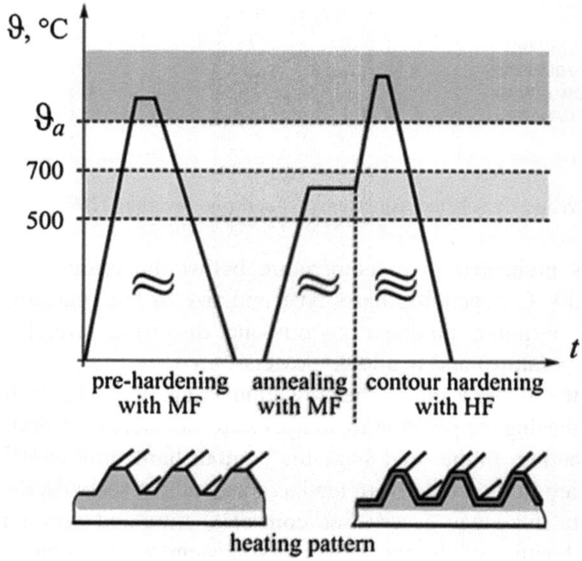

Fig. 4.86 Typical cycle of the dual frequency contour induction hardening with pre-quenching and tempering stage [98]

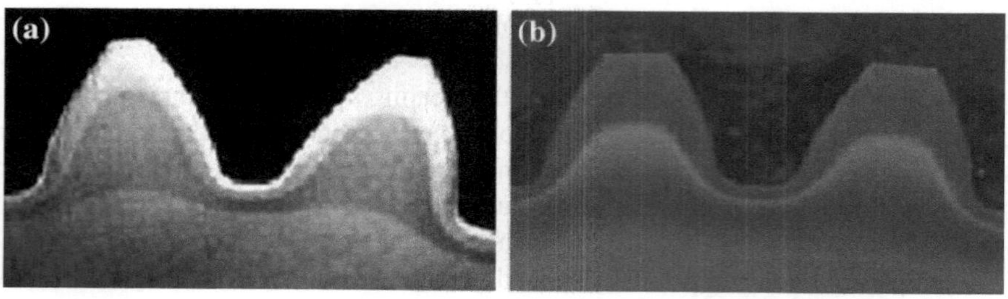

Fig. 4.87 Hardened profiles obtained with dual frequency processes **a** with MF pre-heating stage (OD: 63 mm; thickness: 15 mm; module: 2.5; material: Ck45/1045); **b** with pre-quenching and tempering (OD: 125 mm; thickness: 30 mm; module: 3.0; material: XC48/1045) [98]

In this way it is possible to achieve uniform case tooth profile without influencing core material in heating times less than 0.5 s, using a power ratio in the root of the tooth to the tooth flanks of approximately 2:1.

The power supply of a simultaneous dual frequency induction hardening unit is constituted by two resonant MF and HF converters, whose output are combined in a single output transformer that supplies the load. As shown in the schematic circuit of Fig. 4.89, a HF filter prevents that the HF converter affects the MF circuit. The typical cycle of a Simultaneous Dual-Frequency (SDF) process is shown in Fig. 4.90; the preheating stage, usually applied in non simultaneous dual frequency processes, in this case is not always used.

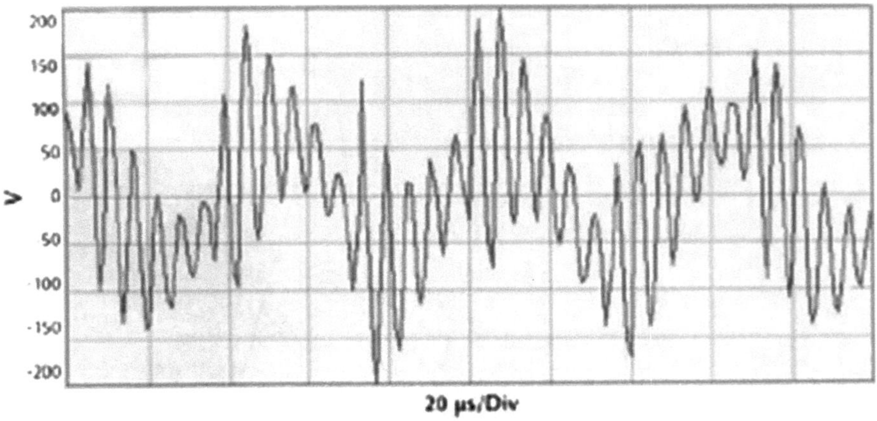

Fig. 4.88 Voltage waveform of a dual frequency generator: HF is superimposed on MF

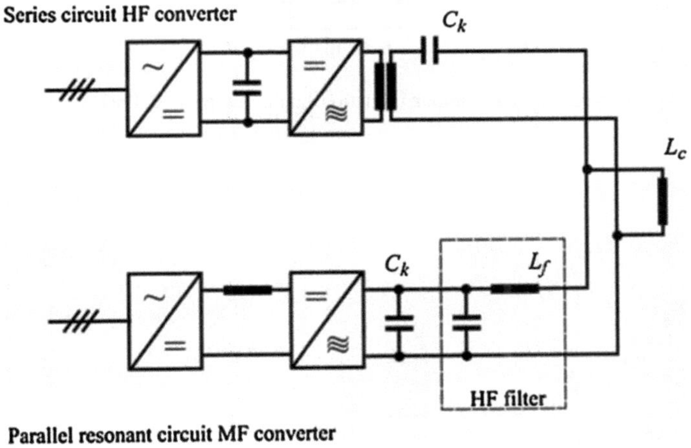

Fig. 4.89 Schematic of the power supply of a SDF induction hardening system (C_k—compensating capacitor; L_f—frequency filter inductance; L_c—inductor coil inductance)

Figure 4.91 show some examples of hardened profiles obtained with the SDF process with different MF to HF ratios. They show that the SDF method can be used for contour hardening different types of workpieces (gear, sprockets, cams, etc.) made of various materials, like steels, metal powders, cast irons.

The experience has shown that a correct choice of heating time and MF/HF power ratio allows to achieve an evenly heated contour profile; for gears the power ratio is usually 70 % MF in the range 3–15 kHz and 30 % HF power at 150–300 kHz.

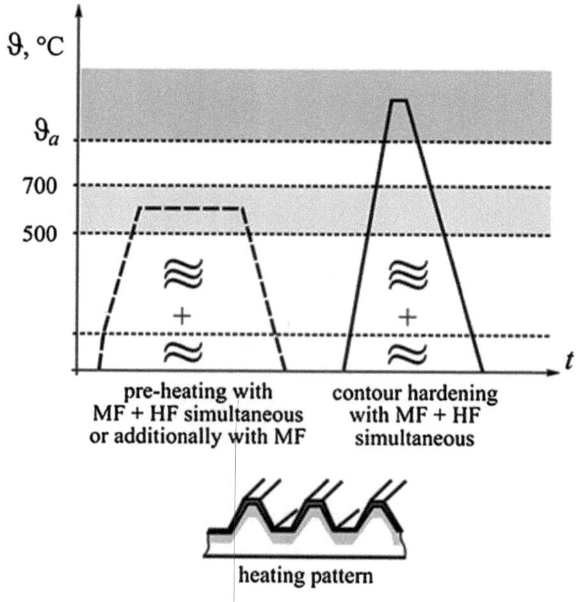

Fig. 4.90 Typical cycle of SDF contour induction hardening process [98]

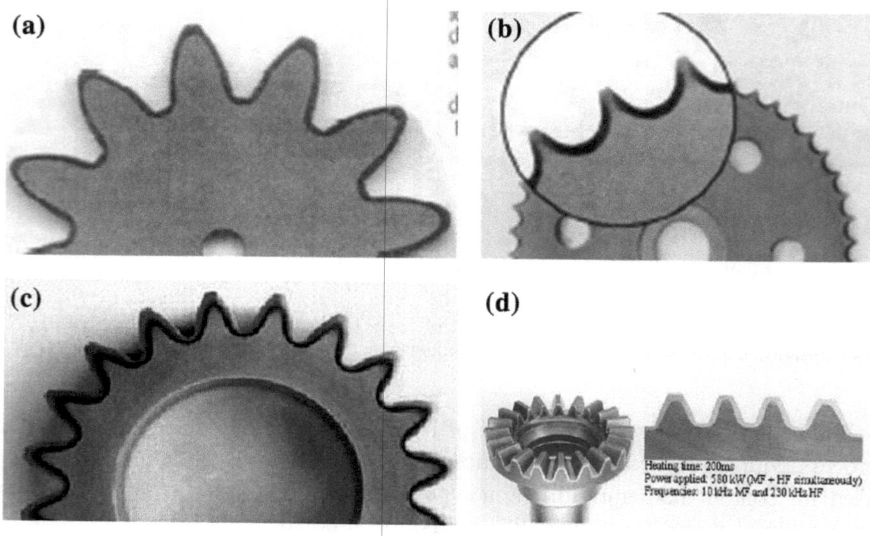

Fig. 4.91 Different complex workpieces contour hardened with SDF process **a** Gear of mid-grade carbon steel: tip diameter (OD) 112 mm; root diameter 76 mm; thickness 6 mm; heating time 1.8 s; MF/HF 80/20 %; SDF unit power supply 75 kW; **b** Powder metal sprocket: OD 120 mm; thickness 4 mm; heating time 0.3 s; MF/HF 74/26 %; unit power supply 150 kW; **c** Gear: OD 83 mm; thickness 38 mm; module 4.0 mm; material Ck53; **d** Bevel gear: heating time 200 ms; simultaneous MF + HF power applied 580 kW; frequencies 10 kHz MF, 230 kHz HF (Courtesy of EMAG Eldec Induction GmbH)

4.6.4 Gear Spin Hardening: Main Factors Influencing the Process

In addition to the parameters mentioned in the previous paragraphs, other factors are of fundamental importance for the hardening result and the process repeatability.

- *Material initial structure*—As known, carbon content of unalloyed steels recommended for induction hardening should not be lower than 0.40 %. Moreover, due to the short austenitizing time available in dual-frequency processes, a homogeneous fine-grained pre-quenched and tempered structure with hardness of 30–34 HRC is the best suited for fast and consistent metallurgical response, with low distortion of the workpiece, minimum amount of grain growth and without crack formation. This type of prior microstructure gives deeper hardened depth and higher hardness compared with an initial ferritic/pearlitic microstructure.
- *Gear dimensions and shape*—The simultaneous dual frequency process is best suited for modules M ≤ 4–5 and gear diameters d ≤ 250 mm. The limitation on diameters is a consequence of the high power densities, in the range 6–12 kW/cm^2, required for heating the gear in very short time for avoiding through hardening of the tooth. The limitation on the maximum module depends mainly on the surface area to be hardened and thus the power that must be delivered from the power supply. In fact the required power density values are obtained with frequency converters which can reach rated power of several hundreds of kW up to some MW.

As regards the shape of the hardened layer, special attention must be paid to the axial end regions of the gear where, due to the concentration of induced power in correspondence to sharp corners, at high frequency there is a tendency to overheat the tooth edges. A possible solution of this problem is a convenient shaping of the inductor or the use of magneto-dielectric yokes that modify the induced current density distribution

Another problem occurs when hardening helical gears: towards the front side of the gears, an asymmetrical hardness pattern occurs due to the tendency of the current to flow along the path of lowest impedance. On one side of each tooth will appear a soft zone, called "*finger nail*" due to its shape. Figure 4.92 illustrates this phenomenon which produces overheating of the sharp edge and insufficient heating of the obtuse edge.

Figure 4.93 shows the resulting asymmetric hardness pattern of the finger nail; the thickness B increases with the helical angle of the gear.

Possible solutions consist in the use of special inductors, with exciting currents at 90° angle to the teeth flanks (so-called "*saw tooth inductor*"), or field concentrators with a shape adapted to the gear. Figure 4.94 shows respectively a special saw tooth inductor and the arrangement of a helicoidal gear with upper and lower field concentrators made of magneto-dielectric material.

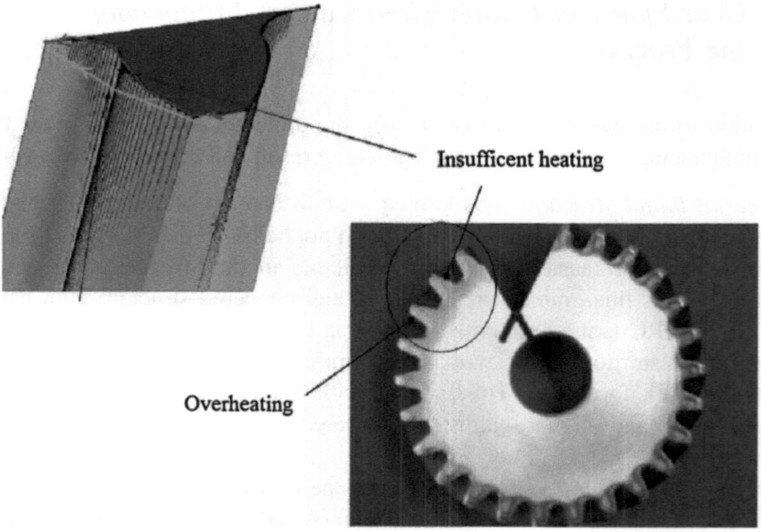

Fig. 4.92 Asymmetric hardness pattern toward front side of helical gears: preferred current path in the front side of a tooth (Courtesy of Eldec Induction GmbH)

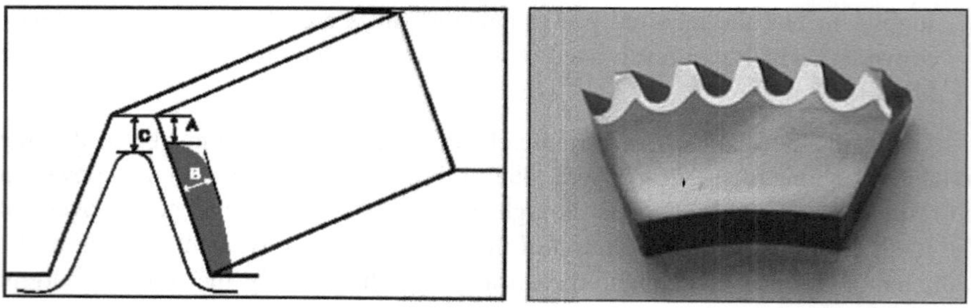

Fig. 4.93 Asymmetric "finger nail" shape of the hardness pattern at the front side of helical gears [98]

However, the design of inductors capable to eliminate the finger nail zone are still under development.

Relative position of gear and coil—concentricity of the gear within the inductor coil strongly affects the hardness pattern; to avoid this problem gears are rotated during heating to ensure an even distribution of the energy along their perimeter. A suitable rotation rate must be chosen according to the process characteristics, e.g. 600–800 rpm during the heating phase, for heating times in the range of few hundreds of milliseconds.

- *Quenching*—quenching from the austenitizing temperature can be done either by submerging or spraying the workpiece with an adequate coolant medium.

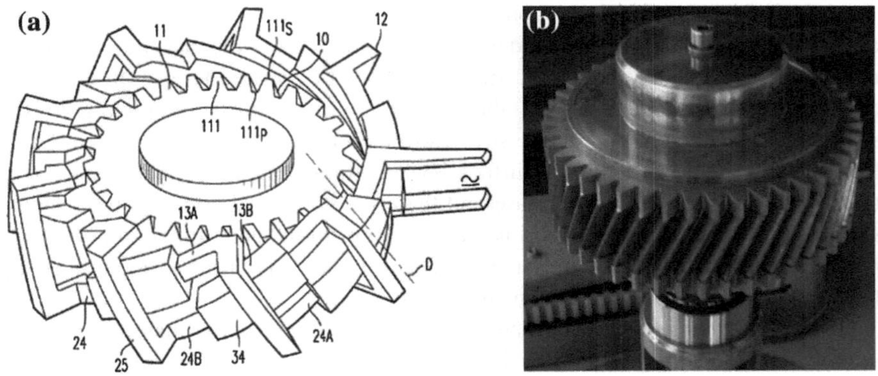

Fig. 4.94 Arrangements for hardening helical crankshaft gears **a** saw tooth inductor, Courtesy of Saet Group SpA; **b** system with upper and lower field concentrators, Courtesy of Eldec Induction GmbH

The first method is applicable to large-size gears, the second to medium and small ones. Quenching should be made with a suitable quenching agent, immediately after the final heating stage, at rotation speed of the workpiece below 50 rpm in order to avoid the so-called "*shadow effect*" on the flank opposite to the direction of rotation. The quenchant is usually a mixture of water and polymer: the higher is the concentration of polymer, the softer will be the quenching effect (i.e., high concentration of polymer leads to low cooling velocity).

In case of very short heating times (in the range of 200–300 ms), a minimal amount of energy is transferred to the workpiece and quenching can be obtained by a combination of self- and spray-quenching, which is the most favorable solution for achieving low residual stresses.

Finally, it must be taken into consideration that quenching is the most critical step of the process because the fast cooling is usually responsible for distortion and crack formation. Low distortion in the quenching process can be obtained by assuring a homogeneous quenching over the whole component, which in turn will produce the lowest temperature gradients required to produce the desired hardness pattern.

- *Residual Stresses and Distortion*—The components treated with surface induction hardening are usually characterized by high residual stresses and low distortions. These are two very important aspects because compressive residual stresses improve the fatigue strength under cyclic loading, whereas low distortion reduces the cost of subsequent machining or even makes it unnecessary. These two phenomena are linked together because residual stresses can produce unpredictable distortion of the component. The distortion of the component occurs during the final heat treatment because, in this stage, the residual stresses induced during the manufacturing are released and may lead to plastic deformation if the stresses exceed the elasticity limit of the material at high-temperature.

As known, in surface induction hardening, only the outer surface layer of the component is heated to the austenitization temperature, whereas the underlying material remains “cold”. Once the surface reaches the desired temperature and the component is quenched rapidly, the surface layer contracts, producing tensile stresses within it and compressive stresses in the underlying transition zone. Upon further cooling, the structure of surface layer is transformed into martensite. Martensite will cause a volume increase and therefore additional compressive residual stresses. At the same time, the shrinkage of the transition zone further contributes to produce in the surface compressive residual stresses which are compensated by tensile residual stresses in the transition zone.

The driving force producing compressive residual stresses is therefore the volume increase due to the martensitic transformation, which depends on carbon content and plastic elongation of the surface. The plastic elongation of the surface is influenced by the core temperature which, in turn, can be controlled by the process parameters.

Preheating of the component is a mean to prevent cracks occurring during quenching for the high tensile stresses caused by the contraction of the surface layer.

The main advantage of surface pulse induction hardening is the relatively small volume of material which undergoes martensitic transformation, thus producing lower dimensional variations and consequent lower stresses and distortion. Although temperature gradients during the pulse heating are much higher than during usual induction hardening, the resulting plastic strain is not a real issue because the soft layer is supported by the untreated core of the tooth and therefore the plastic strains cause only small deformations of the tooth profile. In conclusion, the smaller distortion produced by surface pulse induction hardening can be attributed to the quite small volume that is heated and to the much smaller amount of dissolved carbon.

After carburizing, the temperature of the component is lowered to approximately 800–850 °C in order to decrease the possibility of distortions as well as the retained austenite. On the other hand, the goal of reducing distortion by lowering the temperature during surface induction hardening is achieved by the so called “*quench delay*”. The length of such time delay depends on the material and particularly on its alloying elements. The quench delay produces a certain degree of temperature equalization which allows to store more heat in the transition zone with beneficial effects on the residual stress distribution.

4.6.5 Numerical Simulations and Results

Surface induction hardening represents a rather complex multi physical problem and a satisfactory solution is still non available with numerical methods. Major improvements have been made in the last years, nevertheless, the simulation of induction hardening process of complex components is still a matter of research. In fact a comprehensive model of hardening problem must take into account all the phenomena that lead to the metallurgical result: electromagnetism, heat conduction, convection and radiation losses, metallurgical transformations and mechanical stress.

The basic simulation of the induction heating process consists of calculating the currents and the heat sources induced in the workpiece. The heat sources produce a temperature increase with non-uniform distribution within the workpiece, which in turn modifies the local values of resistivity and magnetic permeability and finally the distribution of the heat sources.

Other phenomena should be considered in the numerical model to fully describe the process effects. In particular thermal stresses, which always occur in fast induction heating hardening processes, may cause plastic deformation. In addition the phase transformation of the microstructure, due to the heating itself, must also be taken into account in the numerical model, since plasticity transformation may play an important role in producing distortion as well as in the residual stress distribution [98, 109–112].

A further problem is due to the non-linearity of material properties. In particular, the magnetic permeability, which depends on temperature and local magnetic field intensity, requires iterative solutions of the electromagnetic problem which increases substantially the calculation time. Moreover, availability of correct material data is of fundamental importance, since the results of simulation are strongly dependent on the accuracy of the input data. To this end, it is important to know the chemical composition of the steel.

Figure 4.95 gives an overview of the above complex physical interactions occurring during surface induction hardening processes.

These interactions are more strong in pulse hardening where all physical phenomena occur in very short times.

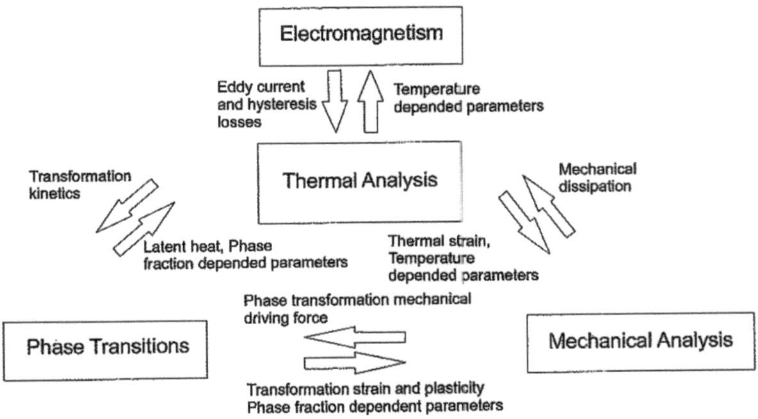

Fig. 4.95 Coupled physical phenomena occurring in the simulation of surface induction hardening

To the best of our knowledge, there are no commercial numerical codes able to simulate the whole induction hardening process taking into account all the involved physical phenomena.

Numerical simulations of the hardening process are still object of academic researches and only large private companies can sustain the high costs required for hardware and software facilities as well as qualified man power.

However, it can be easily predicted that in the future also small companies will take advantage of these tools, particularly when fully coupled solutions will be available on a unique platform. Numerical simulations give some remarkable advantages, like the reduction of design costs and time for experimental testing. They allow at the same time to achieve a deeper understanding of phenomena involved in the process.

Anyway it must be taken into account that even if the design has been carried out through numerical simulations, some experimental tests are always needed in order to fine tuning the hardening process.

In the following, some examples of numerical simulations of induction hardening processes are presented.

4.6.5.1 Coupled Electromagnetic-Thermal Solutions

These solutions, for single- and dual-frequency pulse processes, have been developed in the last years at University level, in particular at Padua and Hanover Universities.

Some models have been developed for the simulation of the full 3D non-stationary heating process. These models, based on the commercial packages ANSYS or FLUX, can take into account all dependencies of material characteristics (electrical resistivity, magnetic permeability, thermal conductivity, specific heat capacity) with temperature, and for magnetic permeability with local magnetic field intensity [102, 113–115].

Figure 4.96 shows a typical scheme of calculation in which the following steps are iteratively performed.

First a harmonic electro-magnetic calculation is done with resistivity and permeability values temperature (and local magnetic field) dependent. The result of this calculation is the induced power density distribution in the workpiece, which is used as input data for the subsequent thermal analysis, performed again with temperature dependent physical parameters.

Each thermal calculation is done for a given time interval Δt.

Result of the calculation is the 3D temperature distribution in the workpiece at the end of the time interval considered.

If the final assigned heating time t_{max} or the maximum final process temperature ϑ_{max} are not reached, another electromagnetic calculation starts, considering new updated temperature dependent values of electrical parameters.

After this, the thermal calculation starts again beginning from the temperature distribution of the last global time-step.

Fig. 4.96 Flow-chart of electromagnetic-thermal calculation

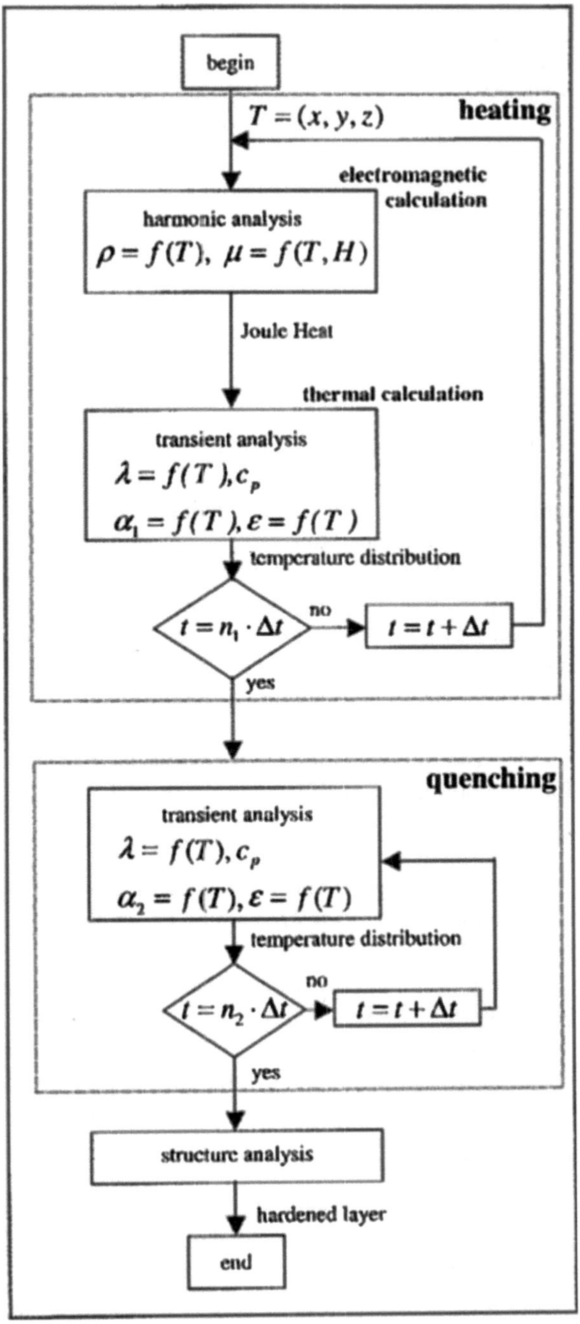

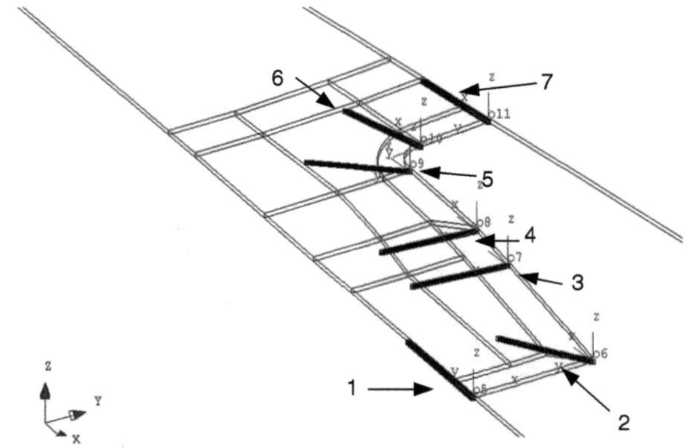

Fig. 4.97 Positions of paths along tooth contour for evaluation of austenitisation depth (AD)

These models can be used for the evaluation along the teeth contour of the "*austenitisation depth*" (AD), defined as the depth of the surface layer where the temperature at the end of the heating transient is equal to or higher than the set austenitisation temperature.

The results given in the following as typical examples of the current calculation capabilities, refer to the hardening of a gear wheel with module M = 4 mm, number of tooth Z = 21, tip diameter 92 mm and root diameter 74.6 mm. The heating parameters (power density and heating time) have been chosen for reaching a maximum surface temperature of about 1,050 °C, with austenitisation temperature set at 850 °C [102].

Several positions along teeth contour have been defined, as shown in Fig. 4.97, for evaluating the austenitisation depth in the direction normal to the tooth surface.

- *Single-shot single frequency process*

For the above gear wheel, the optimal frequency and heating time suggested by the diagrams of Fig. 4.79, should be in the range 19–38 kHz and 0.6–0.7 s respectively.

As Fig. 4.98 shows, these process data lead to a very uneven distribution of temperature and AD.

However, as shown in Fig. 4.99, a more convenient choice of process parameters, i.e. frequency, heating time and surface magnetic field intensity, allows to obtain a reasonably uniform distribution of temperature and AD.

- *Full simultaneous dual-frequency (SFD) process*

As previously described, the simultaneous dual frequency hardening process is performed feeding the same inductor with two different power supplies, working at different frequencies.

In this case the total induced power can be evaluated as the sum of the power induced at medium frequency and the one induced by the high frequency exciting field. The simultaneous double frequency computation can be carried out by subdividing the thermal transient into a number of constant time steps and solving the

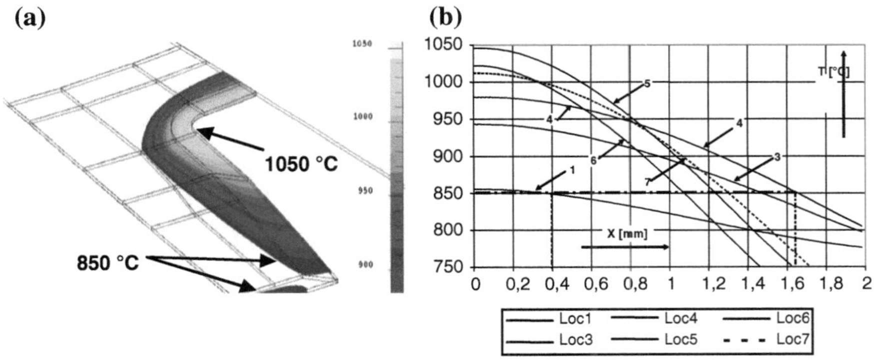

Fig. 4.98 Final temperature distributions (**a**) and AD at different positions along tooth contour (**b**) (frequency: 25 kHz; maximum surface temperature: ϑ_{max} = 1,050 °C; heating time: 0.66 s; *x* distance from the surface; *1–7* paths of Fig. 4.97

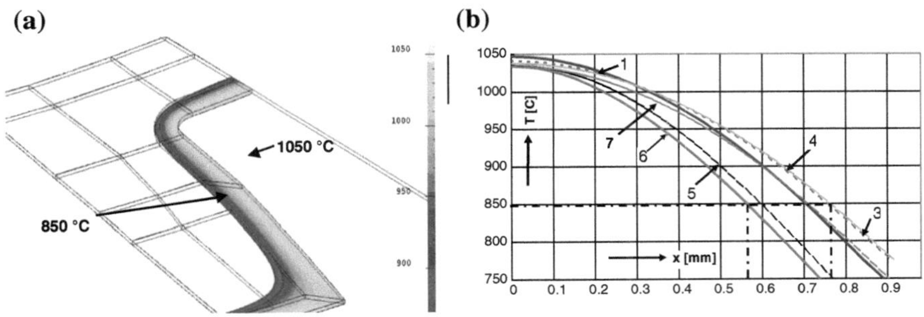

Fig. 4.99 Final temperature distributions above 850 °C (**a**) and AD at different positions along tooth surface (**b**) (f = 100 kHz; H_0 = 240 kA/m; ϑ_{max} = 1,050 °C; heating time: 0.25 s)

EM and thermal problems alternatively for the medium (MF) and the high frequency (HF).

In particular, in one step the MF electromagnetic problem is solved and the calculated induced power sources are used as input for the thermal problem solution; in the subsequent step the same scheme of calculation is repeated applying as source only the HF exciting magnetic field. Attention must be paid, however, to the fact that the induced power sources values used as input for the thermal problem, must be the double of those obtained as solution of the electromagnetic one, since in the real process the energy is delivered continuously to the workpiece and not intermittently as in the described calculation scheme.

The computation accuracy depends on a correct choice of the time step: the results presented here have been obtained by subdividing the heating period into 25 steps for the single frequency process and 50 steps for the double frequency one. These values have been experienced to represent a good compromise between computation accuracy and calculation time. The proposed simulation approach does

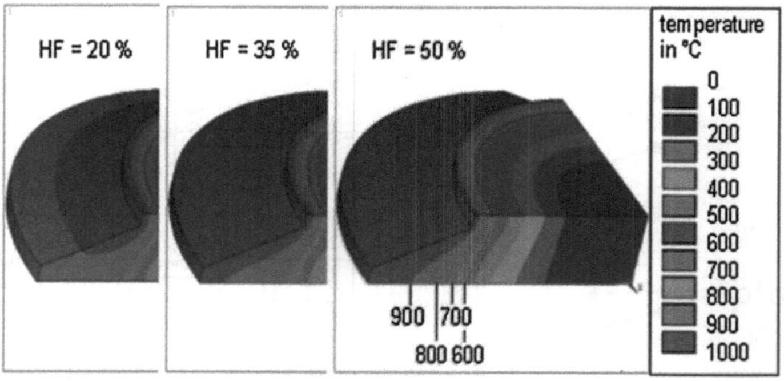

Fig. 4.100 Influence of HF to MF power ratio on final temperature distribution in the SDF process [113]

not consider a correct value of the magnetic field intensity, that is given at any time by the sum of the instantaneous values of MF and HF field, so the computed relative permeability value are affected by a systematic error. Some attempts to overcome this issue have been proposed in Ref. [116].

In SDF processes, the final result is strongly influenced by the ratio of the HF and MF power, as illustrated by the example shown in Fig. 4.100.

As previously mentioned, the experience suggests as optimal recipe 30–35 % of HF power, corresponding to surface power density values about double on the tooth flanks in comparison with the root when the steel is magnetic, and to about equal values above Curie point, in the non-magnetic phase.

An example of calculation for the SDF process is given in Fig. 4.101. The diagrams show that in this example the AD is not sufficiently uniform and that the process parameters should be further optimized.

- *Dual-frequency process with MF pre-heating*

Another example of calculation of a dual frequency process is given in Fig. 4.102; it refers to a gear wheel of modulus M = 1.59 mm, number of teeth Z = 47, tip diameter 77.8 mm, root diameter 70.9 mm and process with MF pre-heating.

As the figure shows, here again a uniform case depth has not been reached, but the metallurgical results are sufficiently good for industrial application.

The comparison of calculation with the experimental result, shown in Fig. 4.102d, points out a fairly good agreement and confirms the reliability of numerical solutions as a tool for predicting the results of the heat treatment, as well for the selection of process parameters.

In particular, the analysis of the results shows that in this case (1) the thickness of the layer heated above 850 °C in the root is thicker than at flanks and tip; (2) simulation and experiment confirm that the tip is not sufficiently heated, and further optimization is needed.

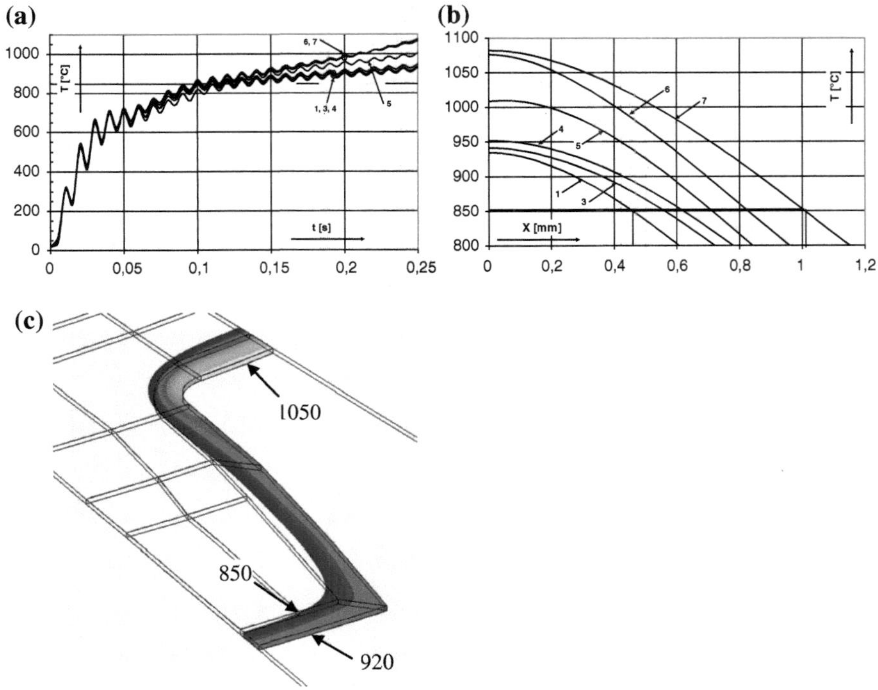

Fig. 4.101 SDF process: **a** transient temperature distribution during the process at surface points of Fig. 4.97; **b** AD at different positions along tooth contour (paths of Fig. 4.97); **c** final temperature pattern above 850 °C; (M = 4; H_{MF} = 500 kA/m-3 kHz; H_{HF} = 160 kA/m–300 kHz)

However, as already stated, the result fulfils the industrial requirements since root and flanks are the parts more mechanically stressed.

The numerical models are also useful for optimization of the inductor, by profiling the copper conductor or by using additional magnetic yokes for guiding the field, in order to limit temperature inhomogeneity which occurs at the edge of the tooth in the axial direction of the gear. An example is given in Fig. 4.103.

4.6.5.2 Metallurgical Models

Computer models are available today for electromagnetic—thermal—metallurgical simulation of the pulse induction hardening process (see Fig. 4.95) [117, 110].

Through these models, temperature and austenite distributions after the heat treatment can be evaluated and the influence of cooling processes with different cooling media can be analyzed. These models, used in connection with the relevant CCC curves, give metallurgical results which allow to predict the final distributions of microstructure, hardness profile and influence of the heating and cooling cycles on the end-product properties and quality.

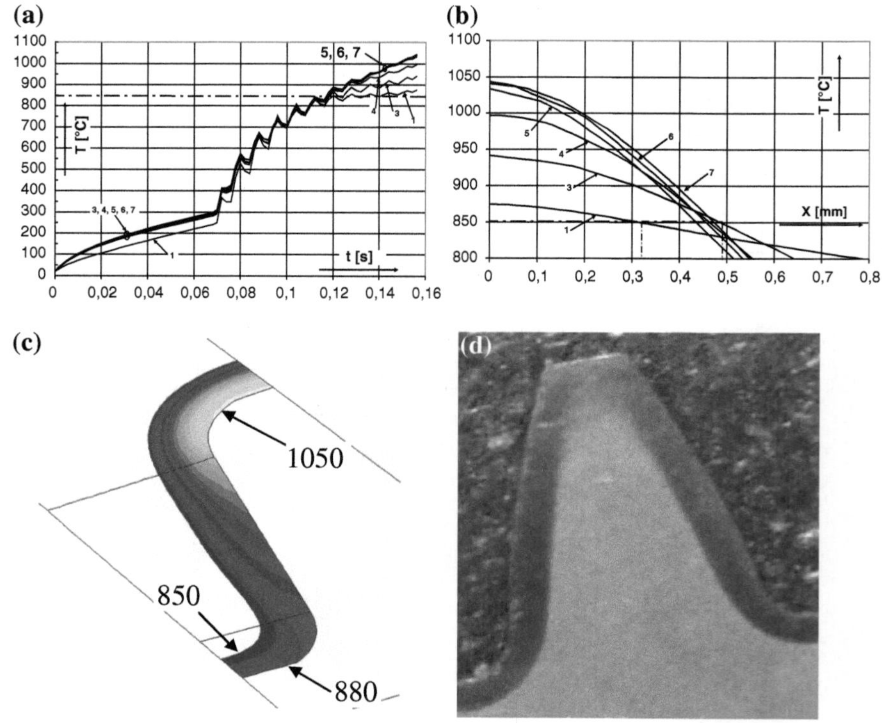

Fig. 4.102 Dual-frequency process with MF pre-heating: **a** transient temperature distribution during process at surface points of Fig. 4.97; **b** AD at different positions along tooth contour; **c** final temperature pattern above 850 °C; **d** experimental hardened pattern for the same heating conditions ($M = 1.6$; $H_{MF} = 390$ kA/m–11 kHz; $H_{HF} = 190$ kA/m–236 kHz; Pre-heating: 70 ms; SDF heating: 86 ms)

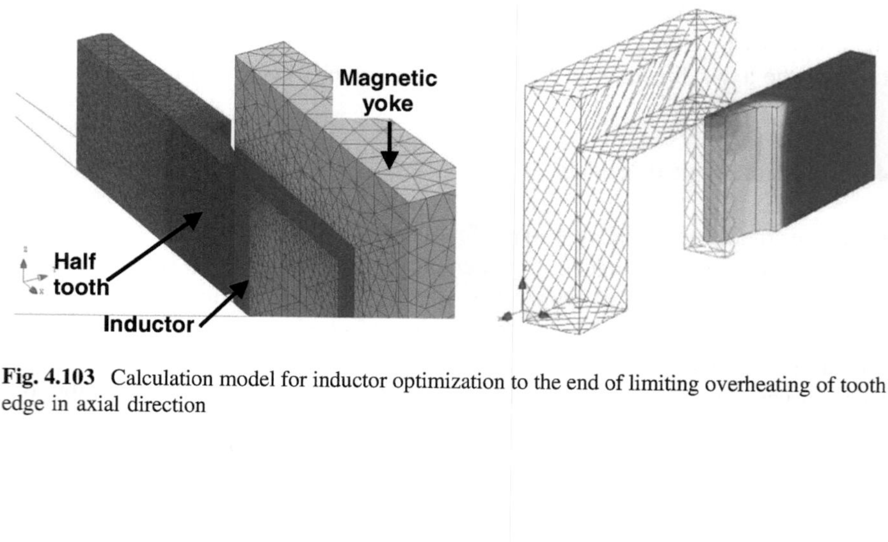

Fig. 4.103 Calculation model for inductor optimization to the end of limiting overheating of tooth edge in axial direction

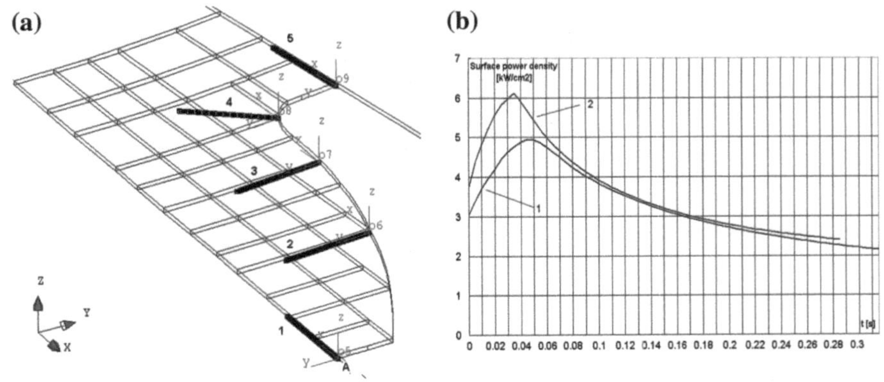

Fig. 4.104 **a** Schematic of paths considered for analysis of temperature, microstructure and hardness profiles; **b** surface power density during pulses (*1* 80 kHz, 315 ms; *2* 120 kHz, 285 ms)

In order to show some potentialities of these models, in the following are given two examples which refer to typical pulse hardening cycles and the schematic of the tooth shown in Fig. 4.104.

The heating and cooling conditions of the two processes are:

- Process 1—frequency 80 kHz, heating time: 315 ms, surface power density: given by curve (1) of Fig. 4.104, maximal temperature: 1,050 °C
- Process 2—frequency 120 kHz, heating time: 285 ms, surface power density: given by curve (2) of the same figure, maximal temperature: 1,085 °C
- Cooling: indicated in the following with the letters A and B, 'A' denoting quenching with polymer fluid and 'B' natural cooling.

The calculated temperature patterns at the end of the heating transients are shown in Fig. 4.105 (diagrams *a*); in the same figure the corresponding austenite percent distributions are shown by diagrams (*b*).

The diagrams show that in both processes the formation of austenite is incomplete. A more uniform distribution and higher percentage of austenite are achieved in process 2, which is characterized by a higher maximum temperature, higher frequency and shorter heating time.

From the final heating pattern and cooling phase it is also possible to draw the curves of the temperature distributions along the paths defined in Fig. 4.104a and the corresponding hardness profiles. They are shown as an example in the Fig. 4.106b, c with reference to process 2.

The curves show that:

- the shape of the curves and the maximum values of the hardness, allow to predict the hardened depth and to analyze the influence of the cooling conditions;

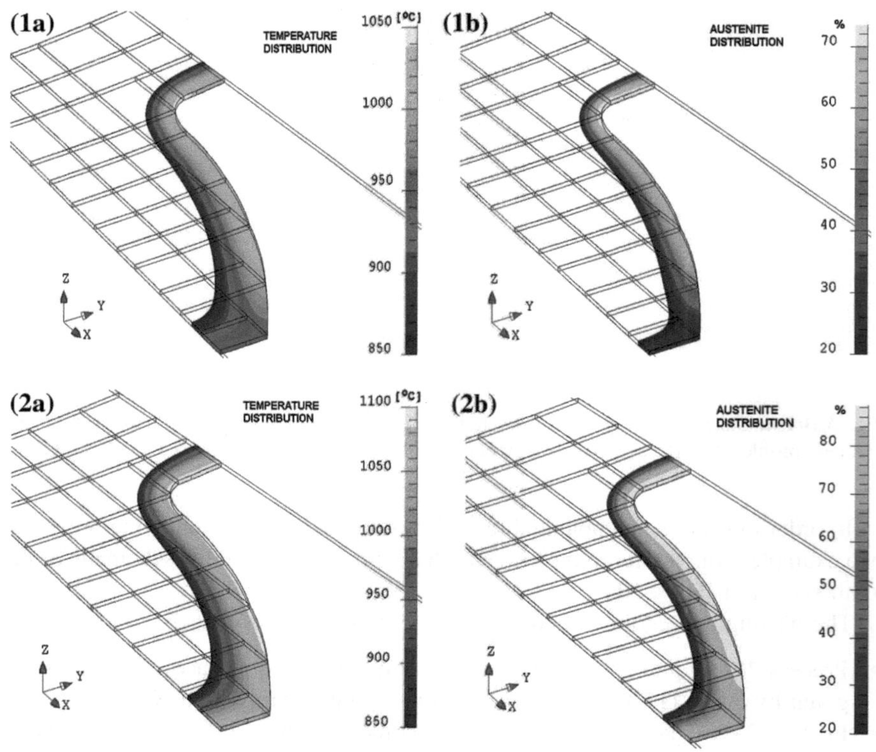

Fig. 4.105 Temperature (**a**) and per cent austenite (**b**) distributions at the end of the heating phase (1—80 kHz, 315 ms, 2—120 kHz, 285 ms)

- the relatively low maximum values of hardness can be attributed to the low percent distribution of austenite at the end of heating and to the characteristics of the steel to which the simulation refers.
- there is a strong influence of the cooling conditions on the final distribution of martensite and the consequent hardness profile.

Finally, the martensite and hardness distributions are given in Fig. 4.107 at the end of process 2 with polymer cooling.

The calculated distributions confirm that, from the metallurgical point of view, good hardness profiles can be obtained in pulse hardening of gear wheels by a suitable choice of process parameters and quenching means.

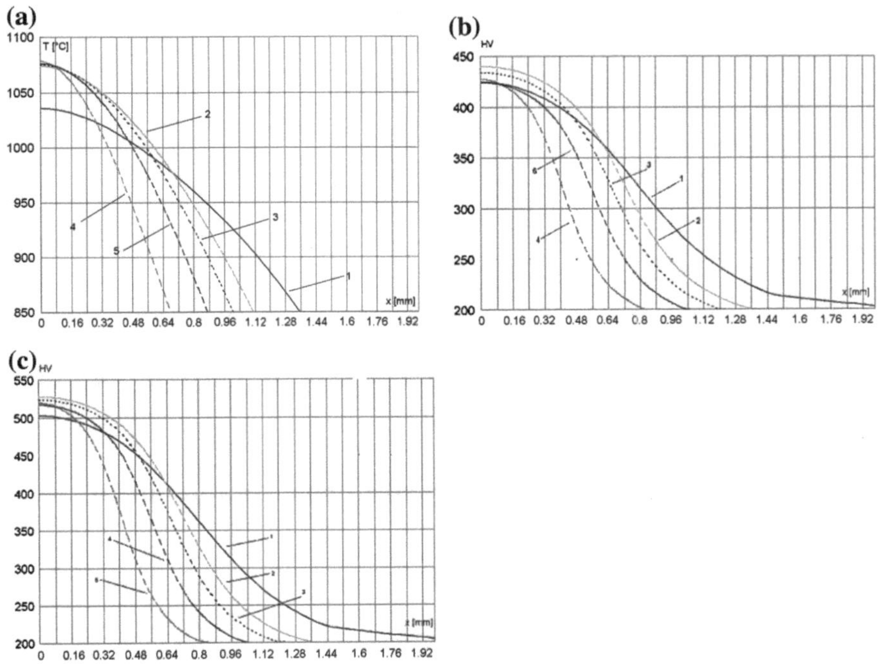

Fig. 4.106 Process 2—120 kHz, 285 ms: **a** Temperature profiles at the end of heating (along the paths of Fig. 4.104**a**, **b** and **c** Hardness profiles at the end of cooling along the same paths, with natural cooling (**b**) and with polymer cooling (**c**)

4.6.6 Conclusions

The results obtained by the numerical non-linear FEM models available for modelling single and double-frequency hardening of gear-wheels have shown the possibility of electromagnetic, thermal and metallurgical analysis of the different hardening processes. It has been demonstrated the possibility of reaching sufficiently uniform austenitising depths along the gear contour with the single- and dual-frequency processes by the choice of optimum heating parameters.

The dual-frequency heat treatments, with or without MF pre-heating period, offer a great process flexibility for the possibility of choosing independently the heating parameters, like the MF and HF power ratio and the frequency values.

However, both for single and dual-frequency processes, the high calculation times make presently very difficult to perform a real process optimization. Further developments are still under development for including in the multi-physics computation the prediction not only of heating and metallurgical transformations, but also of residual stresses and deformations in order to allow a systematic analysis and optimization of the whole process and a correct choice of the quenching medium.

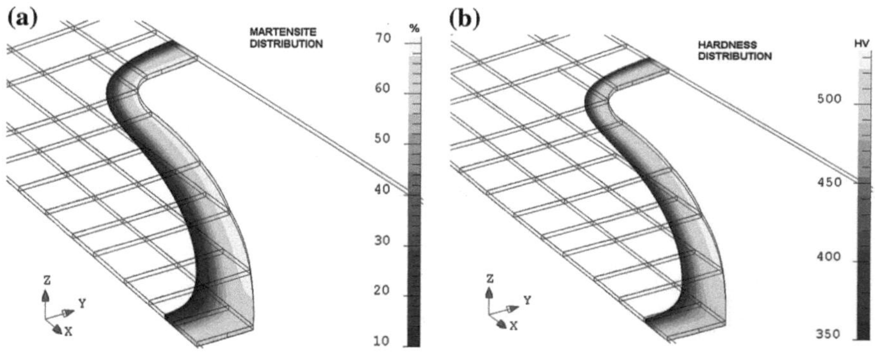

Fig. 4.107 Final martensite (**a**) and hardness (**b**) distributions Process 2—120 kHz, 285 ms and of cooling with polymer

4.7 Induction Heating of Cylindrical Billets Rotating in DC Magnetic Field

4.7.1 Introduction

Electromagnetic induction is widely applied in the industry for heating metals before hot working, like forging, forming, extrusion. In this process the workpiece must be heated to a temperature in the plastic deformation range of the metal, providing uniformity in the temperature distribution within the body, since temperature uniformity greatly affects the formability of the material and the lifetime of dies. The process is fast and clean and, in many cases, very energy efficient.

However, when heating low resistivity nonmagnetic metals, such as aluminum, copper and brass, the efficiency of the process becomes low.

In aluminum extrusion plants 50/60 Hz longitudinal flux induction heaters are used to preheat solid cylinders (billets) from room temperature to around 500 °C before extrusion.

At the beginning of 2000 in Europe there were about 500 aluminum extrusion plants. Each plant has at least one extrusion press with typically two billet preheaters working in parallel. In total the number of aluminum billet heaters in Europe exceeds 1,000. Modern induction heaters with multi-layer coils operate with efficiencies in the 55–60 % range. This means that in large heaters having typical power rating of 1 MW, 400–450 kW are dissipated in the water-cooled copper conductors of the inductor coils and then dispersed into useless cooling water at low temperature. With 4,000 operating hours per year this means an energy loss of about 1.5 MWh/year in each installation [118].

Many theoretical studies have been done for improving the energy efficiency of this process.

Some of them have investigated the possibility of using the Litz-wire technology for fabrication of the induction coils; theoretically this solution improves coil efficiency up to about 65 %, but has the main drawback of the difficulty of providing reliable cooling of Litz-wire cables capable of carrying high currents [106].

Other studies in the late 1970s and early 80s developed multilayer inductors working at line frequency. They have shown that, for power ratings over 1 MW, properly designed single-phase multilayer coils allow to reach efficiency values in the 55–60 % range [119–122].

At the beginning of this century an innovative induction technique has been proposed for reducing inductor losses, by the use of DC superconductive windings, able to produce high intensity magnetic fields in large volumes. In this approach the billet to be heated is forced to rotate inside a transverse DC magnetic field by means of an electric motor drive (Fig. 4.108a). Due to the change of the magnetic flux linkage, an induced current distribution reacts to the driving torque during the

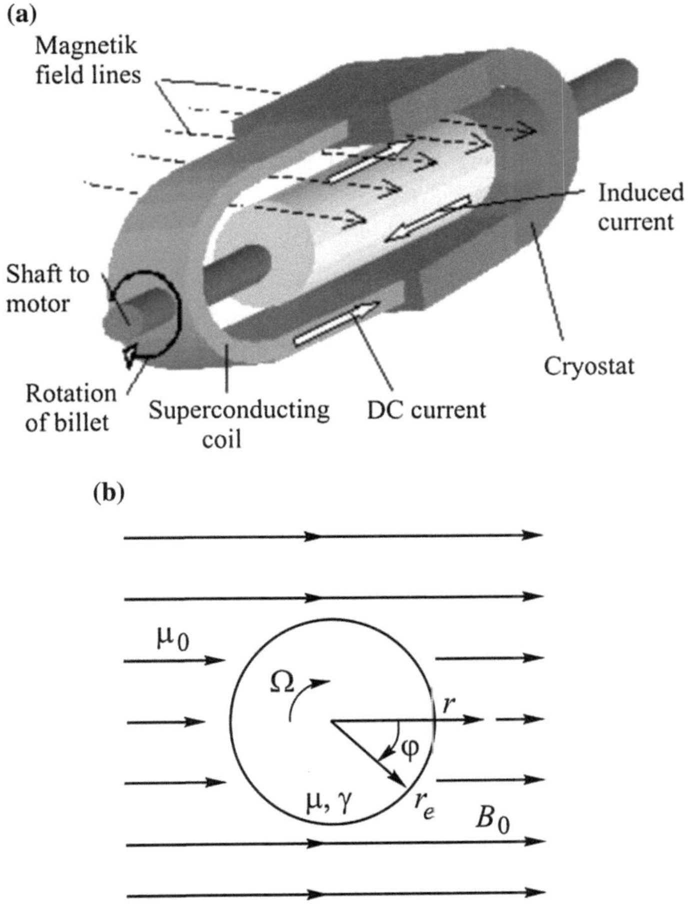

Fig. 4.108 **a** Principle superconducting induction heating; **b** coordinates system for analysis of a circular infinitely long cylindrical workpiece rotating in uniform transverse DC magnetic field (r -radius of billet; Ω—angular velocity; B_0—external DC magnetic field)

rotation and generates thermal power within the billet. The overall efficiency of the process is coincident with that of the motor drive, which in the high power range is above 90 %.

Several theoretical works have considered the energetic and superconductive aspects of the device [123–127] while the analysis of the electromagnetic and thermal phenomena taking place inside the billet has been developed in [128].

4.7.2 *Electromagnetic Solution for the Infinitely Long Geometry*

The electromagnetic solution is obtained with reference to the solid cylinder of Fig. 4.108b, rotating in an Eulerian reference frame in a DC stationary magnetic field.

The cylinder (*billet*) has radius r_e [m], electrical conductivity σ [$\Omega^{-1}m^{-1}$] and uniform constant relative permeability μ. The cylinder is assumed to be infinitely long along z (the axis normal to Fig. 4.108b), surrounded by a non-conducting medium with permeability $\mu_0 = 4\pi\cdot 10^{-7}$ [H/m], and rotates in an external uniform magnetic field B_0 [T] with angular velocity $\Omega = n\cdot(2\pi/60)$, with n in [r/min].

Introducing the quantities:

$\delta = \frac{1}{\sqrt{\Omega\mu_0\mu_r\sigma}}$ –diffusion length [m]; $m = \frac{r_e}{\delta}$; $\xi = \frac{r}{r_e}$; $j = \sqrt{-1}$;

$\Re = m^2 = \left(\frac{r_e}{\delta}\right)^2$—magnetic Reynolds number

the radial and tangential components of the magnetic field inside (suffix i) and outside the cylinder (suffix e) can be written as follows [129]:

$$B_{ri} = \Re e\left\{\frac{B_0\cdot(1+\beta^2)\cdot r_e\cdot J_1(m\xi\sqrt{j})\cdot e^{-j\phi}}{\xi\cdot J_1(m\sqrt{j})}\right\} \tag{4.44a}$$

$$B_{re} = \Re e\left\{B_0\cdot\left(1+\frac{\beta^2}{\xi^2}\right)\cdot e^{-j\varphi}\right\}$$

$$B_{\varphi i} = \Re e\Big\{j\cdot B_0\cdot(1+\beta^2)\cdot \frac{[J_1(m\sqrt{j}) - m\xi\sqrt{j}\cdot J_0(m\xi\sqrt{j})]\cdot e^{-j\phi}}{\xi\cdot J_1(m\sqrt{j})}\Big\} \tag{4.44b}$$

$$B_{\varphi e} = \Re e\left\{-j\cdot B_0\cdot\left(1+\frac{\beta^2}{\xi^2}\right)\cdot e^{-j\varphi}\right\}$$

with:

$$\beta^2 = \frac{(\mu_r+1)\cdot J_1(m\sqrt{j}) - m\sqrt{j}\cdot J_0(m\sqrt{j})}{(\mu_r-1)\cdot J_1(m\sqrt{j}) + m\sqrt{j}\cdot J_0(m\sqrt{j})} \tag{4.45}$$

and J_o, J_1—Bessel functions.

An electric field, also called motional EMF, electromotive force, is induced in a conductive body moving inside a magnetic field. The motional EMF depends by the vector product of the induction field and velocity vectors, so the induced electric field is normal to both the magnetic field and velocity direction.

Since for a generic internal point of the billet, of coordinates r and φ, velocity and radial components of the magnetic induction are at right angle, from Eqs. (4.44a, 4.44b) and (4.45) can be easily derived the induced current density J and the induced power per unit volume w:

$$J(\xi,\varphi) = \sigma E_i = \sigma B_{ri}\frac{2\pi r_e \xi}{60}n \tag{4.46}$$

$$w(\xi,\varphi) = \sigma\left(B_{ri}\cdot\frac{2\pi r_e}{60}n\cdot\xi\right)^2 \tag{4.47}$$

As shown in Fig. 4.108, the induced currents flow along paths parallel to the billet axis, and—as the induced power—have different values along billet radius and azimuthal position.

The diagrams of Fig. 4.109 give the distributions of the induced power density as a function of the azimuthal angle φ, at different radial positions. In particular, they show the following:

- the dramatic influence of parameter m on magnitude and radial distributions of w;
- the shift of the maximum of the curves in the direction of the billet rotation which increases with the value of m and—for the same m—is different at various radial positions.

Figure 4.110 shows a typical induced power density pattern, which confirms the above statement.

It must be noticed that the distribution of the induced currents and consequent thermal power, do not exhibit instantaneous axial symmetry. However, due to the high rotation speed of the billet, each point located at the same radial position undergoes the same electrical and thermal history; axial symmetry therefore exists in average sense and no net thermal flux occurs in azimuthal direction.

By integration of the local specific induced power along the azimuthal coordinate φ, we can derive its mean value as a function of the position ξ and the relative induced power radial distribution; this distribution is given in Fig. 4.111 for different values of m (i.e., for a certain billet, at different rotation rates).

It can be easily verified that these distributions are the same used in the classical induction heating of a static cylindrical billet in a longitudinal exciting AC flux, which are given by the relationship [93]:

$$\frac{w_{m\xi}}{w_{me}} = \frac{ber'^2(m\xi) + bei'^2(m\xi)}{ber'^2(m) + bei'^2(m)} \tag{4.48}$$

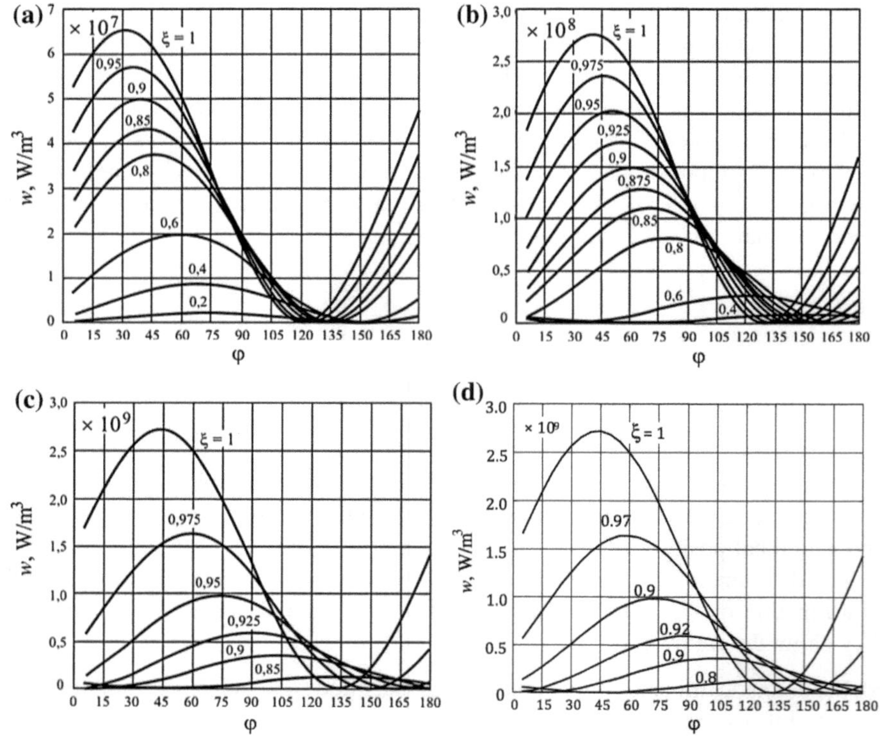

Fig. 4.109 Distributions of power density w as a function of the angle φ, for different radial positions ξ and values of m = 2.5–5–7.5–15 in (**a**), (**b**), (**c**), (**d**) respectively (Aluminium billet: r_e = 200 mm; $\sigma = 2 \cdot 10^7\ \Omega^{-1}\text{m}^{-1}$)

Fig. 4.110 Typical induced power density pattern

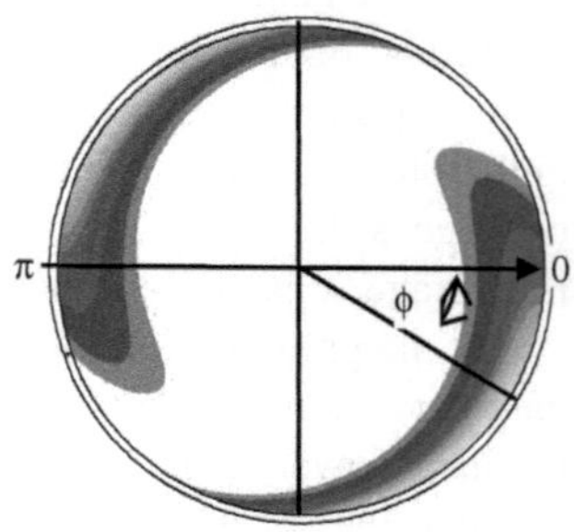

By integration along the radius of the mean specific induced power w_m, the total induced power per unit billet length P_1 is obtained; its value can be calculated by the relationship:

$$P_1 \approx 4 \cdot 10^{12} \cdot \rho \cdot B_0^2 \cdot m^2 \cdot A \tag{4.49}$$

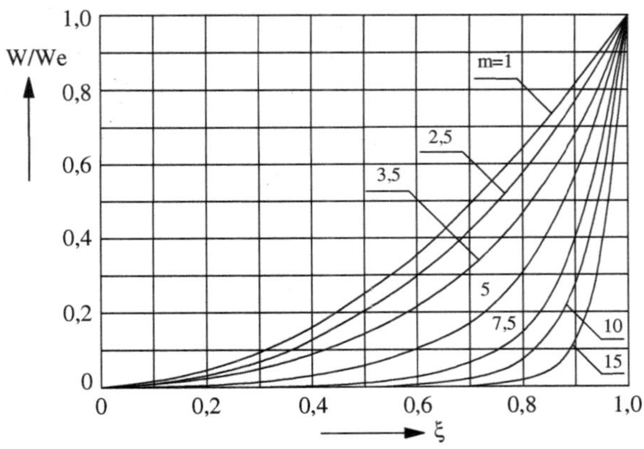

Fig. 4.111 Distributions of mean specific power w induced in the billet as a function of the radial position ξ, for different values of m

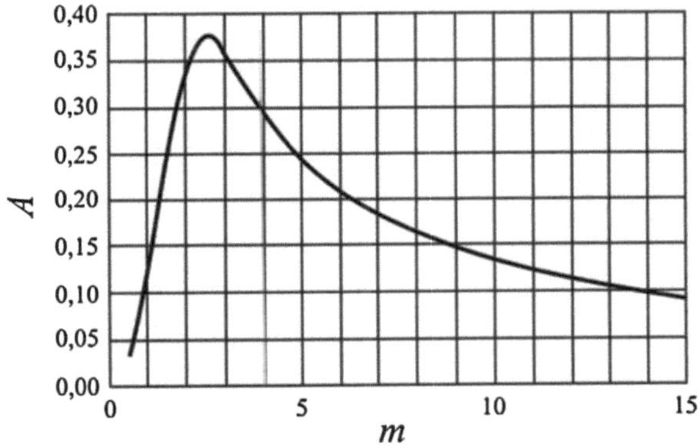

Fig. 4.112 Coefficient of the total power induced in the billet per unit length

with:P_1 [w/m], $\rho = 1/\sigma$ [Ω·m], B_O [T], and the coefficient A is a function of *m*, given by the diagram of Fig. 4.112.

The absolute value w_{me} of the mean specific power induced at the workpiece surface is linked to the total induced power P_1 by the relationship:

$$P_1 = 2\pi r_e^2 \cdot w_{me} \cdot F(m) \tag{4.50}$$

with:

$$F(m) = \frac{berm \cdot ber'm + beim \cdot bei'm}{m(ber'^2m + bei'^2m)}$$

The values of *F*(*m*) are given, as a function of *m*, in the diagram of Fig. 4.113.

As example, the values of the total power P_1 induced in the billet per unit length at different rotation rates are given in Fig. 4.114 for a "long" aluminum billet, 200 mm diameter, heated from 20 to 500 °C. The diagrams show the influence of the variations of resistivity with temperature during the heating transient.

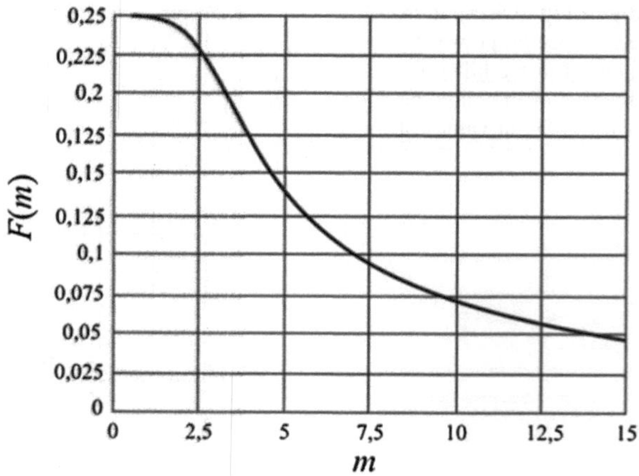

Fig. 4.113 Values of $F(m)$ as a function of m

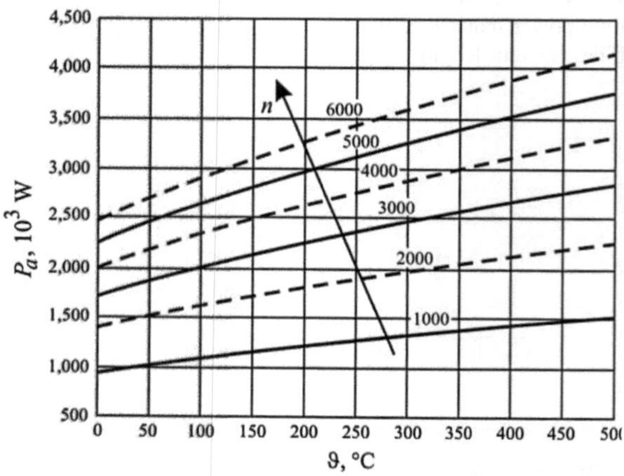

Fig. 4.114 Induced power P_1 as a function of billet mean temperature, at different rotation rates (Al billet: r_e=0.1 m; ℓ = 1 m; B_0 = 1 T; n = 1,000–2,000–3,000–4,000–5,000–6,000 r/min)

4.7.3 Heating Parameters

Taking into account the radial distribution of the mean induced power, given by Eq. (4.48), and the value of the total induced power by Eq. (4.50), the evaluation of heating transients and process parameters can been done with classical axis-symmetric 1-D and 2-D numerical programs, analogous to those used for longitudinal flux heating.

One significant difference is the value of the heat transmission coefficient α from the billet surface to the surrounding ambient due to the high rotation speed. A reliable value of α can be evaluated by the following empirical relationship, which has been deduced experimentally at the Laboratory for Electroheat of Padua University [128]:

$$\alpha \approx 20 \cdot v^{0.54} \tag{4.51}$$

with: α [W/m^2 °C] and v—peripheral speed [m/s].

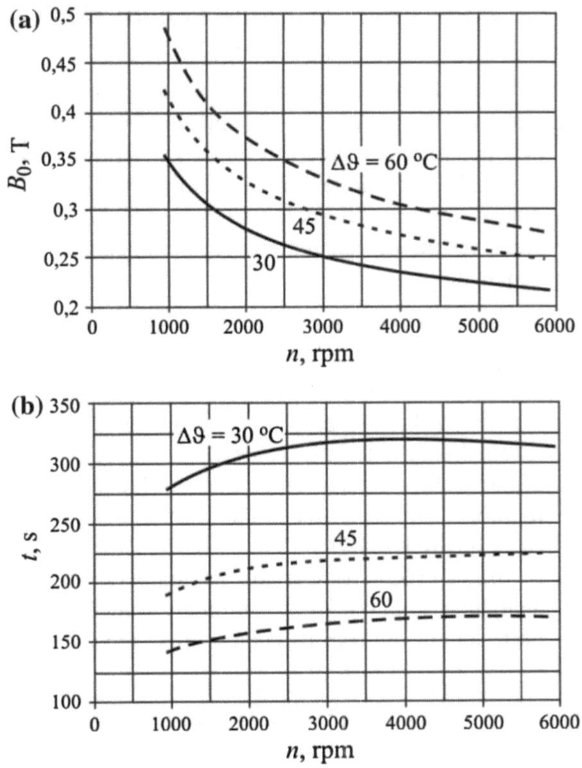

Fig. 4.115 Magnetic flux density B_0 **a** and heating times **b** as a function of rotation rate for heating to 500 °C an aluminum billet, 200 mm diameter, with final axis-to-surface differential of 30, 45, 60 °C ($\rho = 5 \cdot 10^{-8} \Omega{\cdot}\text{m}$)

For a preliminary evaluation of the process parameters, a 1D solution has been used considering a constant mean resistivity value in the temperature range of the process, equal to $\rho = 5 \cdot 10^{-8} [\Omega{\cdot}\text{m}]$ for heating aluminium from room temperature to 500 °C. The analytical solution allows to evaluate, as a function of the rotation rate n [r/min], the total average induced power P_1 and the induced specific mean power distribution w_m by means of the equations of the previous paragraph.

Assuming that this distribution remains unchanged during the heating transient, the heating time and power to achieve a final surface temperature of 500 °C with an assigned surface-to-axis temperature differential can be calculated as a function of the rotation speed.

The results of these calculations are summarised in Fig. 4.115.

Figure 4.115a gives the exciting magnetic flux density required at different rotation rates for reaching the final surface temperature of 500 °C with an axis-to-surface temperature difference of 30, 45 and 60 °C respectively; Fig. 4.115b gives the corresponding duration of the heating process. They allow a quick approximate estimation of process parameters.

4.7.4 Edge Effect in Finite Length Billet

In previous paragraphs the study has been carried out making reference to a billet of infinite axial length rotating in uniform DC magnetic field.

In order to investigate the influence of the billet edge effect, numerical calculations have been done for an aluminium billet 200 mm diameter, 1 m length, rotating in uniform DC field and heated to 500 °C [130, 131].

Some results of calculations are summarised in the diagrams of Fig. 4.116; in particular they give:

- Figure 4.116a—average induced power values during heating transient as a function of rotation rate, for external magnetic field values $B_0 = 0.2–0.3–0.4–0.5$ T;
- Figure 4.116b—heating times as a function of the rotation speed, for the same external magnetic field values;
- Figure 4.116c—final temperature differential in the billet, as a function of the rotation speed, for the same external magnetic field values $B_0 = 0.2–0.3–0.4–0.5$ T.
- Figure 4.116d—comparison of temperature differentials calculated with the analytical solution for the infinitely long billet and the numerical ones for the finite length billet in its central cross-section.

An example of the influence of the billet edge effect on the final temperature pattern at different rotation rates is shown in Fig. 4.117, which refers to a final maximum temperature inside the billet of 500 °C, and external magnetic field $B_0 = 0.3$ T.

4.7.5 Influence of Shape and Position of Superconducting Coils

The comparison of diagrams of Fig. 4.116a, b and temperature distributions of Fig. 4.117, shows that the billet edge effect has a considerable influence on the final temperature differential.

The final temperature pattern strongly depends also on the magnetic field distribution produced by the superconducting coil and, in turn, by the shape of the coil itself.

For a deeper understanding of the combined effect of the magnetic field distribution and the billet edge effect, fully coupled 3D numerical electromagnetic and thermal calculations are necessary [132].

However, useful information can be obtained also from simplified solutions.

A 3-D time dependant finite element electromagnetic model has allowed to calculate the induced currents and specific power distributions in the billet, with reference to the system geometry of Fig. 4.118 with an aluminium billet, 200 mm

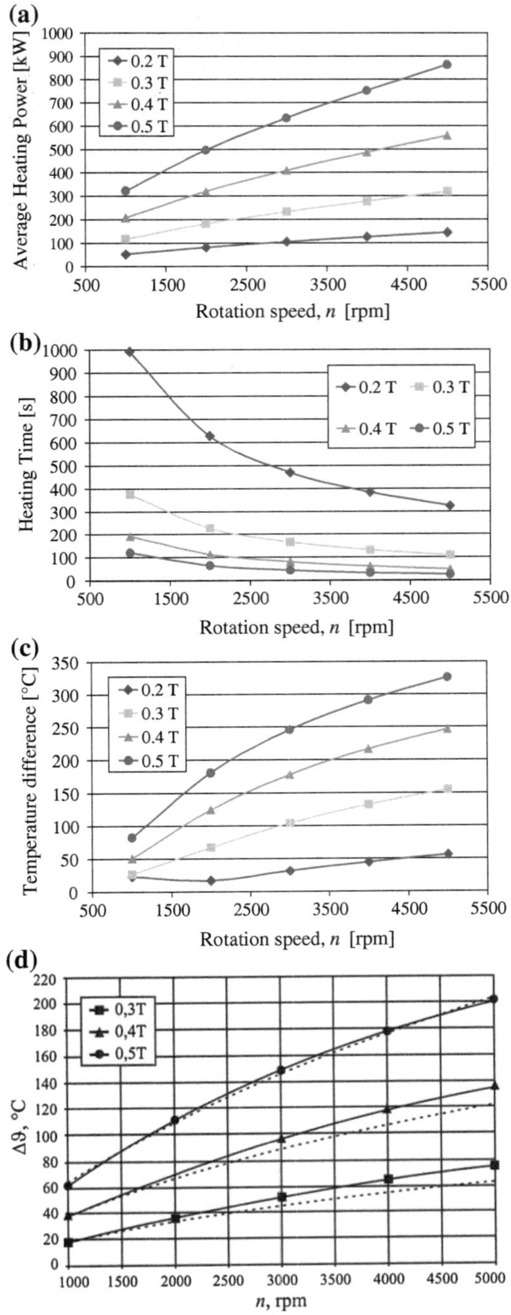

Fig. 4.116 Process parameters for heating to 500 °C a 200 mm diameter billet, 1 m length, as a function of rotation rate, with magnetic field values $B_0 = 0.3$—0.4—0.5 T: **a** total induced power at end of heating; **b** heating times; **c** final temperature differential; **d** comparison of final temperature differential in the billet central cross-section calculated with analytical solution and FEM numerical method

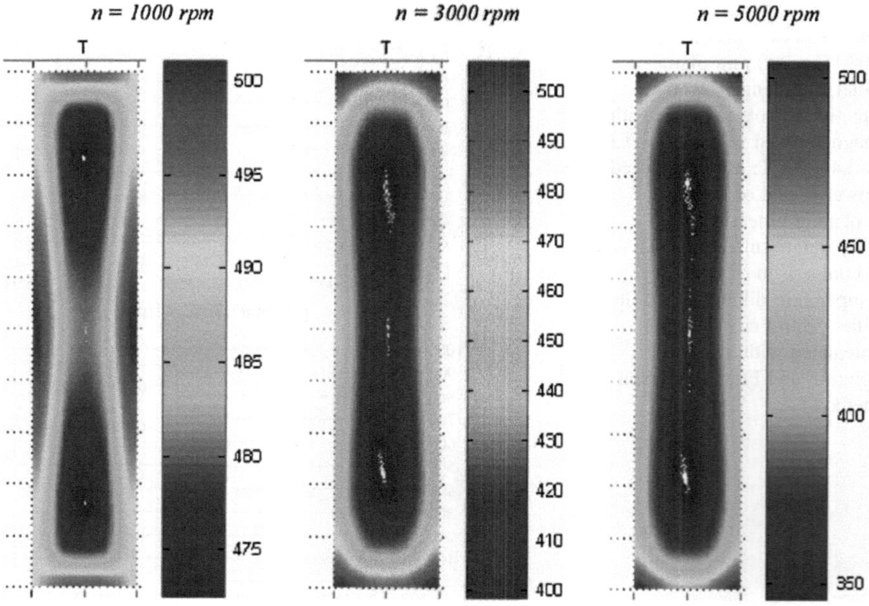

Fig. 4.117 Temperature distributions in the billet cross-section at the end of heating process with external magnetic field B_0 = 0.3 T, final maximum temperature 500 °C and rotation rates 1,000–3,000–5,000 rpm

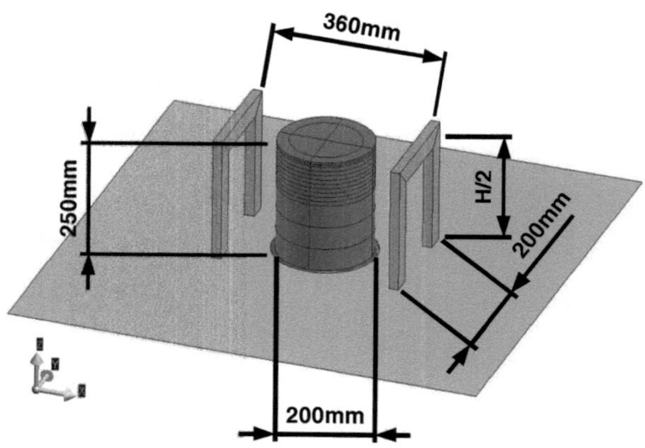

Fig. 4.118 Schematic of position of rotating billet and superconducting coils (half geometry)

diameter, 500 mm height, rotating in the central region of the rectangular superconducting coils shown in the figure [133].

The calculations have been done with constant value of electrical resistivity, equal to 5×10^{-8} Ω·m, corresponding for aluminum to the mean value in the range 20–500 °C.

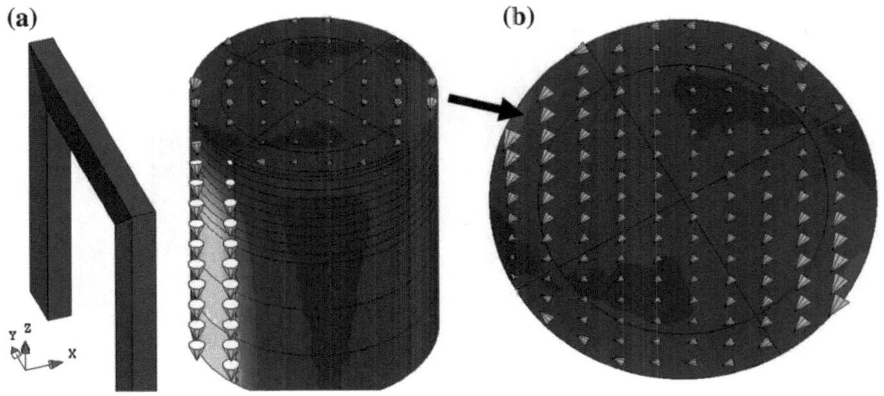

Fig. 4.119 Induced current density paths **a** intensity: *colour shade*, direction: *arrows*; **b** edge surface

The 3-D eddy current distribution in the billet, given in Fig. 4.119, shows that the induced current lines change direction near the billet edge; in particular, in the central part they flow mainly along the axial direction, accordingly with the 2D model, but near the edge of the billet, their paths are characterized by components normal to the billet axis.

Other examples of induced power distributions are shown in Fig. 4.120a, b for different rotation rates and inductor heights.

These distributions suggest that the billet can be ideally subdivided into three main heating zones along the axis. The first zone, close to the symmetry plane, is the *regular zone* where the directions and distributions of induced currents and power densities are similar to the ones computed on 2D models, the second one is a *transition zone* where the current paths change direction, the third zone is the *edge zone*, close to edge faces, where the induced currents are mainly normal to the billet axis.

Moreover, the figures show that, as the speed increases, the axial extension of the transition zone is reduced and the currents paths change direction closer to the billet edge.

The induced power density distribution varies both along billet axis and azimuthal position. However, the variation of induced power along the azimuth has practically no consequence on the temperature distribution because of the high rotation speed of the billet.

On the contrary, the power variation along the axial position can lead to inhomogeneous temperature distributions near the edge of the billet.

The unevenness of the induced power along the billet axis is illustrated in Fig. 4.121a–d.

The diagrams give the induced power values per unit length, obtained by integration of the local volume power density on several cross-sections of the billet, and plotting the results as a function of the axial position.

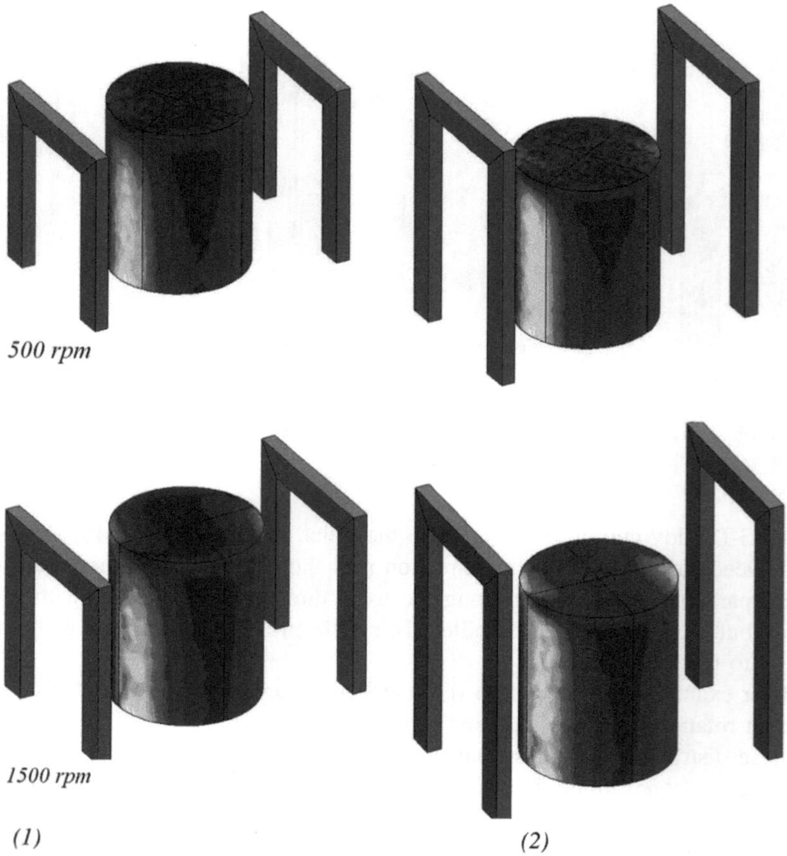

Fig. 4.120 Induced power distributions at different rotation rates and inductor heights. *H* [(*1*) $H = 500$, (*2*) $H = 700$ mm]

The diagrams show again that it is possible to single out three main regions along the axial direction.

In the central "regular" zone the power distribution is not affected by the height of the inductor coils and is almost flat for every rotation speed.

In the "transition zone" the induced power value is lower in comparison with the central zone and the difference between them decreases as rotation speed increases or when the height of the inductor is bigger.

The maximum of the induced power occurs on the billet edge for all the cases analysed and its peak value depends both on rotation rate and height of exciting coils.

However, it must be taken into account that, in order to obtain a uniform final temperature distribution, at the edge it is usually required a higher induced power, because in the edge region there is a wider surface where thermal exchanges take

Fig. 4.121 Induced power distribution along billet length, as a function of distance from the billet middle cross-section, with different inductor heights (inductor: 170 kA·turns; resistivity of billet: 5×10^{-8} Ω·m; **a** 500 rpm; **b** 1,000 rpm; **c** 3,000 rpm)

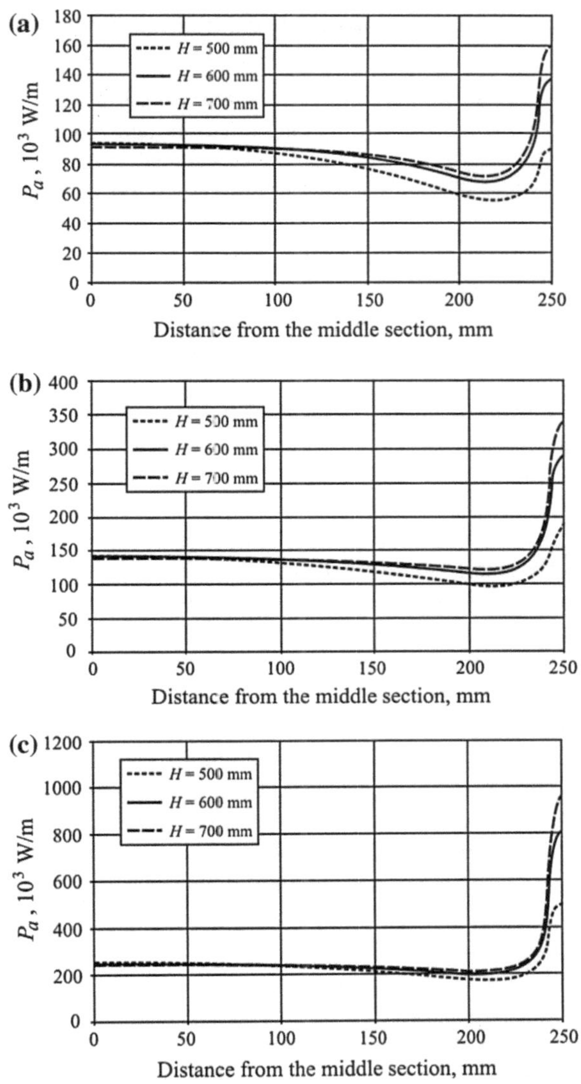

place and consequently more power is needed for compensating the higher thermal losses. Moreover, more power must be supplied to the edge region also to compensate the relative lack of power in the transition zone.

4.7.6 Optimisation of the Heating Process

In order to analyze how the previous power distributions influence the heating results, some coupled magneto-thermal analysis have been made by the numerical model described in [130], for the evaluation of the final heating patterns when the billet has reached the maximum temperature of 500 °C.

The calculations have been done for rotational speed of 500 and 1,500 rpm and inductor height equal to 700 mm.

These velocities have been chosen because the induced power distributions are significantly different in these two cases: at the lower speed, the power in the transition zone is much lower than in the regular one and the peak at the edge is the smallest one.

On the contrary, at the higher speed, in the transition region the variation of induced power is very low, while the peak at the edge is the biggest one.

The average induced power in a revolution and the final heating patterns are shown in Fig. 4.122.

At the higher speed, the temperature differences along the billet axis are relatively small, while the main differences are between core and surface. These differences can be minimized by increasing the heating time.

At the lower speed, the temperature distribution on the cross-section is more even and the main differences are experienced along the billet axis.

From the previous analysis it can be concluded that there are several means for controlling the heating transient and the final temperature pattern in the billet.

The first one is a convenient choice of the values of the induction B_0 and the rotation speed n according to the diagrams of Fig. 4.116. As an example a heating time of about 300 s—a typical value for heating a 200 mm diameter aluminum billet before extrusion—can be obtained with $B_0 = 0.3$ T and $n = 1{,}500$ rpm with a maximum temperature differential inside the billet of about 50 °C.

The second one consists in the control of the rotation speed during the heating transient, with consequent variation of the power density distribution inside the billet. The third is to 'shape' the DC magnetic field along the billet surface either by the use of magnetic cores for direct or concentrate the field on specific zones of the billet or by a convenient shape of the superconductive coil.

An example of this last possibility is given in Fig. 4.123. Figure 4.123a shows the final temperature distribution in a billet, 200 mm diameter, rotating at 250 rpm inside a uniform transverse field of 0.7 T.

Figure 4.123b gives the temperature distribution obtained by applying, at the same rotation rate on the same billet, a DC field having the shape shown in Fig. 4.124.

This example confirms that the temperature uniformity can be greatly improved by a convenient profile of the DC magnetic field. Figure 4.124 shows also a pair of saddle coils that can be used to produce the required field profile [131].

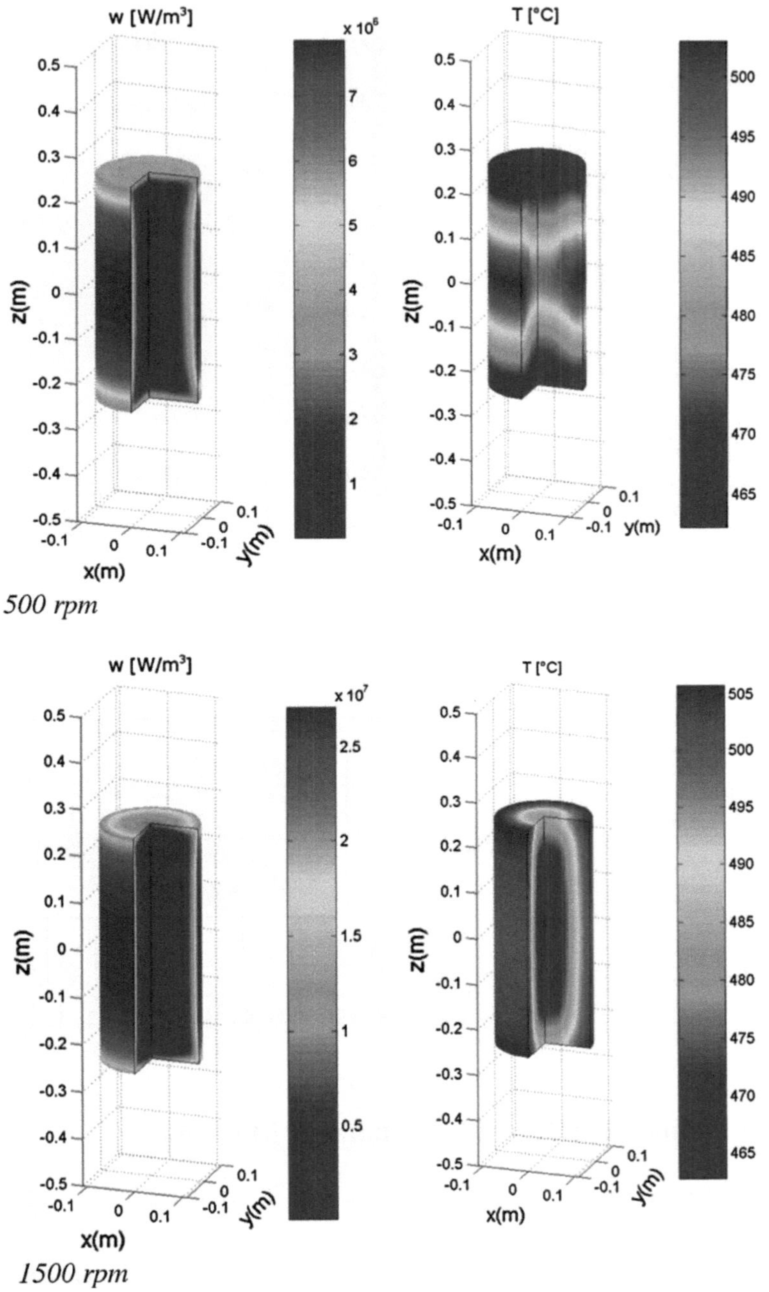

Fig. 4.122 Mean induced power in a revolution (on the *left*) and final temperature distribution in the billet (on the *right*) (inductor: H = 700 mm; 170 kA·turns; billet: $c\gamma$ = 2.61e6 W·s/m^3·K; λ = 238 W/m·K; heating time: at 500 rpm, 560 s; at 1,500 rpm, 230 s)

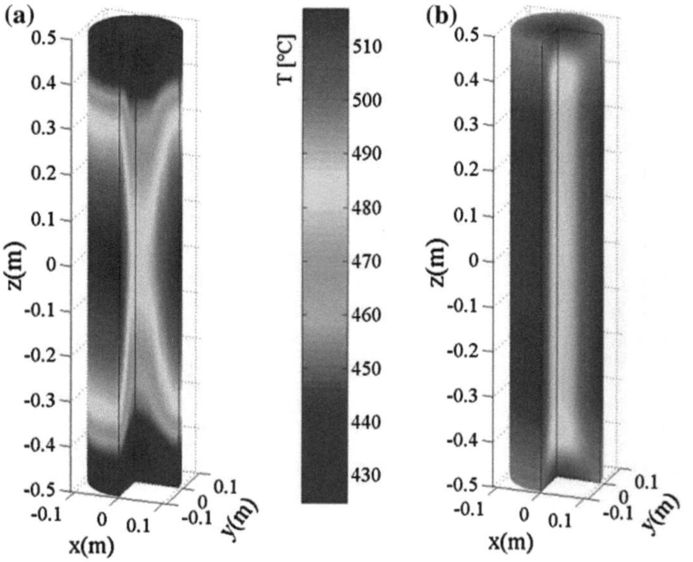

Fig. 4.123 Temperature distributions at the end of heating a 200 mm diameter billet rotating in uniform magnetic field (**a**) and in a conveniently shaped DC field (**b**)

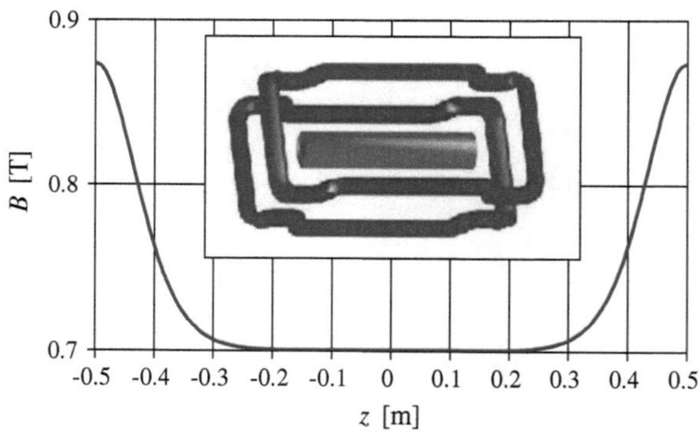

Fig. 4.124 Profile of flux density B along billet length and shape of the superconducting coil producing that field

4.8 Induction Heating with Permanent Magnets

4.8.1 Introduction

High performance hard magnetic materials, like 'NdFeB' or 'SmCo' are commonly applied in electrical devices. A relative movement of a Permanent Magnets (PM), i.e. a relative movement of magnetic field, induces eddy currents in a conductive body. The motional electric field and, in turn, the induced power depend on several design parameters, like the velocity and the type and characteristics of *PM*. In this process,

Fig. 4.125 Industrial scale prototype installed at the Laboratory for Electroheat of Padua University **a** prototype; **b** arrangement of permanent magnets inside a steel rotating frame

mechanical power is transformed into heat with no losses and this allows to heat by induction with high efficiency also high conductive materials, like aluminum or copper.

Several research groups have investigated this type of process, in Academies and industrial companies [134–139].

Here is presented the application developed in the Laboratory for Electroheat of Padua University (LEP) since the first years of 2000 [135, 136].

A PMH—Permanent Magnets Heater is a heating system based upon the rotation of permanent magnets near or around a stationary billet. This solution looks very promising because it allows to achieve high efficiency, which depends mostly on the efficiency of the motor drive, without the need of expensive superconductive systems [131].

An industrial scale prototype (Fig. 4.125) has been realized in 2012 and several tests have been carried out. The prototype is designed to heat aluminum billets, 200 mm diameter, 500 mm length, approximate weight 42 kg. The motor drive is rated power 52 kW, and the magnetic field is produced by SmCo rare earth permanent magnets.

The heating of aluminum billet is obtained thanks to the eddy currents induced by the rotation around the billet itself of the induction field *B* produced by a system of permanent magnets. Two possible sketches of the system are shown in Fig. 4.126: the first one shows ¼ of a 4 poles system while the second one the geometry of a 8 poles system.

The design of the prototype has been carried out using a commercial FEM package and an ad hoc analytical model. The analytical solution requires very short computational time and can be conveniently applied to an automatic optimization procedure. The FEM model uses a transient magnetic solution that reaches the steady state condition only after a high number time steps.

The solution requires the computation of the field of induction B produced by the permanent magnets inside the billet and the induced currents produced by the motion:

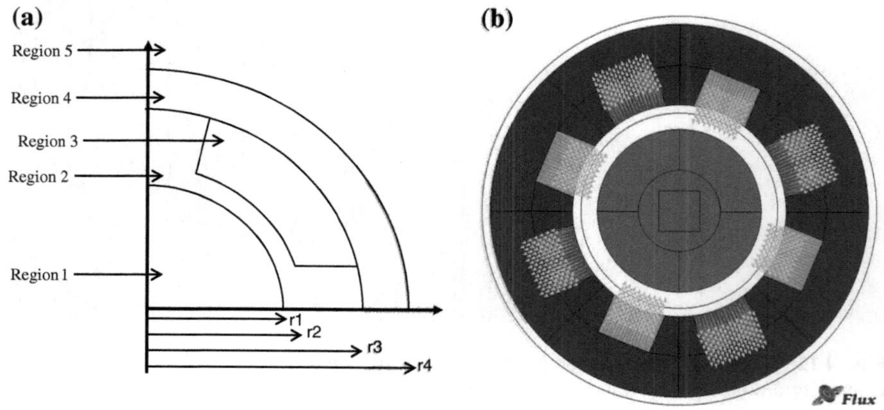

Fig. 4.126 System geometry. **a** 4 poles system: 1—aluminum billet; 2—air-gap (comprising thermal insulation to protect permanent magnets); 3—permanent magnets; 4—iron frame; 5—air; **b** 8 poles system

$$\overline{E}_i = \omega r \overline{u}_\varphi \times \overline{B} \tag{4.52}$$

where $\overline{u}_\varphi$is the azimuthal unit vector.

In the hypothesis of a rotational speed ω in the range 500–3,000 rpm, the magnetic Reynolds number $\Re_M$:

$$\Re_M = \sigma \mu v_0 L_0 \tag{4.53}$$

(with σ—electrical conductivity; μ—magnetic permeability; v_0 and L_O—characteristic velocity and length) is much bigger than the unit and, the motional term must be considered.

4.8.2 FEM Solution

The FEM model is solved taking into account the rotation of permanent magnets by means of a transient magnetic solution.

The conditions used in the computation are the following:

- state variable are time dependent, i.e. $d/dt \neq 0$
- computation only of fields $\overline{B}$, $\overline{H}$ and $\overline{E}$, while $\overline{D}$ is not computed.

The 2D electromagnetic formulation is based on the following PDE:

$$\text{rot}\left(\frac{1}{\mu_r \mu_0}\text{rot}\,\overline{A} - \overline{H}_c\right) + \sigma\left(\frac{\partial \overline{A}}{\partial t} + \text{grad}\ V\right) = 0 \tag{4.54}$$

where μ_r—relative permeability; $\overline{A}$—vector potential, Wb/m; $\overline{H}_c$—coercive magnetic field of the permanent magnets, kA/m; σ—conductivity, S; V—electric scalar potential.

The constitutive equations for the magnetic materials are:

$$B = \mu H \Rightarrow \mu_r \mu_0 H \tag{4.55}$$

for soft magnetic materials, e.g. region 4 of Fig. 4.126

$$B = \mu_r \mu_0 H + B_r = \mu_r \mu_0 H + \mu_0 M \tag{4.56}$$

for hard magnetic materials, e.g. region 3 of the same figure, with B_r—remanence, T, and M—magnetization vector.

Since the displacement of the rotating part modifies the geometry of the modelled device, the computation domain must be re-meshed at each time step.

The state variables, the magnetic vector potential $\overline{A}$ and the electric scalar potential $V$are initialized to zero at $t = 0$. They are time dependant and the differential equations of first order with respect to time:

$$\frac{dX}{dt} = F(X, t) \tag{4.57}$$

are integrated by means of the Euler explicit method in the time domain. The time domain is sub-divided into small time steps Δt, in which the unknowns are supposed to vary linearly with time. The numerical code applies then an explicit method where the derivative is computed using the values of the current and the previous steps:

$$\frac{dX}{dt} = \frac{X_{i+1} - X_i}{\Delta t} \tag{4.58}$$

The precision of this method strongly depends upon the choice of the time step. Its value must be chosen taking into account the rotational speed ω as well as the number of magnetic poles. In practice it is convenient to decrease tentatively the time step until the result is not affected by this choice.

4.8.3 Analytical Solution

The analytical solution has been developed using a cylindrical coordinates system (r, φ, z) with the origin in the center of the billet. In the following formulas, the index of regions makes reference to the schematic of Fig. 4.126.

Maxwell's equations are considered neglecting displacement currents:

$$rot\,\overline{H} = \overline{J} \tag{4.59}$$

with the constitutive equation:

$$\overline{J} = \sigma\left(\overline{E} + \overline{v} \times \overline{B}\right) \tag{4.60}$$

where: $\overline{v} = \Omega \cdot r \cdot \overline{u}_{\varphi}$

Computation of Magnetization M

In the permanent magnets (region 3), the constitutive equation is:

$$\overline{B}_3 = \mu_3 \overline{H}_3 + \mu_0 \overline{M} \tag{4.61}$$

Considering magnets having in the second quadrant a linear demagnetization characteristic and remanence B_r, the amplitude of $\overline{M}$ is:

$$M = \frac{B_r}{\mu_0} \tag{4.62}$$

The direction of the vector $\overline{M}$ depends on magnetization of the magnets; in polar coordinates is:

$$\overline{M} = M_r \overline{u}_r + M_{\varphi} \overline{u}_{\varphi} \tag{4.63}$$

M_r and M_{φ} can be expressed by Fourier series:

$$M_r = \sum_{n=1,3,5,\ldots}^{\infty} M_{rn} \cos(np\varphi) \tag{4.64a}$$

$$M_{\varphi} = \sum_{n=1,3,5,\ldots}^{\infty} M_{\varphi n} \sin(np\varphi) \tag{4.64b}$$

where $2p$ is the number of poles.

Using magnets with parallel magnetization, M_{rn} and $M_{\varphi n}$ are:

$$M_{rn} = \frac{B_r}{\mu_0} \alpha_p (\xi_{1n} + \xi_{2n}) \tag{4.65a}$$

$$M_{\varphi n} = \frac{B_r}{\mu_0} \alpha_p (\xi_{1n} - \xi_{2n}) \tag{4.65b}$$

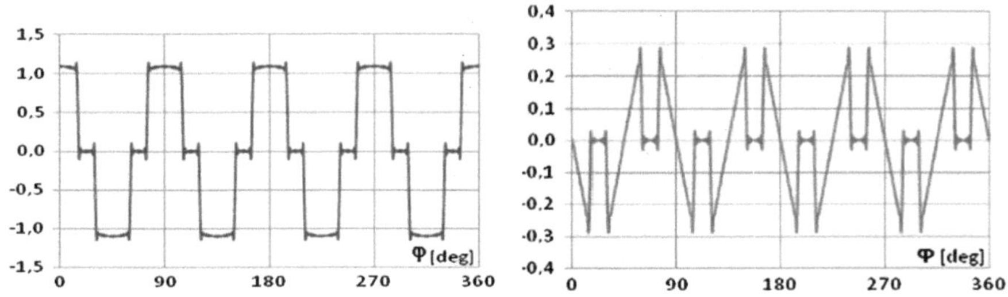

Fig. 4.127 Radial and azimuthal components M_r and M_φ of magnetization

where α_p is the magnet pole-arc to pole-pitch ratio and ξ_{1n}, ξ_{2n}, for 2p greater than 2, are:

$$\xi_{1n} = \frac{sin\left[(np+1)\alpha_p \frac{\pi}{2p}\right]}{(np+1)\alpha_p \frac{\pi}{2p}} \tag{4.66a}$$

$$\xi_{2n} = \frac{sin\left[(np-1)\alpha_p \frac{\pi}{2p}\right]}{(np-1)\alpha_p \frac{\pi}{2p}} \tag{4.66b}$$

The two components of magnetization M_r and M_φ are shown in Fig. 4.127a, b, for a 8 poles system.

General Solution: Diffusion Equation

The solution is obtained in terms of the magnetic vector potential $\overline{A}$:

$$\overline{B} = \text{rot } \overline{A} \tag{4.67}$$

$$\overline{E} = -\frac{\partial \overline{A}}{\partial t} \tag{4.68}$$

Substituting (4.56) and (4.60) in (4.59) we get:

$$rot \frac{1}{\mu} \cdot \left(\overline{B} - \overline{M}\right) = \sigma\left(\overline{E} + \overline{v} \times \overline{B}\right) \tag{4.69}$$

Substituting (4.67) in (4.69), in the hypothesis that no source current exists, is $\overline{E} = 0$ and (4.65a, b) becomes:

$$rot \frac{1}{\mu} \cdot \left(\text{rot } \overline{A} - \overline{M}\right) = \sigma\left(\overline{v} \times \text{rot } \vec{A}\right) \tag{4.70}$$

Equation 4.73 can be written as:

$$\text{rot}\left(\text{rot}\,\overline{A}\right) = \sigma \cdot \mu\left(\overline{v} \times \text{rot}\,\overline{A}\right) + rot\overline{M} \tag{4.71}$$

Considering the Coulomb gauge $\left(\text{div}\,\overline{A} = 0\right)$ we obtain:

$$\nabla^2 \overline{A} = \sigma \cdot \mu\left(-\overline{v} \times \text{rot}\,\overline{A}\right) - rot\overline{M} \tag{4.72}$$

Since the magnetic vector potential has only one component along z, Eq. (4.72) becomes:

$$\frac{\partial^2 A}{\partial r^2} + \frac{1}{r}\frac{\partial A}{\partial r} + \frac{1}{r^2}\frac{\partial^2 A}{\partial^2 \varphi} = \sigma\mu\Omega\frac{\partial A}{\partial \varphi} + \frac{1}{r}\frac{\partial M_r}{\partial \varphi} - \frac{1}{r}M_\varphi - \frac{\partial M_\varphi}{\partial r} \tag{4.73}$$

The general solution given by Eq. (4.73) can be written now for each region of the model.

General Solution Applied to Each Region of the Model

The general solution of Eq. (4.73) is:

$$A_i = \sum_{n=1,3,5,\ldots}^{\infty} A_{ni}\sin(np\varphi + \alpha) \tag{4.74a}$$

In the complex domain, Eqs. (4.74a) can be written as:

$$A_i = Im\left\{\sum_{n=1,3,5,\ldots}^{\infty} \dot{A}_{ni}e^{jnp\alpha}\right\} \tag{4.74b}$$

where: $Im\{\ldots\}$—is the imaginary part of a complex number.

By applying the method of separation of variables, from Eq. (4.73) the solution in the various regions becomes:

- *Region 1* (aluminum billet):

$$\frac{\partial^2 \dot{A}_1}{\partial r^2} + \frac{1}{r}\frac{\partial \dot{A}_1}{\partial r} - \left(p^2 + jk^2 r^2\right)\dot{A}_1 = 0 \tag{4.75}$$

with: $k^2 = \sigma_1 \mu_1 np\Omega$ and the solution is:

$$\underline{A}_1 = \sum_{n=1,3,5,\ldots}^{\infty} \left[C_{1np} J_{np}\left(j^{\frac{3}{2}} kr\right)\right] \tag{4.76}$$

with J_{np}—Bessel function of first kind and order np and $j^2 = -1$

- *Region 2* (air-gap):

$$\dot{A}_2 = \sum_{n=1,3,5,\ldots}^{\infty} \left[C_{2np} r^{np} + D_{2np} r^{-np} \right] \tag{4.77}$$

- *Region 3* (permanent magnets):

$$\dot{A}_3 = \sum_{n=1,3,5,\ldots}^{\infty} \left[C_{3np} r^{np} + D_{3np} r^{-np} + \frac{\mu_0 M_n r}{(np^2 - 1)} \right] \tag{4.78}$$

with:$M_n = npM_{rn} + M_{\varphi n}$

- *Region 4* (iron): Eq. (4.73) in the complex domain has the same terms of Eq. (4.77)
- *Region 5* (external air): the governing equation contains only the r^{-np} term.

Boundary Conditions

The integration constants $\dot{C}_{1np}$, $\dot{C}_{2np}$, $\dot{D}_{2np}$, $\dot{C}_{3np}$, $\dot{D}_{3np}$, $\dot{C}_{4np}$, $\dot{D}_{4np}$, $\dot{D}_{5np}$ are determined by the boundary conditions at the interfaces between regions:

$$\dot{A}_i\Big|_{r=r_{ij}} = \dot{A}_j\Big|_{r=r_{ij}}, \ \dot{H}_{\varphi i}\Big|_{r=r_{ij}} = \dot{H}_{\varphi j}\Big|_{r=r_{ij}} \tag{4.79}$$

where i, j are the indexes of two regions, r_{ij} is the radius at the interface.

Eddy Currents and Power Density

The induced current density in the complex domain is:

$$\dot{J}_1 = -j\frac{k^2}{\mu_1}\dot{A}_1 \tag{4.80a}$$

Thus the current density induced in the billet is:

$$J_1(r, \varphi) = \mathrm{Im}\{\dot{J}_1 e^{jnp\varphi}\} \tag{4.80b}$$

and the power density is given by:

$$w(r, \varphi) = \frac{J_1^2(r, \varphi)}{\sigma} \tag{4.81}$$

Therefore, the instantaneous value of the active power, integrated over the volume of the billet, is:

$$P = \int_0^r \int_0^{2\pi} \frac{J_1^2(r, \varphi)}{\sigma} r d\varphi dr \tag{4.82}$$

4.8.4 Calculation, Experimental Results and Design Optimization

Some comparisons between the results obtained by the analytical formulation and the ones resulting from the FEM solution are shown in Fig. 4.128 for a 8 poles system, rotating at different velocities. The main geometric dimension of the system, with reference to Fig. 4.126, are: r_1 = 100 mm; r_2 = 130 mm; r_3 = 180 mm; r_4 = 230 mm; α_p = 30/45 = 0.666 (magnet pole-arc to pole-pitch ratio); axial length = 0.5 m.

The rare earth permanent magnets have remanence B_r = 1.1 T and coercive field H_c = 838 kA/m.

Analytical and numerical method must be applied in the design of a Permanent Magnet Heater. The main parameters to be defined are number of magnetic poles, dimension of permanent magnets and thickness of air gap.

The electromagnetic design of this system is entirely different from the one used for classical induction heating systems. In this case, it is impossible to define inductor efficiency, i.e. the ratio between the power induced in the workpiece and the total power fed to the inductor clamps, since permanent magnets produce a magnetic field without energy consumption.

In order to define a parameter that quantifies the quality of the design, we have chosen as quantity to be maximized the ratio χ between induced power at a given speed (n, rpm) and weight of the permanent magnets:

$$\chi = \frac{P_{induced}(n)}{PMWeight}, \quad \left[\frac{\text{kW}}{\text{Kg}}\right] \tag{4.83a}$$

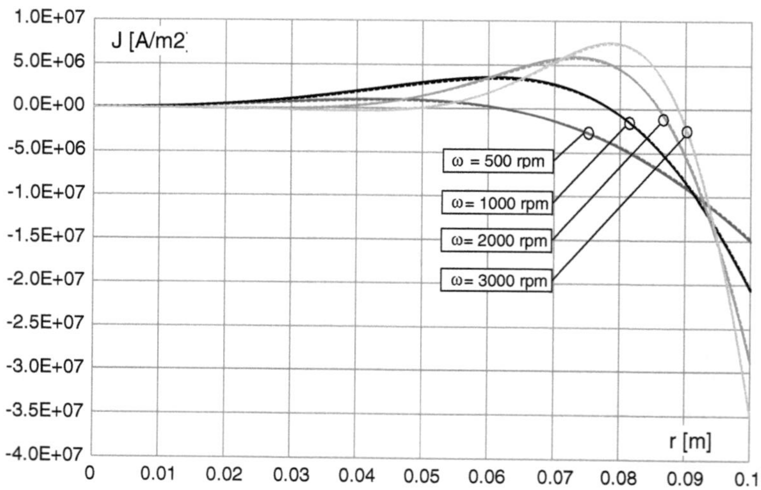

Fig. 4.128 Current density distribution along the radius, for different rotational speeds: FEM solution—*dotted lines*; analytical solution—*continuous lines*

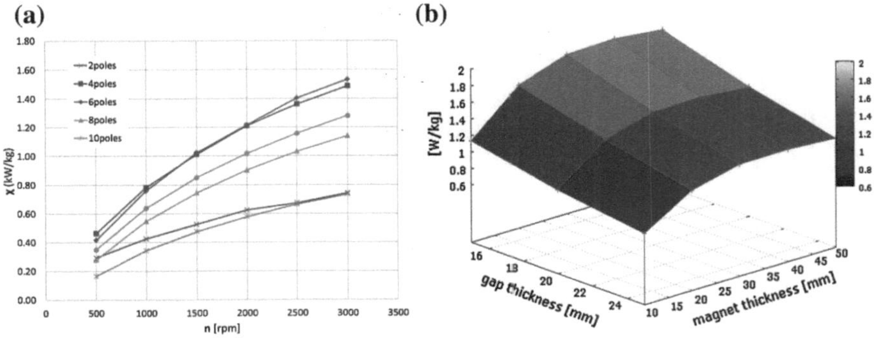

Fig. 4.129 Parametric and DOE analysis for the PMH prototype design. **a** quality parameter χ as a function of rotational speed n, for different number of poles; **b** surface response of DOE analysis of a 6 poles system for optimizing air gap and thickness of magnets

The possible solutions are limited to a finite number of cases: the number of poles can be only an even number, 2, 4, 6, 8, 10, and the azimuthal length of the magnetic poles depends upon billet diameter and minimum admissible air gap, which are chosen to fulfill mechanical constraints (e.g., minimum air gap 12 mm).

In the following, some results of the design optimization of the PMH prototype are presented. For the design optimization a DOE analysis has been carried out.

Figure 4.129 gives the values of χ, computed with different rotation speeds and number of poles (*a*) and the surface response of DOE analysis (*b*) for the optimization of the air gap and the thickness of permanent magnets. The DOE analysis has suggested to realize a 6 poles system, with air gap 12 mm and thickness of magnet 15 mm.

Experimental Results

Figure 4.130 gives the comparison of experimental and FEM numerical results.

The comparison shows a fairly good agreement: the almost constant difference between computed and measured power is explained by the power losses in the real system (mostly losses in the driving motor system and mechanical transmission), which are not taken into account in the electromagnetic transient FEM analysis.

It can be concluded that the prototype of *PMH* has demonstrated the feasibility of this innovative heating process. The measured overall efficiency is in the range 75–80 %, in accordance with the efficiency values of the electrical drive (inverter and induction motor).

Moreover, the measured data have shown that numerical or analytical models are robust and reliable tools for designing this type of heating systems.

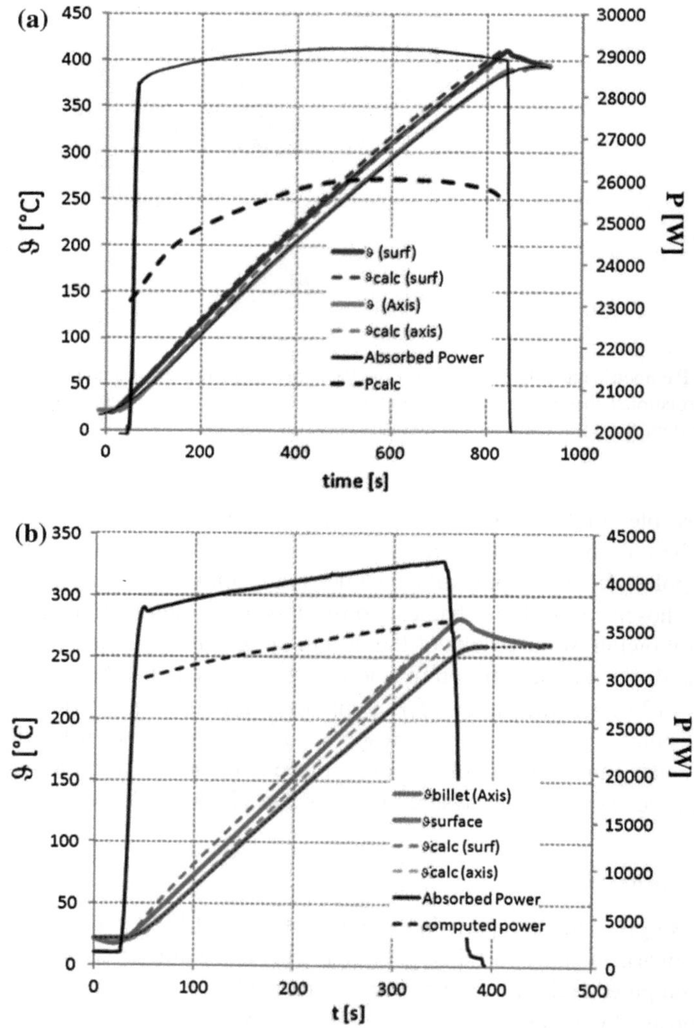

Fig. 4.130 Experimental and numerical results of active power, *W*; surface temperature, °C; axis temperature, °C, at different rotation rates: **a** 500 rpm; **b** 750 rpm *continuous lines* measurements; *dashed lines* computations

4.9 Inductors for Induction Heating of Internal Cylindrical Surfaces

Cylindrical inductors, used in induction heating of interior surfaces of hollow workpieces, are typically solenoid single layer coils with few turns made mostly of rectangular cross-section copper tube, coupled as closely as possible with the internal surface [140, 141].

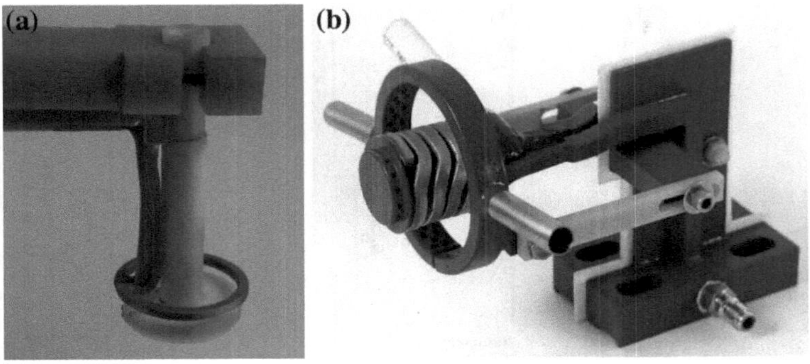

Fig. 4.131 Examples of internal cylindrical inductors **a** courtesy of SAET (Italy); **b** courtesy of CIT (USA)

The inductor external surface (i.e. its outside diameter) and the inner surface of the workpiece are separated by an air gap.

Since the return leg of the inductor usually goes through the centre of the coil, the smallest outside diameter of the inductor cannot be smaller than 20–25 mm.

Magnetic cores (flux concentrators) are often used inside the inductor coil to improve electromagnetic coupling and to shift the highest coil current density closer to the heated workpiece.

Typical examples of internal inductors with and without flux concentrators are shown in Fig. 4.131.

Because of their particular geometry, the equivalent resistance of this type of inductors is strongly influenced by the combined action of skin-, proximity- and ring-effects. In fact, it depends on system geometry, frequency of exciting current, number of inductor turns and the presence or absence of the magnetic yoke in the inner cavity.

4.9.1 Inductor Equivalent Resistance

System Geometry

The resistance of the inductor depends on the current density distribution in the cross-section of the water cooled copper tube.

The calculation of such distribution in the inductor turns has been made for the coil design shown in Fig. 4.132, in which the influence on the magnetic field of the axial current flowing in the return leg of the inductor has been neglected [142, 143].

The analysis refers to inductors with one, two or four turns, made of copper tube, 10×10 mm cross-section, wall thickness $s = 1$ mm, and axial distance between turns $t = 2.5$ mm.

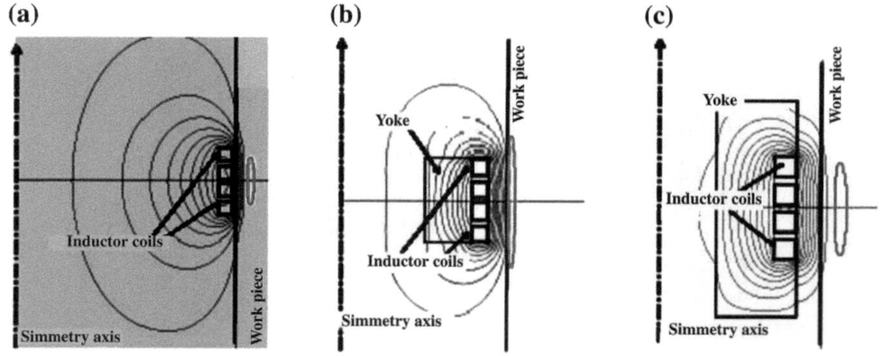

Fig. 4.132 Models of inductor-load system **a** without yoke; **b** with yoke; **c** with yoke and poles

The load is constituted by non-magnetic material with resistivity $\rho_2 = 100 \times 10^{-8}$ Ω·m. The inner radius r_i of the load was varied in the range 25–150 mm; the gap Δ between inductor and workpiece surface between 1 and 10 mm, and the frequency f of the inductor exciting current between 50 Hz and 500 kHz.

The schematic of the inductor without internal magnetic yoke and the heated workpiece is shown in Fig. 4.132a.

Other calculations have considered the presence of magnetic yokes inside the inductor, in order to understand their influence on the current density distribution in the cross-section of the inductor and its equivalent resistance. The geometric layout of magnetic yoke, inductor and workpiece system is shown in the Fig. 4.132b, c. They refer to the following data: radial thickness of magnetic yoke 25 mm, magnetic yoke material with negligible losses and constant relative magnetic permeability $\mu_r = 10$.

4.9.2 Results of Numerical Analysis

With the criteria and in the range of parameters defined in the previous section, a number of finite element numerical simulations have been done, using the software package FLUX, for evaluating the influence of design parameters on the resistance of typical inductors used in industrial installations [142, 143].

The numerical analysis refers to 2D axis-symmetric models representing a heated hollow workpiece of infinite axial length and infinite radial thickness, and an inductor placed in the inner cylindrical cavity of the workpiece.

The FEM model correctly represents the distribution of field quantities both in the thickness of the inductor tube and in the load up to a frequency of 500 kHz. Moreover, in order to describe the series connection of inductor turns, the model has been coupled with an external electrical circuit assuring the same amplitude and phase of the current flowing in each turn.

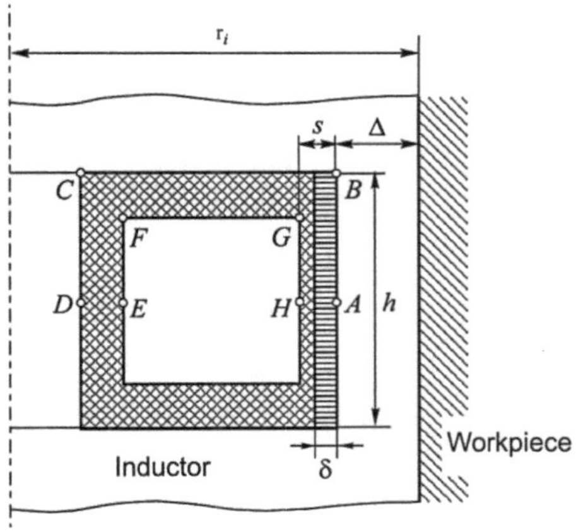

Fig. 4.133 Cross-section of an inductor turn and its position in relation to the load

The equivalent series resistance of the inductor coil is calculated as the ratio of the total active losses in the copper coil and the square of the inductor current.

The results are conveniently represented as the ratio of the AC resistance R_{ac} of the inductor tube to the DC resistance $R_{dc\delta}$ of a layer of thickness equal to the penetration depth δ_{cu} in the copper coil

$$k_R = \frac{R_{ac}}{R_{dc\delta}}, R_{dc\delta} = N \cdot \frac{\rho \pi d_e}{h \delta_{cu}} \tag{4.83b}$$

where ρ—electrical resistivity of the water cooled inductor copper (2×10^{-8} Ω·m); $d_e = 2(r_i - \Delta)$—external diameter of the inductor coil; Δ—air gap between inductor and workpiece; h—axial dimension of inductor turns (= 10 mm); N—number of inductor turns; s—wall thickness of inductor tube; $m' = s/\delta$.

Figure 4.133 represents schematically the relative positions of a single inductor turn and the hollow workpiece.

The diagrams of Fig. 4.134 show the dependence of coefficient k_R as a function of m' for inductors with one, two or four turns, gaps between inductor and load Δ = 1–2–10 mm, and two different radii r_i of the cavity of the workpiece, r_i = 25 and 50 mm.

The curves show that k_R has a minimum at a frequency of approximately 10 kHz. Moreover, for Δ = 10 mm and frequency f > 10 kHz, the coefficient k_R remains nearly constant and close to the minimum.

Figure 4.135a–d give the diagrams of the coefficient k_R as a function of m', for N = 4, with r_i = 75, 100, 125 and 150 mm, respectively.

The analysis of the diagrams allows to draw the following conclusions:

- for inductors without internal magnetic yoke (Fig. 4.132a), the value of k_R becomes smaller when the gap Δ increases. Moreover, for the smaller values of the radius r_i, when the parameter m' increases are observed lower values of the minimum of k_R;

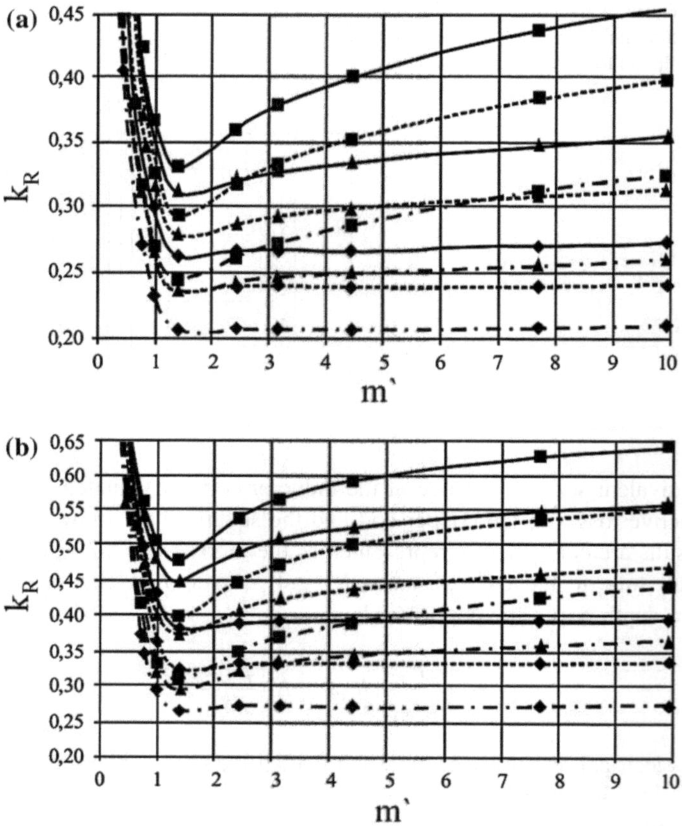

Fig. 4.134 Coefficient k_R as a function of the parameter $m' = s/\delta$ for inductors without magnetic yoke: **a** $r_i = 25$ mm; **b** $r_i = 50$ mm; *solid lines* $N = 4$; *dotted lines* $N = 2$; *dashed lines* $N = 1$; ■ − $\Delta = 1$ mm; ▲ − $\Delta = 2$ mm; ◆ − $\Delta = 10$ mm

- the presence of the internal magnetic yoke reduces the magnetic reluctance $\Re_m$ of the return path of the flux in the inner cavity of the inductor and, as a consequence, increases considerably, in comparison with an inductor without magnetic core, the induction field on the inner surface of the workpiece. These remarks are consistent with the results given in Ref. [141].

In conclusions, the characteristics of the induction system can be significantly improved with the use of magnetic yokes. This is due to the higher value of k_R, shown in Fig. 4.135 (case B), in comparison with the corresponding values of k_R in inductors without magnetic yoke (Case A).

Since the reluctance of the return circuit of magnetic flux is higher for inductors of small diameter (because of the smaller cross-section of the inner cavity of the inductor), while the magnetic impedance $\mathbb{Z}_m$ of the "active zone" (workpiece and air gap between inductor and load) remains almost unchanged, the improvements

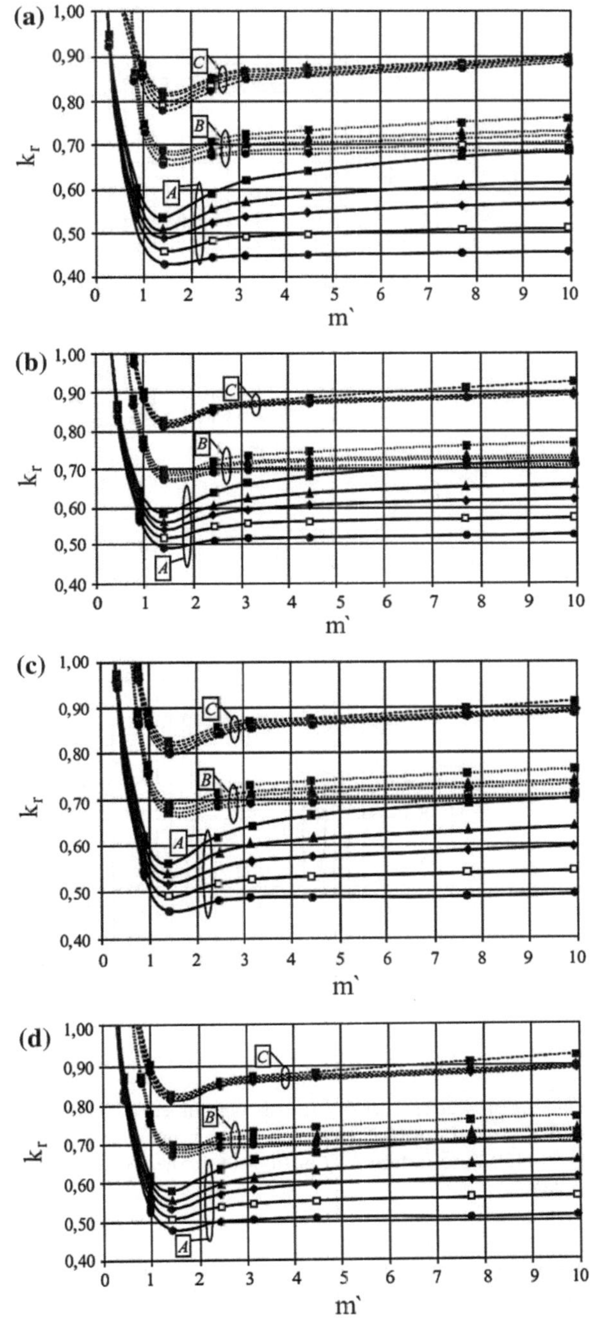

Fig. 4.135 **a**, **b**—Coefficient k_R for internal inductors with four-turns ($N = 4$): **a** $r_i = 75$ mm; **b** $r_i = 100$ mm; *A*—without magnetic core; *B*—with rectilinear magnetic core (as in Fig. 4.132b); *C*—with C-shaped magnetic core (as in Fig. 4.132-c); ■ − $\Delta = 1$ mm; ▲ − $\Delta = 2$ mm; ◆ − $\Delta = 3$ mm; □ − 5 mm; ● − $\Delta = 10$ mm. **c**, **d**—Coefficient k_R for internal inductors of a four-turns ($N = 4$): **c** $r_i = 125$ mm; **d** $r_i = 150$ mm; *A*—without magnetic core; *B*—with rectilinear magnetic core (as in Fig. 4.132b); *C*—with *C*-shaped magnetic core (as in Fig. 4.132c); ■ − $\Delta = 1$ mm; ▲ − $\Delta = 2$ mm; ◆ − $\Delta = 3$ mm; □ − 5 mm; ● − $\Delta = 10$ mm

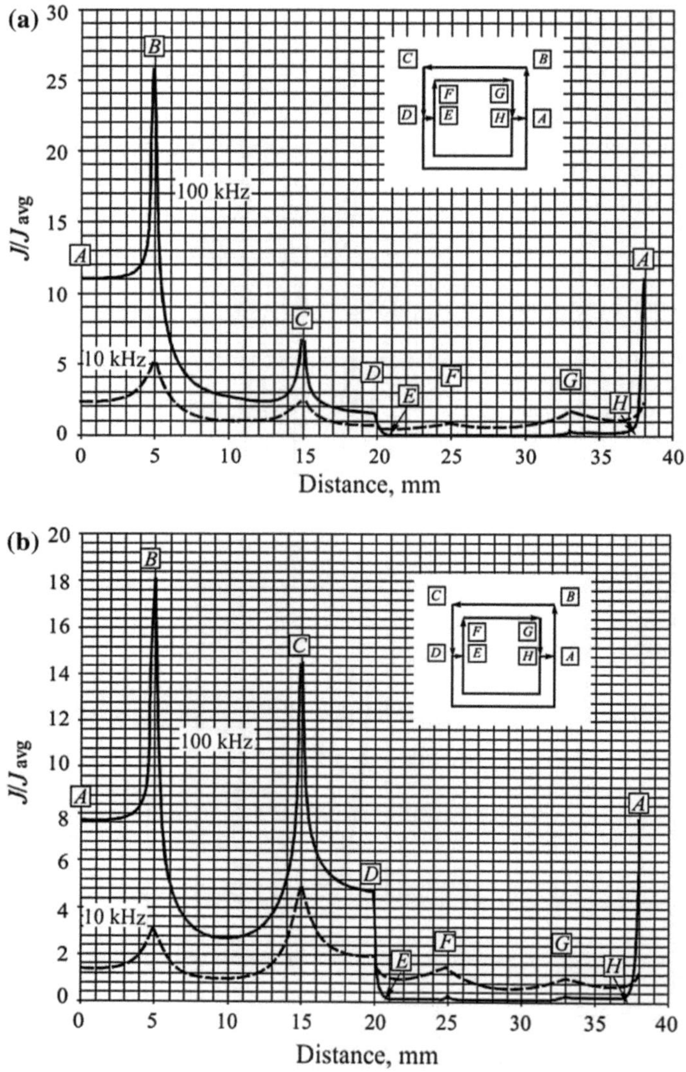

Fig. 4.136 Current density distribution in cross-section of an inductor turn, along the line *A–B–C–D–E–F–G–H–A*, with gap between inductor and workpiece $\Delta = 1$ mm: **a** $r_i = 25$ mm; **b** $r_i = 150$ mm; surface *A–B* in front of the heated workpiece; surface *C–D* towards the axis of system

due to the use of a magnetic yoke are more relevant for inductors of small diameter. The positive influence of the magnetic yoke is more pronounced for high values of the gap Δ.

For inductors with a C-shaped magnetic circuit, as in Fig. 4.132c, the coefficient k_R is much greater than the corresponding values with simple rectilinear magnetic core or without core. At the same time, the reluctance of the return path of the magnetic flux decreases. Moreover, with C-shaped magnetic cores, the effect of variations of the gap Δ and the radius r_i is relatively weak.

This behaviour can be better explained by the diagrams of Fig. 4.136 which show the distribution of current density on the cross-section of the inductor's conductor.

They show that the increase of frequency leads to more pronounced skin, ring and proximity (between inductor and workpiece) effects, which produces a more uneven distribution of the current density and a consequent increase of the AC resistance (see Fig. 4.136a).

For the same frequency and same gap between inductor and workpiece, bigger inductor radii lead to a growing influence of the proximity effect in comparison with the ring-effect. Therefore, with the increase of radius, a more uniform distribution of the current density in the cross-section occurs giving—as a consequence—lower values of the equivalent resistance.

4.9.3 Conclusions

In this paragraph the influence on the equivalent resistance of internal inductors with or without magnetic concentrators has been illustrated.

The analysis has shown the strong influence on the AC resistance of the skin-, proximity- and ring-effects for a wide range of geometrical parameters of inductors used in industrial processes.

These results give to the designer of inductors useful information about the advantages of the use of magnetic flux concentrators for the induction hardening of inner surfaces of hollow workpieces.

4.10 Electromagnetic Forces

4.10.1 Introduction

In induction heating applications the interaction of the exciting magnetic field, produced by the inductor coils, and the currents induced in the various components of the inductor-load system, including magnetic yokes and metallic structures, gives rise to electro-dynamic forces acting on such elements. These forces can produce mechanical vibrations, body deformations or displacement of different solid parts of the system and, when acting on molten metals, like in induction melting furnaces, they are responsible of the flow pattern and the shape of the free surface of the melt [144–147].

In this chapter reference is made to a time harmonic excitation of the inductor coils, which is the case of most induction heating applications. We suppose therefore that the coil current has the form:

$$i(t) = I_m \cos(\omega t + \varphi)$$

with: I_m—peak value of coil current, A; ω—angular frequency, rad/sec; $\omega = 2\pi f$; φ—phase angle, rad.

With this assumption and in the hypothesis of linear materials, all field values X have the form:

$$\dot{\bar{X}}(\dot{\bar{r}}, t) = \Re e\left\{\dot{\bar{X}}(\dot{\bar{r}})e^{j\omega t}\right\}$$

There are different methods for the evaluation of forces acting on conductors placed in a magnetic field. Here the most used are described and typical results are given. For a more general treatment of this topic see Ref. [140].

4.10.2 *Evaluation by Lorentz's Law of Forces in "Long" Cylindrical Systems*

The first method, useful for systems without ferromagnetic conductors, is based on the Lorentz's law which states that the specific force per unit volume acting on a conductor is given by the vector product of the current density $\bar{J}(t)$ and the magnetic flux density $\bar{H}$:

$$\bar{f}_v(t) = \bar{J}(t) \times \bar{B}(t) = \mu_0(\bar{J}(t) \times \bar{H}(t)) \tag{4.84}$$

with $\bar{B}(t) = \mu_0 \bar{H}(t)$.

The total force on a conductor of volume v is therefore:

$$\dot{\bar{F}}_v = \int_v (\bar{J}(t) \times \bar{B}(t))dv = \mu_0 \int_v (\bar{J}(t) \times \bar{B}(t))dv \tag{4.85}$$

For sake of simplicity, it is assumed that the current density vector is directed along the x axis while the induction field along the y axis, so Lorentz's force is a vector directed along z axis, since $\bar{u}_x \times \bar{u}_y = \bar{u}_z$. In the case of sinusoidal time variations of field quantities, we can write:

$$\begin{cases} \dot{J}_x = \sqrt{2} J_x e^{j(\omega t + \phi_J)} \cdot \bar{\mathrm{u}}_\mathrm{x} \\ \dot{B}_y = \mu_0 \sqrt{2} H_y e^{j(\omega t + \phi_H)} \cdot \bar{\mathrm{u}}_\mathrm{y} \end{cases} \tag{4.86}$$

and the specific force:

$$f_v \bar{\mathrm{u}}_\mathrm{Z} = \begin{bmatrix} \mu_0 J_x \, \mathrm{H_y} \cos(\phi_H - \phi_J) \\ +\mu_0 J_x \, \mathrm{H_y} \cos(2\omega t + \phi_H + \phi_J) \end{bmatrix} \cdot \bar{\mathrm{u}}_\mathrm{Z} \tag{4.87}$$

The force is therefore characterised by a constant component, independent on time, and a pulsating component oscillating with double frequency.

The constant component is responsible for possible deformations and/or displacements of solid system components (in particular the workpiece) or for a conservative magnetic pressure and a non-conservative vector field of forces able to perform work around a closed path and to cause EM stirring in a molten charge.

The oscillating component at double frequency can give rise to vibration and noise of the solid metallic parts of the installation. The noise can be so strong that, in some cases, has even prevented some induction heating applications.

In liquid metals, since the mechanical time constant is much greater than the period of the exciting field, the molten charge cannot respond to instantaneous forces, except at very low frequency (typically <3 Hz) or in case of low density materials.

In particular, taking into account the relationship $rot\,\dot{\overline{H}} = \nabla \times \dot{\overline{H}} = \dot{\overline{J}}$, the first term of Eq. (4.87) can be written as follows:

$$\overline{f}_m = \mu_0 \Re e\left(\dot{\overline{J}} \times \overline{H}^*\right) = -\frac{\mu_0}{2} \nabla \dot{\overline{H}}\,\overline{H}^* + \frac{\mu_0}{2} \Re e\left(\dot{\overline{H}}\,\nabla\right)\overline{H}^* \tag{4.88}$$

with ∇—Nabla operator.

Here again we can recognize that the first term is a conservative one, opposite to the magnetic pressure $p_m = \frac{\mu_0}{2}\dot{\overline{H}}\,\overline{H}^*$, which produces a confinement force counteracting gravity (e.g. in cold crucible furnaces), while the second term, i.e. the curl of $\dot{\overline{f}}_m$,

$$\nabla \times \overline{f}_m = \frac{\mu_0}{2} \Re e\left(\nabla \times \left(\dot{\overline{H}}\nabla\right)\overline{H}^*\right)$$

provides a measure of the ability of the Lorentz's force field to perform work (e.g. to stir a molten metal).

This method has been applied to the ideal 1D cases of conduction and induction heating of rectilinear molten conductors of infinite length and circular cross-section [146–148].

These cases describe in a very simplified manner forces and pressures acting in the channel of a channel induction furnace or in coreless crucible induction furnaces.

The main results derived for the above two cases are summarised in the following.

(a) *Forces and pressure due to a current flowing in an infinitely long cylindrical conductor*

For the conducting cylinder of Fig. 4.137, in which a sinusoidal current flows, with reference to an azimuthal coordinate system (r, φ, z), the distributions of the magnetic field intensity $\dot{\bar{H}} = H(r) \cdot \dot{\bar{u}}_\phi$ and the current density $\dot{\bar{J}} = J(r) \cdot \dot{\bar{u}}_z$ in the conductor can be expressed in terms of Bessel functions as follows (see Sect. 3.1):

$$\frac{H(r)}{H(r_e)} = \frac{J_1(\sqrt{-j}m\xi)}{J_1(\sqrt{-j}m)} = \frac{ber'\, m\xi + jbei' m\xi}{ber'\, m + jbei'\, m} \tag{4.89}$$

$$\begin{aligned} J(r) &= \sqrt{-j}\frac{m}{2}\left(\frac{I}{\pi r_e^2}\right)\frac{J_0(\sqrt{-j}m\xi)}{J_1(\sqrt{-j}m)} \\ &= -H(r_e)\frac{m}{r_e}\frac{bei\, m\xi - jber m\xi}{ber'\, m + jbei'\, m} \end{aligned} \tag{4.90}$$

with: $\xi = r/r_e$; $m = \sqrt{2}r_e/\delta$; $\delta = \sqrt{2\rho/\omega\mu\mu_0}$; $J_1(u)$—first derivative of the Bessel function $J_0(u)$.

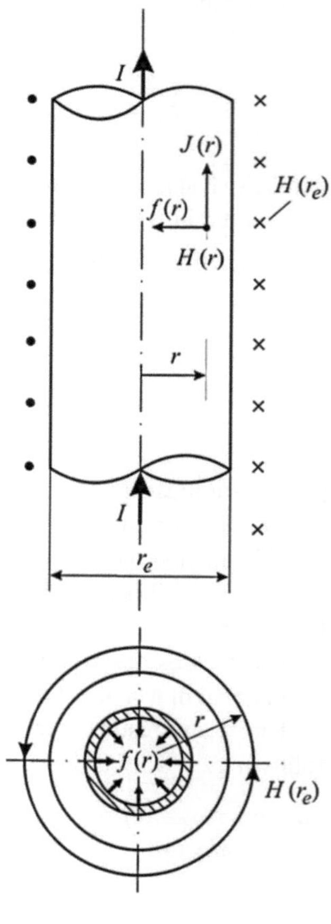

Fig. 4.137 Rectilinear conductor of infinite length and circular cross-section

From Eqs. (4.88) to (4.90) the force per unit volume in the cylinder $\dot{\bar{f}}_m = f(r) \cdot \dot{\bar{u}}_r$ can be written as follows:

$$f(r) = -\frac{\mu_0}{2}|H(r_e)|^2 \frac{2\,m}{r_e} \frac{ber m\xi \cdot bei'\, m\xi - bei\, m\xi \cdot ber'\, m\xi}{(ber'\, m)^2 + (bei'\, m)^2} \quad (4.91a)$$

The maximum force density on the external surface layer of the cylinder (i.e. for $\xi = 1$) is therefore:

$$f(r_e) = -\frac{\mu_0}{2r_e}|H(r_e)|^2 \Psi(m) \quad (4.91b)$$

with:

$$\Psi(m) = 2m \cdot \frac{ber\, m \cdot bei'\, m - bei\, m \cdot ber'\, m}{(ber'\, m)^2 + (bei'\, m)^2} \quad (4.92a)$$

The ratio:

$$\frac{f(r)}{f(r_e)} = \frac{ber\, m\xi \cdot bei'\, m\xi - bei\, m\xi \cdot ber'\, m\xi}{ber\, m \cdot bei'\, m - bei\, m \cdot ber'\, m} \quad (4.92b)$$

is given, as a function of m, by the curves on the right in Fig. 4.138.

By integration of the force density, the time average value of the electrodynamic pressure at radius r can be obtained by the following relationship:

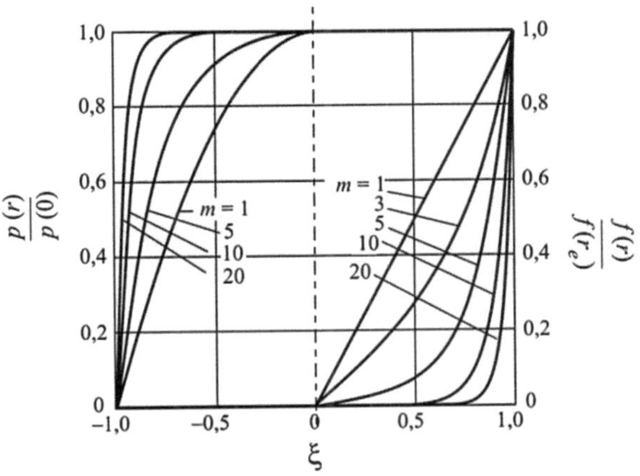

Fig. 4.138 Relative radial distributions of force density (*right*) and pressure (*left*) as a function of $\xi = r/r_e$, for different values of $m = \sqrt{2}\, r_e/\delta$

$$p(r) = \int_r^{r_e} dp = \mu_0 \int_r^{r_e} \Re e\left(\dot{\bar{J}} \times \bar{H}^*\right) dr$$
$$= \mu_0 H(r_e)^2 \frac{\int_\xi^1 (ber m\xi \cdot bei' m\xi - bei\, m\xi \cdot ber' m\xi) d\xi}{(ber' m)^2 + (bei' m)^2} \qquad (4.93a)$$

The maximum value of pressure, which takes place on the axis of the cylinder, can be therefore written in the form:

$$p(0) = \mu_0 H(r_e)^2 \frac{\int_0^1 (ber\, m\xi \cdot bei' m\xi - bei\, m\xi \cdot ber' m\xi) d\xi}{(ber' m)^2 + (bei' m)^2} = \qquad (4.93b)$$
$$= \mu_0 H(r_e)^2 \cdot \Phi(m)$$

with the pressure factor $\Phi(m)$ given, as a function of m, in Fig. 4.139.

The curve of $\Phi(m)$ shows that the maximum pressure occurs with direct current, for $m = 1$, where $\Phi(m) = 1$. Increasing the frequency, or 'm', the pressure continuously reduces towards a limit at very high frequency, where $\Phi(m) \to 1/2$.

The curves on the left in the diagram of Fig. 4.138 give, for different value of m, the radial distribution of the ratio:

$$\frac{p(r)}{p(0)} = 1 - \frac{\int_0^\xi (ber\, m\xi bei' m\xi - bei\, m\xi ber' m\xi) d\xi}{\int_0^1 (ber\, m\xi bei' m\xi - bei\, m\xi ber' m\xi) d\xi} \qquad (4.94)$$

It's easy to express the above equations as a function of the current flowing in the conductor, using the relationship $I = 2\pi r_e \cdot H(r_e)$.

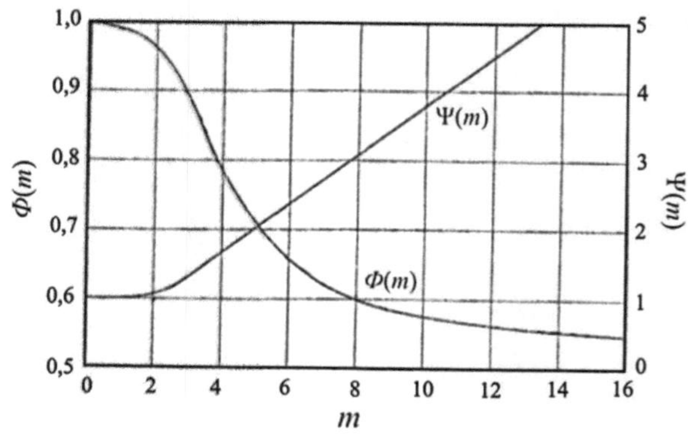

Fig. 4.139 Force density factor $\Psi(m)$ and pressure factor $\Phi(m)$ as a function of $m = \sqrt{2} r_e/\delta$

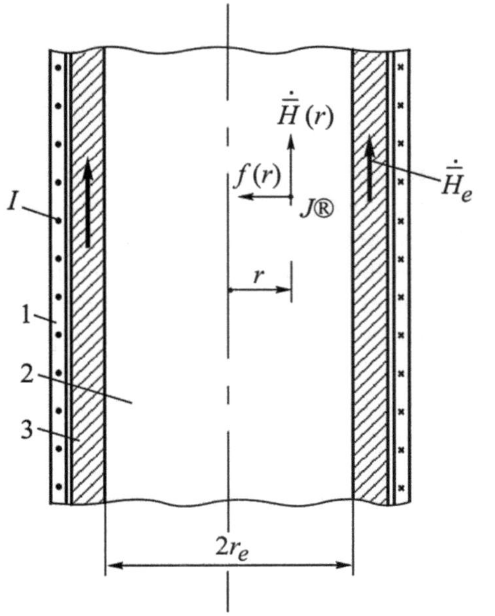

Fig. 4.140 Ideal infinitely long cylindrical induction heating system (*1* inductor; *2* charge; *3* air gap; *I* exciting current)

(b) *Forces and pressure in induction heating of an infinitely "long" cylindrical conductor*

We make reference to the system of Fig. 4.140, which represents a part of an ideal infinitely long induction heating system with an inductor supplied by sinusoidal current I. The distributions of the magnetic field intensity $\dot{\bar{H}} = H(r) \cdot \dot{\bar{u}}_z$ and the induced current density in the charge $\dot{\bar{J}} = J(r) \cdot \dot{\bar{u}}_\phi$ are given by (see Sect. 3.1):

$$\begin{cases} H = H_e \dfrac{J_0(\sqrt{-j}m\xi)}{J_0(\sqrt{-j}m)} = H_e \dfrac{ber\, m\xi + jbei m\xi}{ber\, m + j\, bei\, m} \\ J = \sqrt{-j}H_e \dfrac{\sqrt{2}}{\delta} \cdot \dfrac{J_1(\sqrt{-j}m\xi)}{J_0(\sqrt{-j}\, m)} = \\ \quad = -H_e \dfrac{\sqrt{2}}{\delta} \cdot \dfrac{ber'\, m\xi + jbei' m\xi}{ber\, m + j\, bei\, m} \end{cases} \tag{4.95}$$

From the above equations, the force per unit volume:

$$f = \frac{df}{dv} = \dot{\bar{J}} \times \dot{\bar{B}} = \mu_0 \left[\dot{\bar{J}} \times \overset{*}{\bar{H}} \right], \; [\mathrm{N/m^3}]$$

can be written as follows:

$$f(\xi) = -\frac{\mu_0}{\rho} p_e \frac{ber\, m\xi \cdot ber'\, m\xi + bei\, m\xi \cdot bei'\, m\xi}{ber\, m \cdot ber'\, m + bei\, m \cdot bei'\, m} \tag{4.96a}$$

where: $p_e = H_e^2 \frac{\rho}{\delta} F_{cyl}$—specific induced power per unit external surface of the cylinder [w/m^2]; F_{cyl}—coefficient of the active power induced in the cylinder defined in the Sect. 3.1 (see Fig. 3.5).

This force is directed radially towards the axis of the cylinder and varies from the maximum value on the surface (for $\xi = 1$) to zero on the axis ($\xi = 0$); moreover, as shown by Eq. (4.96a), the force is proportional to the surface specific induced power.

From Eq. (4.96a), it's easy to derive the ratio of the force at the coordinate ξ to the maximum surface value:

$$\frac{f(\xi)}{f(1)} = \frac{ber\, m\xi \cdot ber'\, m\xi + bei\, m\xi \cdot bei' m\xi}{ber\, m \cdot ber'\, m + bei\, m \cdot bei'\, m} \tag{4.96b}$$

which is given, as a function of ξ and for different values of m, by the curves on the right in the diagram of Fig. 4.141.

The pinching pressure caused by these forces increases from the surface towards the axis according to the equation:

$$p(\xi) = \frac{1}{2} \mu_0 H_e^2 \left[1 - \frac{ber^2(m\xi) + bei^2\,(m\xi)}{ber^2\,(m) + bei^2\,(m)} \right] \tag{4.97a}$$

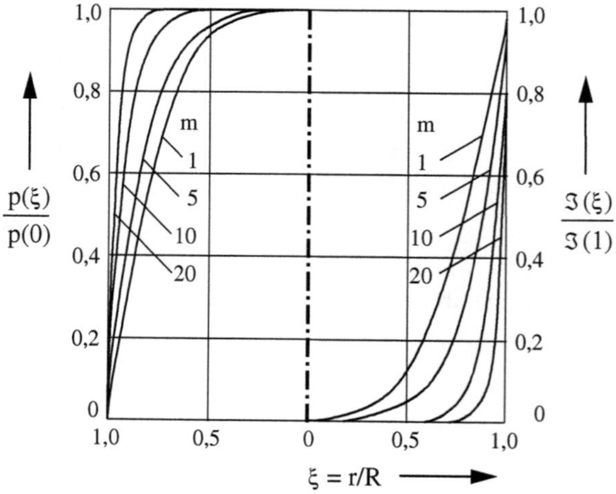

Fig. 4.141 Relative radial distributions of force per unit volume and pressure in a cylinder of infinite axial length

Its maximum value for $\xi = 0$, i.e. on the axis of the cylinder, is given by:

$$\begin{aligned} p(0) &= \frac{1}{2}\mu_0 H_e^2\left[1 - \frac{1}{ber^2(m) + bei^2(m)}\right] \\ &= \frac{\mu_0}{\rho}\frac{\delta}{2} p_e Ж(m) = 3.16 \cdot 10^{-4}\frac{p_e}{\sqrt{\rho f}} Ж(m) \end{aligned} \quad (4.97b)$$

with:

$$Ж(m) = \frac{ber^2(m) + bei^2(m) - 1}{\sqrt{2}[ber\, m \cdot ber'\, m + bei\, m \cdot bei'\, m]}$$

In the ideal case of an infinitely long inductor-load system the pressure acts to compress radially the load, with constant intensity along z-axis.

The values of the ratios $p(\xi)/p(0)$ and the pressure coefficient ж(m) are given as a function of ξ and m by the curves of the Figs. 4.141 (left) and 4.142 respectively.

The above equations can be used for a rough estimation of the meniscus which characterises the melt in coreless induction furnaces.

With reference to the sketch of a crucible induction furnace of Fig. 4.143, taking into consideration only the time averaged radial component of forces, the height of meniscus h_m can be approximately evaluated by the equilibrium on the axis of the electromagnetic pressure $p(0)$ and the hydrostatic pressure of the molten metal, using the equation:

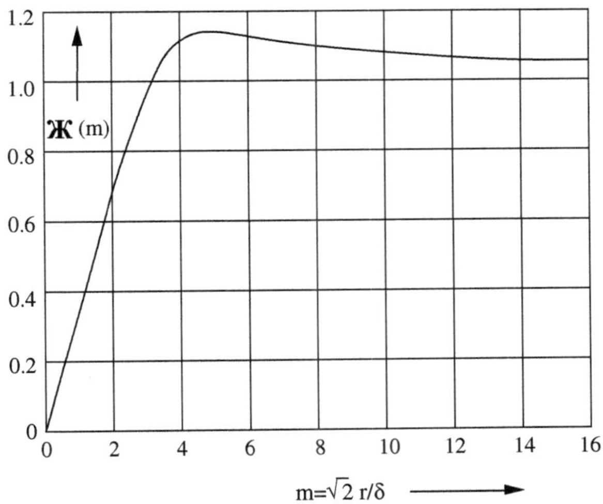

Fig. 4.142 Values of the pressure coefficient ж(m) as a function of m

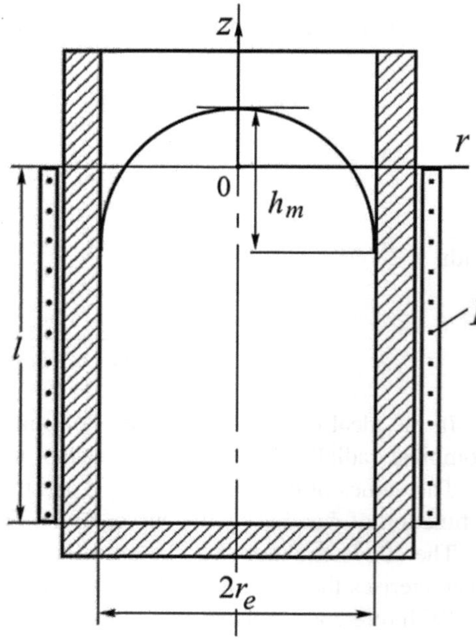

Fig. 4.143 Heigth h_m of meniscus in the molten bath of a crucible induction furnace

$$\begin{aligned} h_m &\approx \frac{p(0)}{g\gamma} = \frac{3.16 \times 10^{-4}}{9.81} \frac{p_e}{\gamma\sqrt{\rho f}} ж(m) \\ &= 0.32 \times 10^{-4} \frac{p_e}{\gamma\sqrt{\rho f}} ж(m) \end{aligned} \tag{4.98}$$

where: γ—molten metal density; ρ—resistivity; f—frequency.

An analytical solution obtained with the same procedure, is available in Ref. [149] also for the ideal case of the induction heating of an infinite long cylinder by means of poly-phase inductors.

4.10.3 Forces in the Induction Heating of Disk Plates with Planar Circular Coils

We refer here to the inductor-load system shown in Fig. 4.144, which is constituted by a circular planar inductor coil with N concentric turns series connected and excited by sinusoidal current I, and by a load unbounded in radial direction, consisting of two homogeneous metal layers of constant resistivity and permeability, placed parallel in front of the inductor [78, 150].

The equations of the vector potential of the magnetic field in the different regions of the system have been derived in Sect. 4.4.

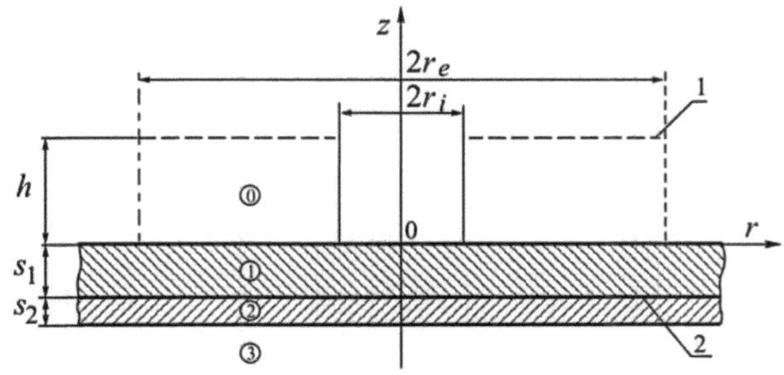

Fig. 4.144 Schematic diagram of the planar inductor-load system (*1* inductor; *2* two layer metal load)

In particular, Eq. (4.24) allows the determination of the electromagnetic field quantities for $z \geq 0$ and, therefore, also at z = h, where are located the idealized inductor turns.

Taking into account that for the axial symmetry of the system the potential vector contains only the azimuthal component, the radial component of the flux density in the points where the inductor is located can be expressed as follows [151]:

$$\dot{B}_r(r,h) = \frac{\mu_0 N\dot{I}}{2(r_e - r_i)} \int\limits_{k=0}^{\infty} kC_0(k)F(k)J_1(kr)e^{-2kh}dk \tag{4.99}$$

with $C_0(k)$ and $F(k)$ given by Eqs. (4.24a) and (4.27a) of Sect. 4.4.

From the above the expression of the time average value of the force acting in z direction on an elementary turn of the inductor of radial thickness dr, located at radius r can be written as:

$$dF_z(r,h) = -\Re_e\left[\frac{N\dot{I}}{(r_e - r_i)}B_r^*(r,h)2\pi r dr\right] \tag{4.100}$$

Substituting from Eq. (4.99) the value of $B_r^*(r,h)$, the pressure p_{zm} corresponding to this force can be written as follows:

$$p_{zm}(r,h) = \frac{1}{2}\mu_0\left(\frac{N\dot{I}}{r_e - r_i}\right)^2 \Re_e\left[\int\limits_{k=0}^{\infty} kC_0(k)F^*(k)J_1(kr)e^{-2kh}dk\right] \tag{4.101}$$

The total average force acting on the coil can be obtained by integration of Eq. (4.100) along radial length:

$$F_z = \int_{r=r_i}^{r_e} dF_z = -\pi\mu_0\left(\frac{N\dot{I}}{r_e - r_i}\right)^2 \Re_e\left[\int_{k=0}^{\infty} k[C_0(k)]^2 F^*(k)e^{-2kh}dk\right] \tag{4.102}$$

As already stated in Sect. 4.4, the most delicate problem for the evaluation of pressure and force is the numerical solution of the integrals appearing in Eqs. (4.101) and (4.102). The solution has been obtained using the Gauss-Laguerre quadrature formulae, as described in Ref. [152].

Some results obtained for the heating of ferromagnetic or bimetallic plates are given below as examples. Other data for this application can be found in Ref. [152].

For ease of analysis and bearing in mind the possibility of applying the electromagnetic modelling laws [see Ref. 152], the results are presented in normalized form by the use of the following coefficients:

$$p_0 = \frac{p_{zm}}{\mu_0 (NI)^2}, \quad f_0 = \frac{F_z}{\mu_0 (NI)^2}$$

which give the values of pressure (in N/m^2) and force (in N) when the currents are expressed in amperes.

In particular, the graphs of Figs. 4.145 and 4.146 give the values of f_0 as a function of the ratio $s_1/\delta_1$for an inductor of 200 mm diameter and load characteristics and distances between inductor and load specified in the captions.

The curves show the influence of the magnetic permeability (Figs. 4.145 and 4.146), the distance between inductor and load (Fig. 4.147) and the frequency.

The corresponding radial distributions of pressure are shown in the graphs of Fig. 4.148a, b and in Fig. 4.149 for different values of s_1/δ_1.

It's interesting to observe that for ferromagnetic plates with constant value of permeability, when the ratio s_1/δ_1 increases the force changes from attractive to repulsive and, for a given value of s_1/δ_1, the force is repulsive for low permeability values while it becomes attractive at high permeability values.

This can be explained by the opposite effects of the induced currents and the material magnetisation: the first one tends to expel the flux lines from the load, the second one to concentrate them inside it.

Fig. 4.145 Coefficient f_0 of force between the inductor and a single layer ferromagnetic plate, as a function of s_1/δ_1, for various values of relative permeability (r_e = 100 mm; r_i = 0; h = 4 mm; s_I = 4 mm; ρ_I = 60 μΩ·cm; μ_{rI} = 1–5–10–50–100)

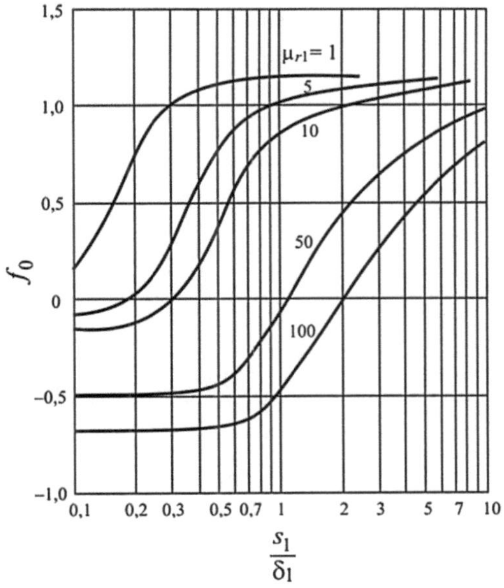

Fig. 4.146 Coefficient f_0 of force between the inductor and a bimetallic plate, as a function of s_1/δ_1, for various values of the relative permeability of the ferromagnetic layer (r_e = 100 mm; r_I = 0; h = 4 mm; s_I = 1 mm; ρ_I = 60 μΩ·cm; μ_{rI} = 1–5–10–50–100; s_2 = 1 mm; ρ_2 = 4 μΩ·cm; μ_{r2} = 1)

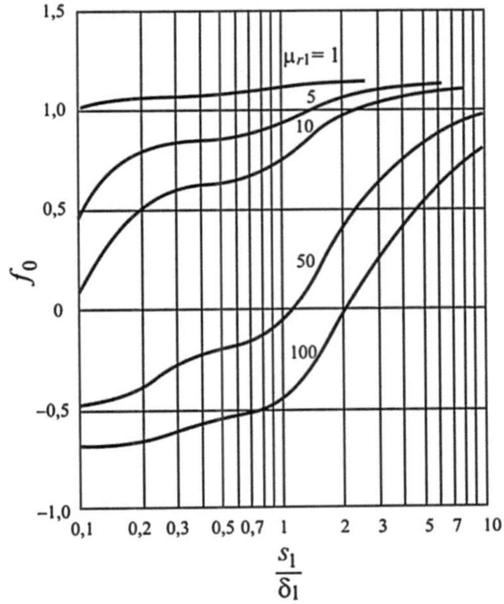

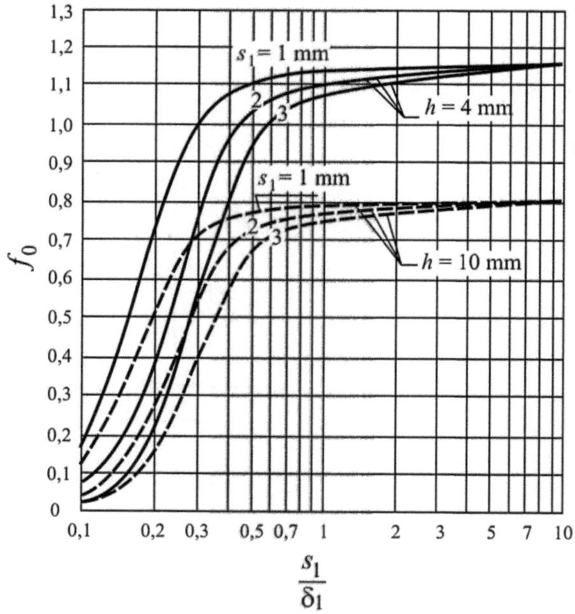

Fig. 4.147 Coefficient f_0 of force between inductor and a non-magnetic bimetallic plate, as a function of s_1/δ_1, for various values of s_1 and h ($r_e = 100$ mm; $r_i = 0$; $h = 4$–10 mm; $s_1 = 1$–2–3 mm; $\rho_1 = 4$ μΩ·cm; $\mu_{r1} = 1$; $s_2 = 1$ mm; $\rho_2 = 120$ μΩ·cm; $\mu_{r2} = 1$)

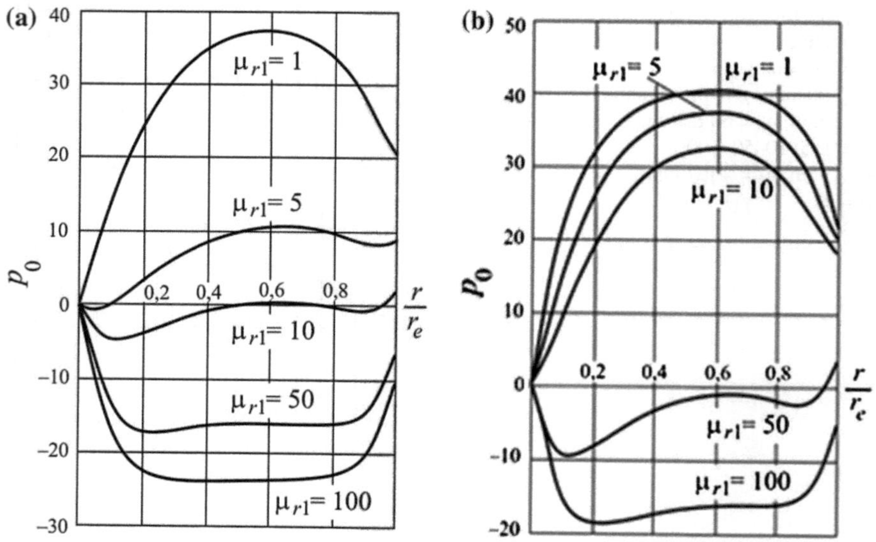

Fig. 4.148 Coefficient p_0 of the pressure acting on the coil in case of ferromagnetic plates, as a function of the radial position, for various values of relative permeability (inductor-load system: as in Fig. 4.145; **a** $s_1/\delta_1 = 0.3$; **b** $s_1/\delta_1 = 1$

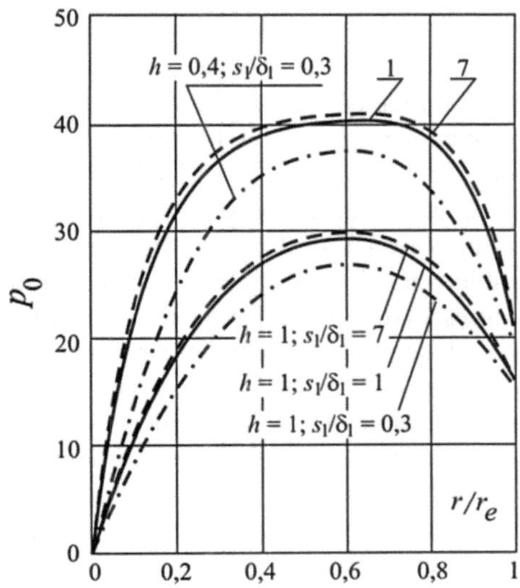

Fig. 4.149 Coefficient p_O of pressure acting on the coil in case of bimetallic non-magnetic plates, as a function of the radial position, for values s_1/δ_1 =0.3–1.0–7.0 and distances h = 4–10 mm r_e = 100 mm; r_I = 0; s_I = 1 mm; ρ_I = 4 μΩ·cm; μ_{rI} = 1; s_2 = 1 mm; ρ_2 = 120 μΩ·cm; μ_{r2} = 1

4.10.4 Evaluation of Forces by Variation of the Field Energy

The second method for the evaluation of forces is based on the principle of virtual works, that here is solved by computing the variation of reactive power Q corresponding to a 'small' displacement of a part of the system. It is the most convenient method for the evaluation of the time mean value of the force; it is also the most used by numerical software packages.

The component of the force along a general coordinate is obtained by taking the derivative of the field energy with respect to this coordinate, keeping constant the current in the inductor, i.e.:

$$F_{mz} = \frac{1}{2\omega}\frac{dQ}{dz} = \frac{I^2}{2}\frac{dL}{dz} \tag{4.103}$$

Since the sum of the forces acting on the system must be zero, the above equation gives either the force acting on the load or (by changing its sign) that on the inductor.

In a linear system comprising n circuits carrying sinusoidal currents at the same frequency, the following expression is therefore valid for the component along the z-axis of the average force acting on the i-th circuit:

$$F_{mi(z)} = \Re_e\left[\dot{I}_i \sum_{k=1}^{n} I_k^* \frac{dM_{ik}}{dz}\right] = \Re_e\left[\dot{I}_i \sum_{k=1}^{n} I_k^* f_{ik}\right] \tag{4.104}$$

The mutual coupled circuits method, which is described in the following at Sect. 5.7, is particularly convenient also for the calculation of forces since it gives as solution the currents $\dot{I}_i$ and I_k^* which appear in Eq. (4.104) and also allows the use of common procedures for the calculation of inductances and forces, with consequent advantages for developing programs and reducing calculation times.

In particular, in the implementation of the mutually coupled circuits method, the force between ideal circuits like circular coaxial filaments, circular coils of negligible radial thickness, etc., can be calculated exactly by simple differentiation of the corresponding formulae for the mutual inductances, expressed as a function of the coordinate along which the component of the force is to be determined [151].

However, the problem becomes more complex, when the dimensions of the cross-section of the subdivision elements are non-negligible compared with the distances between elements and from the elements to the axis of the configuration.

In these cases, or when a high calculation precision is required, it's no longer possible to replace each element by a single filament situated at the geometric centre of the cross-section, and it is necessary to use approximate and more sophisticated solutions [153, 154].

The mutually coupled circuit method, described in Sect. 5.7, has been applied to several practical cases. Some results are shown in the following examples.

4.10.5 *Forces in Induction Heating of the Ends of Non-magnetic Bars*

This case refers to the medium frequency heating of the ends of non-magnetic cylindrical bars or tubes, as shown in the sketch of Fig. 4.150 [155].

The values of the coefficient F_0 given in Fig. 4.151 allow to evaluate the forces acting on the inductor, as a function of the ratio between the bar diameter d and the penetration depth δ of the induced currents, for various positions of the load inside the inductor.

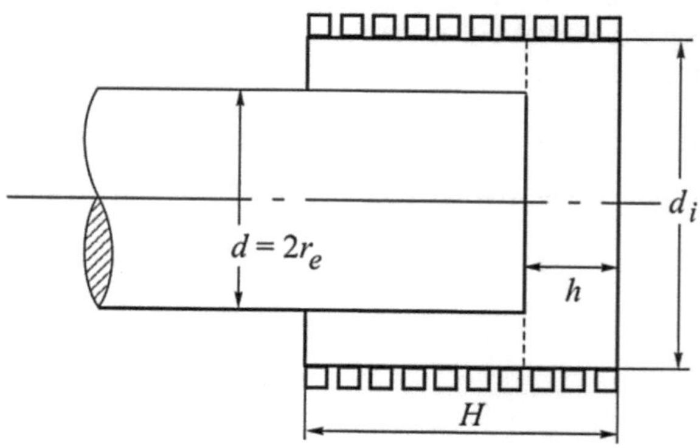

Fig. 4.150 Inductor for heating the end of bars

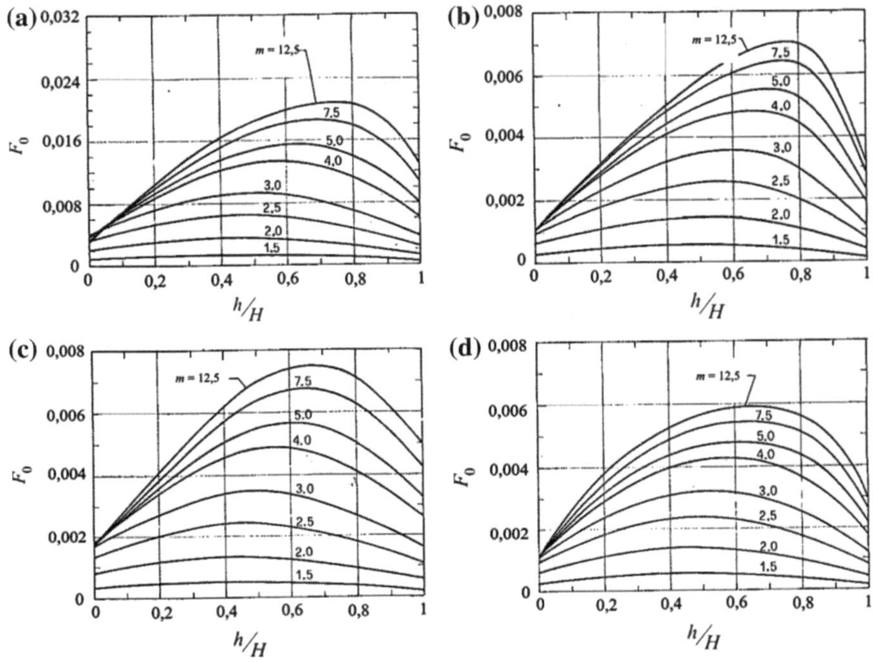

Fig. 4.151 Coefficient F_0 of the force acting between inductor and non-magnetic loads for the geometry of Fig. 4.141, as a function of *h/H* and various values of $m = d/\sqrt{2}\,\delta = \sqrt{2}\,r_e/\delta$ inductor: radial thickness = 1 mm; space factor = 0.8; **a** $r_i/r_e = 1.2$; $H/d = 1$; **b** $r_i/r_e = 1.2$; $H/d = 2$; **c** $r_i/r_e = 1.5$; $H/d = 1$; **d** $r_i/r_e = 1.5; H/d = 2$

The '*z*' component of the force (in Newton) is given by the equation:

$$F_z = 4\pi \times 10^{-7} \cdot 9.81 \cdot (NI)^2 \cdot F_0 \tag{4.105}$$

The curves show that the force is higher when r_e/δ increases and reaches its maximum for different values of *h/H*, which depend on the geometry of the inductor-load system.

The comparison of calculations with experimental data, given in Fig. 4.152, shows a fairly good agreement. The differences can be explained by the simplified representation of the inductor in calculations and the constant value of parameter $d/ = \sqrt{2}\delta = \sqrt{2}r_e/\delta$, which actually varies during heating.

Figure 4.153a shows an oscillogram of the inductor current *I* (peak to peak values) and the force of repulsion *F* taken during experimental tests.

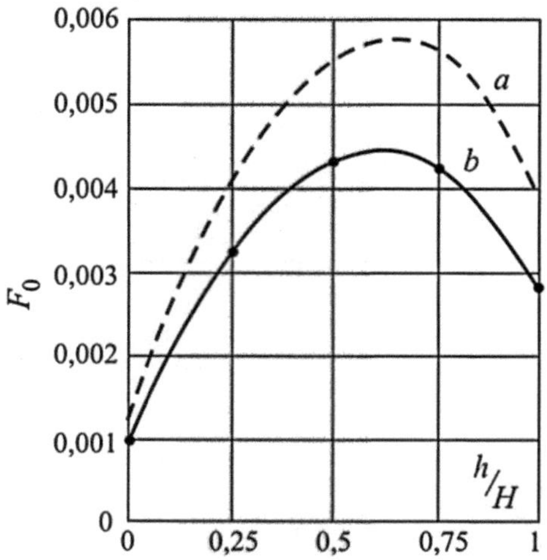

Fig. 4.152 Coefficient F_0 for the system geometry of Fig. 4.150, as a function of the ratio *h/H*, during heating of aluminium bars $d = 40$ mm; $f = 9{,}800$ Hz; $H/d = 2$; $r_i/r_e = 1.5$; $N = 8$ (copper tube 8 × 8 × 1 mm, space factor = 0.8); **a** calculated values; **b** experimental values

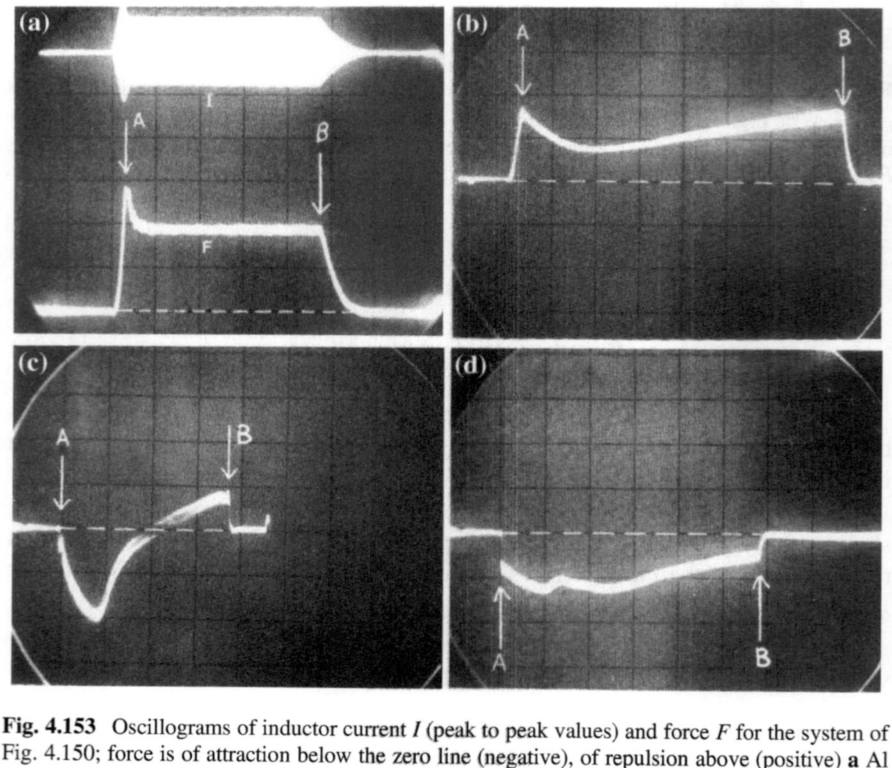

Fig. 4.153 Oscillograms of inductor current *I* (peak to peak values) and force *F* for the system of Fig. 4.150; force is of attraction below the zero line (negative), of repulsion above (positive) **a** Al bar, data as in Fig. 4.152; $h/H = 0.75$; $I = 960$ A; **b, c, d** C40 carbon steel; $f = 9.8$ kHz, **b** inductor I; d = 40 mm, I = 1,600 A; **c** inductor II, d = 33 mm; I = 1,200 A; **d** inductor III, d = 13.3 mm, I = 850 A (see Table 4.4)

Table 4.4 Parameters of inductor-load system used in experimental tests (see Fig. 4.150)

Inductor	I	II	III	IV	V
D_i (mm)	60	50	30	20	15
H (mm)	80	67	40	30	20
N	8	7	5	4	4
Copper tube (mm)	8 × 8 × 1	8 × 8 × 1	6 × 6 × 1	6 × 6 × 1	4 × 4 × 1

4.10.6 *Forces in Induction Heating of the Ends of Magnetic Bars*

For this case, always with reference to the system of Fig. 4.150, a series of experimental data are available; the data were obtained, during heating to about 1,000 °C (at 10 kHz) C40 carbon steel bars with the inductors specified in Table 4.4 [155].

The tests were carried out by taking oscillograms (analogous to the one of Fig. 4.153a) of the inductor current and measuring by a transducer the force acting between the inductor and the bar.

Figure 4.153b–d shows some typical oscillograms measured during tests.

They highlight three different ways in which the force vary during heating: only repulsion (*b*), attraction and repulsion (*c*) or only attraction (*d*), depending on the position of the bar inside the inductor, the magnetic field intensity and the temperature.

The diagrams of Fig. 4.154a, b summarise the results obtained from measurements. They give the values of coefficient F_0 of the maximum force of attraction that can occur during heating of the bar ends as a function of their diameter, for various values of the ampere-turns per unit length of the inductor, under the experimental conditions defined in the captions of the figures.

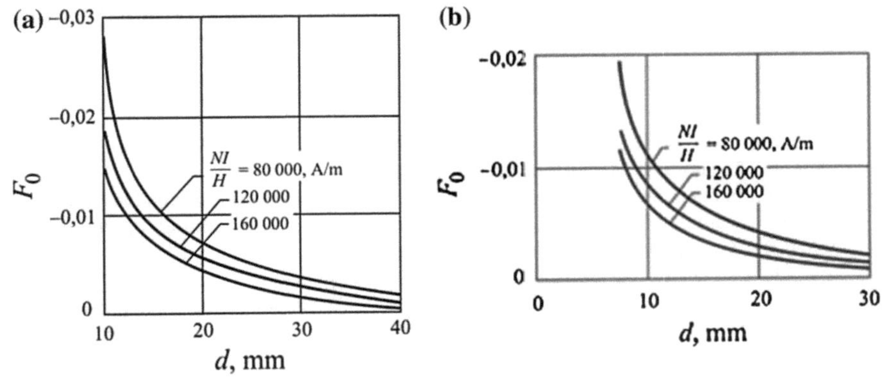

Fig. 4.154 Values of coefficient F_0 of the maximum force of attraction during heating the ends of C40 carbon steel bars (see Fig. 4.150), as a function of bar diameter, for various values of *NI/H*; $f = 9.8$ kHz; inductors I, II, III, IV, V as in Table 4.4; **a** $r_i/r_e = 1.5$, $H/d = 2$; **b** $r_i/r_e = 2.0$, $H/d = 2.66$

4.10.7 Conclusions

In this paragraph the most used methods for the calculation of forces in heating systems have been considered.

Typical calculation results have been presented with data useful for the evaluation of the magnitude of the time average values of forces in some system configurations used in industrial applications.

The treatment of this subject does not pretend to be exhaustive, since many important topics regarding forces haven't been discussed, e.g. evaluation of forces in terms of Maxell stresses, evaluation of the oscillating component and its influence on vibrations, influence of forces in transverse flux induction heating of thin strips, etc. For a discussion of these issues we suggest the reader to consider the following Refs. [156–159].

The purpose of this paragraph is to draw the attention of the reader on the magnitude of these forces and to the fact that they can considerably influence the mechanical design of the installation and, in some special cases, even prevent the possibility of application of the same heating process.

References

1. Aliferov, A., & Lupi, S. (2004). *Direct resistance heating of metals* (223 pp.). Novosibirsk: NGTU. ISBN 5-7782-0475-2 (in Russian).
2. Lupi, S. (2005). *Elettrotermia* (453 pp.). Padova (Italy): Libreria Progetto (in Italian).
3. Lupi, S., & Nunes, M. F. (1990). *Riscaldamento dei metalli mediante conduzione diretta di corrente* (86 pp.). Padova (Italy): CLEUP. ISBN 88-7178-338-7 (in Italian).
4. Romanov, D. I. (1981). *Direct resistance heating of metals* (280 pp). Moscow: Mashinostroenie (in Russian).
5. Trelle, K. (1967). Eigenshaften und Probleme der konduktiven Erwärmung von Stahl. *BBC-Nachrichten, 9,* 425–434 (in German).
6. Crepaz, G., & Lupi, S. (1982). Influence of the installations electrical parameters on the transient temperature distribution in the direct resistance heating of ferromagnetic billets. *Proceedings of Conference on Electroheat for Metals "UIE-BNCE".* Cambridge, England
7. Kalantarov, P., & Centlin, L. (1986). *Inductance calculations.* Leningrad: Energoatomizdat (in Russian).
8. Šneerson, G. A. (1981). *Fields and Transient processes in devices with eddy currents* (200 pp.). Leningrad: Energoatomizdat (in Russian).
9. Lupi, S., Forzan, M., & Aliferov, A. (2008). Characteristics of Installations for direct resistance heating of ferromagnatic bars of square cross-section. *Proceedings of International Scientific Colloquium "Modelling for Electromagnetic Processing,* 27–28 October (pp. 43–49). Hannover (Germany). ISBN 978-3-00-026003-2.
10. Inkin, A. I., Aliferov, A. I., & Blank, A. V. (2007). Calculation of electrical parameters in resistance heating of workpieces with rectangular cross-section. *Elektrichestvo, 8,* 35–38 (in Russian).
11. Inkin, A. I., Aliferov, A. I., & Blank, A. V. (2009). Studies of parameters of ferromagnetic bodies of rectangular cross-section heated by direct resistance or by induction. *Russian Electrical Engineering, 80*(12), 695–700 (in Russian).
12. CEDRAT. http://www.cedrat.com, FLUX users guide.

13. Lupi, S., Forzan, M., & Aliferov, A. I. (2009). Integral parameters of installations for direct resistance heating of ferromagnetic steel bars of square cross-section. *Proceedings of International IEEE Conference for the 150-anniversary of Alexander S. Popov*, 18–23 May. Saint Petersburg, Russia. IEEE Copyrights, IEEE Service Center, Piscataway, Hoes Lane, 2009 (pp. 1581–1587).
14. Sarač, V. (1981). *Influence of current type and characteristics of power supply on the direct heating of steel billets* (20 pp.). Moscow: MEI (in Russian).
15. Aliferov, A., & Lupi, S. (2008). Skin effect in toroidal conductors with circular cross-section. *COMPEL, 27*(2), 408–414.
16. Fok, V. A. (1930). Skin effect in annular circular cross-section. *ZRFHO (Russian Journal of Physical and Chemical society), Part Physics, LXII*(3), 281–297 (in Russian).
17. Tayler, R. J. (1960). The distribution of currents on the surface of a toroidal conductor. *Theoretical Physics Division, Atomic Energy Research Establishment, M 563*, 1–19.
18. Vologdin V. P. (1947). *Surface induction hardening*, (291 pp.). Moscow: Oborongiz (in Russian).
19. Aliferov, A. I. (2002). *Direct resistance electroheat processes in bent tubes* (Vol. 1, pp. 63–73). Researches of NGTU, Novosibirsk: NGTU.
20. Aliferov, A. I., & Lupi, S. (2003). Experimental energetic and thermal characteristics of direct resistance heating of cylindrical bent workpieces. *Proceedings of the of 5th International Conference on Electromechanics, Electrotechnology and Electrical materials, Ukraine, Crimea, Alushta,* (pp. 33–36). Moskow: MEI (in Russian).
21. Aliferov, A. I., & Lupi, S. (2004). Energetic parameters of curvilinear tubes heated by direct resistance heating. *Proceedings of International Symposium HES-04—Heating by Electromagnetic Sources* (pp. 649–654). Padua (Italy): SGEditoriali. ISBN 88-86281-92-7.
22. Aliferov, A. I., Lupi, S., Forzan, M., Vlasov D. S. (2010). *Skin effect in toroidal conductors of round section* (Vol. 4, pp. 73–82). Naucnyi vestnik, Novosibirsk: NGTU (in Russian).
23. Lupi, S., & Aliferov, A. I. (2003). Electromagnetic phenomena in resistance heating of curviliner cylindrical workpieces. *COMPEL—International Journal for Computation and Mathematics in Electrical and Electronic Engineering, 22*(1), 149–157.
24. Aliferov, A. I., & Lupi, S. (2010). *Induction and direct resistance heating of metals, Novosibirsk* (411 pp.). Novosibirsk: NGTU. ISBN 978-5-7782-1622-8 (in Russian).
25. Aliferov, A. I. (2001). Methods of temperature field smoothing in resistance heating of workpieces with arbitrary shape. *Proceedings of HIS-01- International Conference on Heating by Internal Sources* (pp. 207–212). Padua (Italy): SGEditoriali. ISBN 88-86281-64-1.
26. Aliferov, A. I. (1987). *Direct resistance heating of thick-walled tubes of complex configurations*. Optimization of Processes in Electrothermal Installations Using Computers (Vol. 122, pp. 83–87). Moscow: MEI (in Russian).
27. Kovrev G. S. (1975). *Electro-contact heating for processing non-ferrous metals* (312 pp.). Moscow: Metallurgia (in Russian).
28. Aliferov, A. I. (1999). *Effectiveness of different thicknesses of thermal insulation on curved electro-contact heated products* (Vol. 3, pp. 127–132). Environmentally advanced systems and technology, Researches of NGTU, Novosibirsk: NGTU (in Russian).
29. Aliferov, A. I., Selskiĭ, S. V., & Kuvaldin, A. B. (1991). *Electro-contact heating of curved portion of tubular products* (Vol. 31). URSS Patent 1671707.
30. Aliferov, A. I., Selskiĭ, S. V., & Kuvaldin, A. B. (1992). *Method for resistance heating of curved portion of tubular products* (Vol. 19). URSS Patent № 1735387.
31. Sluhočkiĭ, A. E. et al. (1968). *On the application of magnetic yokes in inductors for heating in transverse magnetic field* (pp. 231–235). Researches of High-frequency Currents in Electroheat, Leningrad: Mashinostroenie (in Russian).
32. US Patent 36122809, I N 5 3/00.
33. US Patent 2282384, I 5 3/00.

34. Aliferov, A. I., Dughiero, F., & Forzan, M. (2010). Coupled magneto-thermal FEM Model of direct heating of ferromagnetic bent tubes. *IEEE Transactions on Magnetics, 46*(8), 3217–3220.
35. EDF—Électricité de France. (1996). *Induction, conduction electrique dans l'industrie* (780 pp.). Paris: Electra. ISBN 2-869 95-022-5.
36. Baker, R. M. (1950). Transverse flux induction heating. *AIEE Transactions, 69*, 711–719.
37. Gibson, R. C., Johnson, R. H., & Hilditch, R. W. (1982). High efficiency induction heating as a production tool for heat treatment of continuous strip metal. *BNC-UIE Electroheat for Metals Conference*. Cambridge (UK).
38. Haubitzer, W. (1984). Die Optimierung von Querfeld-induktoren zur kontinuierlichen Blecherwärmung. *Elektrowärme International, 42* (B1), B39–B48.
39. Jackson, W. B. (1972). *Transverse flux induction heating of flat metal products*. 7th U.I.E. (Vol. 206). Warsaw: Congress.
40. Lackner, R. V. (1960). Continuous flash annealing produces fine grained aluminium sheets. *Modern Metals, 9*, 62–64.
41. Lamourdedieu, M. (1951). Continuous heat-treatment of aluminium alloys of the Duraluminium type. *Journal of the Institute of Metals, 80*, 835.
42. Nuns, J., & Poiroux, R. (1981). *Le chauffage de bandes d'aluminium par inducteur a flux tranverse* (27 pp.). EDF—Direction Des Études et Recherches, HE 122 (T. 439) (in French).
43. Peysakhovich, V. A. (1966). Energetic relationships in the heating of metal strips in transverse magnetic field. *Industrial Applications of High-Frequency Currents, Proceedings of VNIITVCH* (Vol. 7, pp. 41–57). Leningrad (in Russian).
44. Ross, N. V., Howell, C. N., Kaltenhauser, R. H., & Walzer, G. A. (1992). *Transverse flux induction heating of steel strips* (pp. 110–119). Montreal (Canada): XII UIE Congress.
45. Scherg, P. (1981). Zur Auslegung von Querfluss-induktions-erwärmungs-aggregaten. *Elektrowärme International, 39*, B84–B90.
46. Schluckebier, D. (1992). Induktive Querfelderwärmung von Flachgut. *Elektrowärme International, 50*, B165–B169.
47. Anderhuber, M. et al. (2002). *Transverse flux induction heating device with magnetic circuit of variable width*. US Patent No US6, 498 328 B2, Dec.24.
48. Battistetti, M., Colaone, F., Dughiero, F., Forzan, M., Lupi S., & Ponchiroli, S. (2000). A 3D analytical tool for design and optimization of transverse flux induction heating systems. *EPM2000—3rd International Symposium on Electromagnetic Processing of Materials*, 3–6 April (pp. 199–204). Nagoya (Japan).
49. Battistetti, M., Dughiero, F., Lupi, S., Farina, M., Di Barba, P., & Savini A. (2001). *COMPEL, 20* (2), 507–522.
50. Bukanin, V., Dughiero, F., Lupi, S., Nemkov, V., & Siega, P. (1995). 3D-FEM Simulation of Transverse Flux Induction Heaters. *IEEE Transactions on Magnetics, 31*(3), 2174–2177.
51. Bukanin, V., Dughiero, F., Forzan, M., Lupi, S., & Nemkov, V. (1996). *Induction heating of flat metal bodies*. XIII UIE Congress on Electricity Applications, Birmingham (UK), MIII 17-25. ISBN 1 86125 007 X.
52. Conraths, H. J. (1993). Rechnergestützte Simulation der induktiven Querfelder-wärmung. *Elektrowärme International, 51*, B4–B12.
53. Dughiero, F., Lupi, S., Mühlbauer, A., & Nikanorov, A. (2001). TFH—Transverse flux induction heating of non-ferrous and precious metal strips—Results of a EU research project. *Proceedings of International Seminar HIS-01—Heating by Internal Sources*, 12–14 September. Padua (Italy), and *COMPEL International Journal 2003, 22*(1), 565–575.
54. Gibson, R. (1996). *The design and performance of multi-megawatt variable width TFX induction heaters for the ultra rapid heat treatment of metal strip*. XIII UIE Congress on Electricity Applications, Birmingham, UK, MIII 69–76. ISBN 1 86125 007 X.

55. Lupi, S., Forzan, M., Dughiero, F., & Zenkov, A. (1999). Comparison of edge-effects of transverse flux and travelling wave induction heating inductors. *INTERMAG 99, Kijoungiu (Korea) and IEEE Transactions on Magnetics, 35*(5), 3556–3558.
56. Mühlbauer, A., Muižnieks, A., & Lessmann, H. J. (1992). Matematische 3D-Modellierung des elektromagnetischen und des thermischen Feldes bei der Querfelderwärmung dünner Bleche. *37th International Wiss Koll Ilmenau, B1* (pp. 100–106).
57. Mühlbauer, A., Muiznieks, A., & Lessmann, H. J. (1994). Berechnung von dreidimensionalen elektromagnetischen Feldern bei der induktiven Erwärmung. *Archiv für Elektrotechnik, 77*, 157–168.
58. Mühlbauer, A., Nauvertat, G., Nikanorov, A., Lupi, S., Colaone, F., & Forzan M., et al. (2000). Numerical tools for optimum design of transverse flux induction heating of non-ferrous metal strip. *Proceedings of EPM2000—3rd International Symposium on Electromagnetic Processing of Materials*, 3–6 April (pp. 205–210). Nagoya, Japan.
59. Nacke, B., Andree, W., Mohring, J. U., & Lessmann, H. J. (1993). Induktive Querfeld-erwärmung—eine flexible und effiziente Erwärmungsmethode für metallische Flachprodukte im Walzwerk. *Elektrowärme International, 51*, B156–B166.
60. Nauvertat, G., Nikanorov, A., & Mühlbauer, A. (2000). Optimiertes Design induktiver Querfeld-Erwärmungsanlagen. *Elektrowärme International, 1*, 22–29.
61. Neau, Y., Paya, B., Anderhuber, M., Griffay, G., Ducloux, F., Hellegouarch, J., & Ren, J. Y. (2003). High Power (3 MW) transverse flux inductor for industrial heating of wide range size strips. *4th International Conference EPM on Electromagnetic Processing of Materials* (pp. 570–575). Lion (France).
62. Nikanorov, A. et al. (2006). *Transverse field heating installation for inductively heating flat objects*. Patent application No: US 2006/0196870 A1, Sept. 7.
63. Nikanorov, A., Nauvertat, G., Schulbe, H., Nacke, B., & Mühlbauer, A. (2001). Investigation, design and optimization of transverse flux induction heaters. *Proceedings of HIS-01—International Seminar Heating by Internal Sources, Padua (Italy) and COMPEL International Journal, 20*(2), 553–558.
64. Nikanorov, A., Schulbe, H., Nacke, B., Zlobina, M., Galunin, S., & Blinov, Y. (2004). Non-linear effects in transverse flux heating systems. *Proceedings of International Symposium HES04—Heating by Electromagnetic Sources*, 22-25 June (pp. 685–691). Padua (Italy): SGEditoriali. ISBN 88-86281-92-7.
65. Rhunke, A. (1999). *Systematische Untersuchungder induktiven Querfeld-Band-erwärmung*. Dissertation, Universität Hannover, Düsseldorf: VDI.
66. Ruhnke, A., Mühlbauer, A., Nikanorov, A., & Demidovich, V. (1998). 3D numerical and experimental investigations of transverse flux induction heating. *Proceedings of HIS-98—International Induction Heating Seminar*. Padua (Italy): SGEditoriali, 109-116. ISBN 88-86281-26-9.
67. Schülbe, H., Nikanorov, A., & Nacke, B. (2004). Flexible transverse flux heaters of metal strip. *Proceedings of International Symposium HES04—Heating by Electromagnetic Sources*, 22–25 June (pp. 293–300). Padua (Italy): SGEditoriali. ISBN 88-86281-92-7.
68. Thorpe, J. C. et al. (2003). *Transverse flux induction heating apparatus*. US Patent No US 6 576 878 B2, June 10, 2003.
69. Tudorache, T., & Fireteanu V. (1998). 3D Numerical modeling of new structures for transverse flux heating of metallic sheets. *Proceedings of HIS-98—International Induction Heating Seminar* (pp. 117–124). Padua (Italy): SGEditoriali. ISBN 88-86281-26-9.
70. Tudorache, T., & Fireteanu, V. (2007). Magneto-thermal-translating motion coupling in 3D FEM analysis of metallic sheets transverse flux induction heating. *Proceedings of International Symposium HES-07—Heating by Electromagnetic Sources*, 19–22 June (pp. 259–266). Padua (Italy): SGEditoriali. ISBN 88-89884-07-X.
71. Walker, R. A., & Hayward, A. (1996). *The application of high power transverse flux induction heating in the metal strip industries*. XIII UIE Congress on Electricity Applications, Birmingham (UK), MIII 1-8. ISBN 1 86125 007 X.

72. Miyagi, D., Saitou, A., Takahashi, N., Uchida, N., & Nozaki, K. (2006). Improvement of zone control induction heating equipment for high-speed processing of semiconductor devices. *IEEE Transactions on Magnetics, 42*(2), 292–294.
73. Fujita, H., Uchida, M., & Nozaki, K. (2007). Zone Controlled Induction Heating (ZCIH) A new concept in induction heating. *PCC '07—Power Conversion Conference* (pp. 1498–1504). Nagoya (Japan).
74. Spagnolo, A., Forzan, M., Lupi, S., Pateau, O., & Paya, B. (2010). Space control optimization of multi-coil transverse flux induction heating of metal strips. *Proceedings of International Symposium HES10—Heating by Electromagnetic Sources*, 18–21 May (pp. 455-46). Padua (Italy): SGEditoriali. ISBN 978-88-89884-13-3
75. Acero, J., Alonso, R., Burdio, J. M., Barragan, L. A., & Puyal, D. (2006). Analytical equivalent impedance for a planar circular induction heating system. *IEEE Transactions on Magnetics, 42*(1), 84–86.
76. Adler M. S. (1974). A field-theoretical approach to magnetic induction heating of thin circular plates. *IEEE Transactions on Magnetics, MAG-10*, 1118–1125.
77. Cheng, D. H. S. (1965). The reflected impedance of a circular coil in the proximity of a semi-infinite medium. *IEEE Transactions, IM-14*, 107–116.
78. Lupi, S. (1978). The calculation of planar circular coils for the induction heating of bimetallic plates. *Archiv für Elektrotechnik, 60*, 313–318.
79. Lupi, S. (1979). Planar circular coils for induction heating. *Elektrowärme International, 37*, B319–B326.
80. Lupi, S., Sieni, E., & Spagnolo, A. (2010). Analytical calculation of planar circular coils induction heating systems. *International Conference HES-10—Heating by Electromagnetic Sources*, 18–21 May (pp. 121–128). Padova (Italy): SGEditoriali. ISBN 978-88-89884-13-3.
81. Mathews, B. E. (1957). Flat work—Coil design. *AIEE Transactions, 11*, 249–256.
82. Smeljanski, A. M., Kuvaldin, A. B., Kolbe, E., & Reiss, W. (1974). *Ein Beitrag zum Problem der induktiven Beheizung durch Fläschenspulen* (pp. 91–96). XIX International Wiss Koll, TH Ilmenau.
83. Zgraja, J. (1975). Shaping of temperature field in induction heated metal plates. *Proceedings of the International Wiss Koll* (pp. 82–87). TH Ilmenau.
84. Zgraja, J. (1996). Non-deforming induction heating of metal discs. *XIII UIE Congress on Electricity Applications*, 16–20 June (MIII, pp. 43–50). Birmingham (UK).
85. Zgraja, J. (1998). Optimization of radial temperature distribution for non-deforming induction heating of metal disks. *HIS-98—International Induction Heating Seminar*, 13–15 May (pp. 87–94). Padua (Italy): SGEditoriali. ISBN 88-86281-26-9.
86. Bucholz, H. (1957). *Elektrische und magnetische potentialfelder*. Berlin: Springer.
87. Lupi, S., & Morini, A. (1971). Induction heating of cylindrical rods using multiple coils. *Elektrowärme International, 29*(12), 663–667.
88. Lupi, S., & Nemkov, V. S. (1977). The calculation of inductors with periodical fields. *Elektrowärme International, 13*(35), 103–109. N-13 978-1-62708-012-5.
89. Lupi, S., & Nemkov, V. S. (1978). Analytical calculation of cylindrical induction systems. *Elektrichesvo, 6*, 43–47 (in Russian).
90. Mahmudov, K. M., & Nemkov V. S. (1975). *Calculation of inductors for heating non-magnetic cylinders*. Leningrad: LETI (in Russian).
91. Nemkov, V. S. (1966). Induction heating of thin-walled tubes in cylindrical inductors. *Proceedings of VNIITVČ* (Vol. 7). Leningrad (in Russian).
92. Onoe, M. (1968). Analysis of a finite solenoid coil near a conductor. *Electrical Engineering, 88*(10) (in Japanese).
93. Sluhockiĭ, A. E., & Ryskiĭ, S. E. (1974). *Inductors for induction heating* (264 pp.). Leningrad: Energia (in Russian).
94. Fikus, F. (1972). Inductionserwärmung mit einem Innenfeld-inductor für netzfreguenz. *Proceedings of VII UIE Congress*, Warsaw (No. 607).

95. Lupi, S., Nemkov, V. S. (1979). Riscaldamento ad induzione delle superfici esterne ed interne di cilindri bimetalici cavi mediante “induttori corti”. *L'Elettrotecnica, LXVI*(2), 141–147 (in Italian).
96. Biasutti, F., Krause, C., & Lupi, S. (2012). Induction hardening of complex geometry and geared parts. *Heat Processing, 3*, 60–70. ISSN 1611-616X.
97. Criqui, B., et al. (1997). Multi-frequency induction hardening of gears to replace carburizing —methodology of development. *1st International Induction Heat Treating Symposium*, 16–18 September (pp. 1–7). Indianapolis, Indiana.
98. Stiele, H., & Marquis, F. (2007). An overview of inductive gear spin hardening. *Conference Proceedings of HES-07—Heating by electromagnetic Sources*, 19–22 June (pp. 183–190). Padua (Italy). ISBN 88-86281-92-7.
99. Babat, G. I. (1946). *Induction heating of metals and its industrial application* (552 pp.). Ed. Gosenergizdat (in Russian).
100. Lozinskiĭ, M. G. (1969). *Industrial applications of induction heating* (672 pp.). London: Pergamon Press Ltd.
101. Lupi, S., & Rubatto, G. (1998). *Tempra di ingranaggi mediante riscaldamento ad induzione con due frequenze*. AIM—Associazione Italiana di Metallurgia, Giornata di studio su Trattamenti superficiali, Milano (Italy), Marzo 31 (in Italian).
102. Lupi, S., Dughiero, F., & Forzan, M. (2006). Modelling single- and double-frequency induction hardening of gear wheels. *Conference on Proceedings of EPM-2006*, 23–27 October (pp. 473–478). Sendai (Japan).
103. Di Pieri, C., Lupi, S., & Crepaz, G. (1988). The design of induction heating machines for single-shot hardening of large ring-gears or gear-wheels. *Electroheat Congress UIE 11*, 3–7 October, n. B7.6. Malaga (Spain).
104. Hassan, M. (1980). Oberflächenhдrtung mit getakteten Hochfrequenzgeneratoren. *Elektrowдrme International, 38*(B2), B84–B92.
105. Langer, E. (1974). *Elektrowaerme—Theorie und Praxis* Essen: Verlag W. Girardet, chapter 4, 578-630. ISBN 3-7736-0355-X.
106. Rudnev, V., Loveless, D., Cook, R., & Black, M. (2003). *Handbook of induction heating* (777 pp.). New York: Marcel Dekker Inc. ISBN 0-8247-0848-2.
107. Rudnev, V., Loveless, D., Cook, R., & Black, M. (2003). *Induction hardening of gears: A review, heat treatment of metals* (Part 1: pp. 97–103, Part 2: pp. 11–15). UK: Wolfson Heat Treatment Centre, Aston University.
108. Stiele, H., & Marquis, F. (2007). Induktive Anwendungen im Flugzeugbau. *Elektrowärme International, 1*, 23–25.
109. Demidovich, V., Kundas, S., Ivanov, D., Lemzikov, A., & Tchmilenko, F. (2007). Simulation of induction heat treatment as coupling of multi-physics phenomena. *Conference on Proceedings of HES-07—Heating by Electromagnetic Sources*, 19–22 June (pp. 719–725). Padua (Italy). ISBN 88-86281-92-7.
110. Dolega, D., & Kurek, K. (2004). The 2D and 3D analysis of the induction hardening process considering changes in the material microstructure. *Conference on Proceedings of HES-04—Heating by Electromagnetic Sources*, 22–25 June (pp. 577–584). Padua (Italy). ISBN 88-86281-92-7.
111. Grum, J. (2004). Influence of induction surface heating and quenching on residual stress profiles, followed by grinding. *Conference on Proceedings of HES-04—Heating by Electromagnetic Sources*, 22–25 June (pp. 585–592). Padua (Italy). ISBN 88-86281-92-7.
112. Schwenk, M., & Schwenk, W. (2007). Substitution/replacement of case hardening gearbox components in the automotive as well as airplane industry through contour true induction hardening. *Conference Proceedings of HES-07—Heating by Electromagnetic Sources*, 19–22 June (pp. 175–182). Padua (Italy). ISBN 88-86 281-92-7.

113. Nacke, B. (2007). Recent developments in induction Heating and future prospects in industrial applications. *Conference on Proceedings of HES-07—Heating by Electromagnetic Sources*, 19–22 June (pp. 243–250). Padua (Italy). ISBN 88-86281-9 2-7.
114. Wrona, E., Nacke, B., & Schwenk, W. (2004). Solving complex induction hardening tasks by numerical methods. *Conference Proceedings of HES-04—Heating by Electromagnetic Sources*, 22–25 June (pp. 593–598). Padua (Italy). ISBN 88-86281-9 2-7.
115. Wrona, E., *Numerisch Simulation des Erwärmung prozesses für das inductive Randschichthaerten komplexer Geometrien*. Goettingen: Ed. Cuvillier.
116. Barglik, J., Smalcerz, A., Przylucki, R., & Doležel, I. (2014). 3D modeling of induction hardening of gear wheels. *Journal of Computational and Applied Mathematics, 270*, 231–240.
117. Bonetto, S., Dolega, D., & Forzan, M. (2007). Metallurgical model for induction pulse hardening of gear wheels. *Conference on Proceedings of HES-07—Heating by Electromagnetic Sources*, 19–22 June (pp. 641–648). Padua (Italy). ISBN 88-86281-9 2-7.
118. Magnusson, N. (1972). Prospects for rotating billet superconducting induction heating. *HES-07—Heating by Electromagnetic Sources Conference Proceedings*, 19–22 June (pp. 479–486). Padua (Italy).
119. Harvey, I. G. (1976). The theory of multi layered windings for induction heating and their application to 1 MW, 50 Hz longitudinal flux billet heater. *Conference Proceedings of VIII UIE Congress,* Liege, 1976, Section IIa, No. 4.
120. Harvey, I. G., & Martin, D. H. (1982). Operating experience with a 450 kW, multi-layer coil induction preheater in a production environment. *BNCE-UIE Electroheat for Metals Conference*, 21–23 September, n. 6.3. Cambridge (England).
121. Magnusson, N., & Runde, M. (2003). Efficiency analysis of a high-temperature superconducting induction heater. *IEEE Transactions on Applied Superconductivity, 13*(2), 1616–1619.
122. Magnusson, N., & Runde, M. (2003). Applying superconductors to electromagnetic processes. *Proceedings of EPM2003—Int. Conference on Electromagnetic Processing of Materials*, 13–17 October (pp. 621–626). Lyon (France).
123. Hiltunen, I., Korpela, A., & Mikkonen, R. (2005). Solenoidal Bi-2223/Ag Induction heater for aluminum and copper billets. *IEEE Transactions on Applied Superconductivity, 15*(2), 2356–2359.
124. Magnusson, N., & Runde, M. (2006). A 200 kW MgB_2 Induction heater project. *Journal of Physics, Conference Series, 43*, 1019–1022.
125. Magnusson, N., Bersås, R., & Runde, M. (2003). Induction heating of aluminium billets using HTS DC coils. *Proceedings of 6th European Conference on Applied Superconductivity*, 14–18 September (pp. 1104–1109). Sorrento (Italy).
126. Runde, M., & Magnusson, N. (2002). Induction heating of aluminium billets using superconducting coil. *Physica C, 372–376*, 1339–1341.
127. Runde, M., & Magnusson, N. (2003). Design, building and testing of a 10 kW superconducting induction heater. *IEEE Transactions on Applied Superconductivity, 13*(2), 1612–1615.
128. Dughiero, F., Forzan, M., & Lupi, S. (2006). Induction heating of aluminium billets rotating in a DC magnetic field. *VIII International Conference on Problems of Control and Modelling Complex Systems*, 24–29 June (pp. 171–176). Samara (Russia).
129. Perry, M. P., & Jones, T. B. (1978). Eddy current induction in a solid conductor cylinder with a transverse magnetic field. *IEEE Transactions on Magnetics, MAG-14*(4), 227–232.
130. Fabbri, M., Morandi, A., & Ribani, P.L. (2007). DC induction heating of aluminium billets by means of superconducting magnets. *Conference on Proceedings of HES-07*, 20–22 June (pp. 505–512). Padua (Italy). ISBN 88-89884-07-X.

131. Fabbri, M., Forzan, M., Lupi, S., Morandi, A., & Ribani, P. L. (2009). Experimental and numerical analysis of DC induction heating of aluminum billets. *IEEE Transactions on Magnetics, 45*(1), 192–200.
132. Nacke, B., Zlobina, M., Nikanorov, A., & Ulferts, A. (2007). Numerical simulation of induction heating of aluminium billets by rotation in DC magnetic field. *Conference Proceedings of HES-07*, 20–22 June (pp. 497–504). Padua (Italy). ISBN 88-89884-07-X.
133. Forzan, M., Lupi, S., Moranti, A., & Spagnolo, A. (2007). Edge effects in aluminium billets heated by rotation in DC magnetic field. *Conference Proceedings of HES-07*, 20–22 June (pp. 317–324). Padua (Italy). ISBN 88-89884-07-X.
134. Aliferov, A., Blank, A., Forzan, M., Inkin, A., & Lupi S. (2012). *Analytical and numerical analysis of electrical parameters of induction heating with permanent magnets.* UIE 2012 Conference St. Petersburg (Russia), May.
135. Dughiero, F., Forzan, M., Lupi, S., Nicoletti, F., & Zerbetto, M. (2011). A new high efficiency technology for the induction heating of non magnetic billets. *COMPEL, 30*(5), 1528–1538.
136. Dughiero, F., Forzan, M., Zerbetto, M. (2013). *Device for induction heating of a billet.* Patent WO2013128241.
137. Mach, F., Karban, P., & Doležel, I. (2012). Induction heating of cylindrical non-magnetic ingots by rotation in static magnetic field generated by permanent magnets. *Journal of Computational and Applied Mathematics, 236*(18), 4732–4744.
138. Mach, F., Kůs, P., Karban, P., & Doležel, I. (2013). Optimization of the system for induction heating of nonmagnetic cylindrical billets in rotating magnetic field produced by permanent magnets. *Computing, 95*(1), 537–552.
139. Suominen, P. (2011). *Method, device and arrangement for heating an object by induction.* Patent WO2010100082.
140. Nemkov, V. S., & Demidovich V. B. (1988). *Theory and calculation of induction heating installations* (280 pp.). Leningrad: Energoatomizdat (in Russian).
141. Nemkov, V., Goldstein, R., & Bukanin, V. (2004). Optimal design of internal induction coils. *Proceedings of HES-04—Heating by Electromagnetic Sources—Padua*, 22–25 June (pp. 489–496). ISBN 88-86281-92-7.
142. Forzan, M., & Lupi, S. (2009). A.C. Resistance and electrical efficiency of inductors for induction heating of internal cylindrical surfaces. *APIH 09—Actual problems of Induction Heating*, 20–22 May (pp. 23–30). St. Petersburg (Russia).
143. Lupi, S., Forzan, M., Aliferov, A., & Melieško A. (2010). Resistance of the inductor in the induction heating of internal cylindrical surfaces. *Elektrotechnika*, (Ed. MEI (МЭИ-ТУ), *5*, 43–47. (in Russian).
144. Lipinski, W., & Sikora, R. (1975). Verluste und Kraftangriff von Wirberströmen im leitenden Halbraum bei Zylindersymmetrischen magnetischen Wechselfeld. *Archiv für Elektrotechnik, 57*, 173–175.
145. Brokmeier, K. H. (1953). Die mechanischen kraftwirkungen beim Induktiosofen. *VDE-Fachber, 17*, S. III/28.
146. Mühlbauer, A. (1967). Über die elektrodynamichen Kräfte in der Schmelze von Induktionsöfen. *Elektrowärme, 25*(12), 461–473.
147. Rummel, T., & Mühlbauer, A. (1968). Kräfte und Strömungen in elektrischen Induktionsöfen. *VI UIE International Electroheat Congress*, 13–18 May, no. 624. Brighton (UK).
148. Rummel, T., & Mühlbauer, A. (1967). Der zeitliche Verlauf elektrodynamicher Kräfte in metallischen Leitern. *International Z. Elektrowärme, 25*, 271.
149. Mühlbauer, A. Kräfte und Strömungen im idealisierten mehrphasig erregten Induktions-Tiegelofen. *XIII International Wiss Koll* (pt. 2, pp. 47–55). TH Ilmenau.

150. Lupi, S., & Orefice, C. (1981). Electrodynamic forces in induction heating of bimetallic plates with planar circular coils. *Archiv für Elektrotechnik, 63*, 213–217.
151. Lipinski, W. (1977). Elektrodynamisch Kraftwirkungen auf Wirbelströme in einer leitenden Platte. *Archiv für Elektrotechnik, 59*, 122–132.
152. Lupi, S., & Orefice, C. (1981). Azioni elettrodinamiche nel riscaldamento ad induzione di piastre ferromagnetiche. *L'Energia Elettrica, 10*, 437–444 (in Italian).
153. Lupi, S., & Ceolini, F. (1978). The mutually-coupled circuits method for calculations regarding inductors for induction heating with special configurations. *IEEE-IAS 1978 Annual Meeting Conference Record* (pp. 1151–1157).
154. Lupi S. (1979, Oct). The numerical calculation of forces in induction heating systems. *IEEE-IA Annual Meeting Conf. Rec., Cleveland (USA)* (pp. 1226–1231).
155. Di Pieri, C., & Lupi, S. (1980). Electromagnetic forces in medium frequency induction heating. *IX Congres International d'Electrothermie*, Cannes, n.IV, Ba1.
156. Dughiero, F., Lupi, S., & Siega, P. (1995). Calculation of forces in travelling wave induction heating systems. *IEEE Transactions on Magazine, 31*(6), 3560–3562.
157. Dughiero, F., Lupi, S., Ruhnke, A., Mühlbauer, A., Nikanorov, A., & Demidovitch, V. (1997). Electromagnetic forces during transverse flux heating of metal strips. *EPM—International Symposium on Electromagnetic Processing of Materials*, 26–29 May (pp. 87–93). Paris (France).
158. Kasjanov, H., Schülbe, H., & Nikanorov, A. (2003). Electrodynamic forces in steel strip during induction heating. *MEP 2003—Modelling for Electromagnetic Processing*, 24–26 March (pp. 337–342). Hannover. ISBN 3-00-011073-9.
159. Neau, Y., Paya, B., Tudorache, T., & Fireteanu, V. (2000). Numerical evaluation and experimental validation of eddy currents and electromagnetic forces in transverse flux induction heating of magnetic steel slabs. *3rd EPM International Symposium on Electromagnetic Processing of Materials* (pp. 211–216), Nagoya (Japan). ISIJ.

Chapter 5
Analytical and Numerical Methods for Calculation of Induction and Conduction Heating Systems

Abstract This chapter is about the analytical and numerical methods that have been successfully applied for modeling induction and direct resistance heating processes. In Sect. 5.1, calculation of induction heating system is presented by applying the equivalent magnetic circuit method. Analytical methods are extensively described in the previous chapters and briefly summarized in the Sect. 5.2 of this chapter. The Sect. 5.3 presents a simple procedure to solve coupled EM and Thermal induction heating problem by means of the finite difference method in 1D axis-symmetric domain. The procedure is fully detailed and can be easily implemented in a code. Typical results produced by a commercial code based on 1D finite difference method are shown in Sect. 5.4. Volume integral methods have been also successfully used in coupled electromagnetic and thermal calculations: these numerical techniques are detailed in the Sect. 5.5. Calculation of parameters of direct resistance heating systems and the importance of the feeding circuit parameters are dealt with in Sect. 5.6 by applying analytical solutions while in Sect. 5.7 the calculation is carried out by resorting again to the 1D finite difference method. Finite element method is the numerical technique most applied in commercial software dedicated to electromagnetic design. Finite element models are nowadays extensively used in electro thermal applications and a short description focused on practical use of finite element is presented in the closing section of the book.

5.1 Calculation of Induction Heating Systems with the Equivalent Magnetic Circuit Method

The calculation of the equivalent magnetic circuit describing the inductor-load system is based on the method of "total flux method" [1, 2].

This method considers the magnetic circuit of the inductor-load system assuming that all inductor windings are linked with the same induction flux. The magnetic circuit is described by means of reluctances, or magnetic resistances, corresponding

S. Lupi et al., *Induction and Direct Resistance Heating*,
DOI 10.1007/978-3-319-03479-9_5

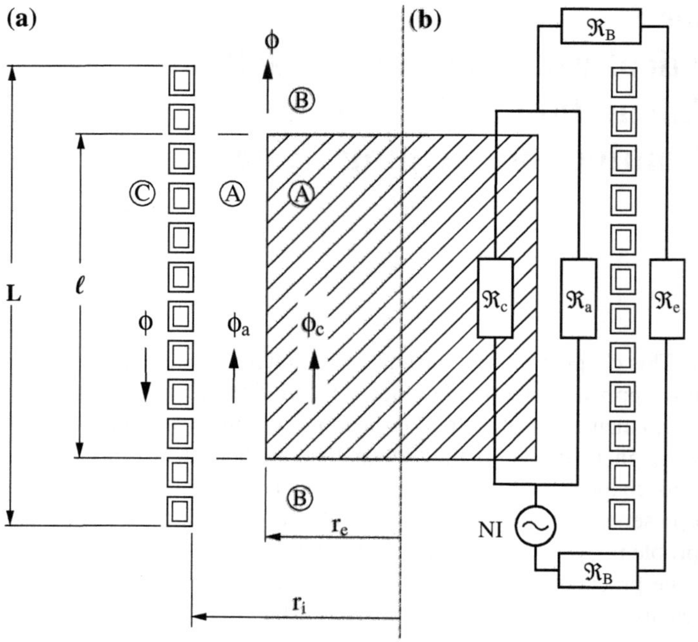

Fig. 5.1 **a** *Left side* inductor-load system, **b** *right side* equivalent magnetic circuit

to finite parts of the magnetic circuit and it is solved through its equivalent electrical circuit. Reluctances are defined as the ratio between the amplitude of magneto-motive force *mmf,* that is the product NI of the current by number of turns of the inductor, and the amplitude of the flux in the magnetic circuit. Magnetic circuits are solved by analogy with electrical circuit, where voltage is analogous to *mmf*, current to magnetic flux and impedance to reluctance.

For calculation of reluctances, the magnetic circuit is subdivided into three "sections": the heating section (region A of Fig. 5.1) and the sections corresponding to the return path of magnetic flux (regions B inside the coil, and C outside it).

In section A, the total flux Φ is subdivided in two components: one Φ_a in the space between inductor and load, the second one Φ_c inside the heated workpiece.

Finally, it is assumed an ideal distribution of the magnetic field, such that all field lines in section A are parallel to the axis, i.e. region A is considered a segment of an infinitely long system. On the contrary, the configuration of the field in sections B and C is assumed to be the same of that of the "empty" inductor (i.e. the system without load).

With the above assumptions, to the layout of Fig. 5.1a corresponds the magnetic circuit of Fig. 5.1b. The total *mmf* NI produced by the inductor is the source of the total flux Φ. The total flux depends on the reluctances $\Re_c$ and $\Re_a$ of section A and the reluctance $\Re_0$ of the "return path".

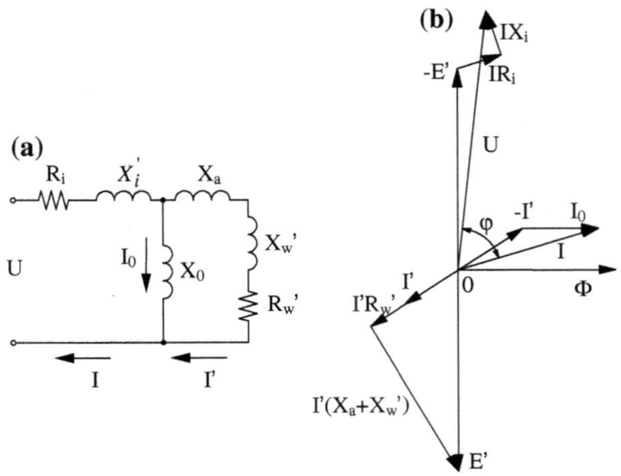

Fig. 5.2 **a** Electric circuit equivalent to the magnetic scheme, **b** phasor diagram

The electrical circuit of Fig. 5.2a and the phasor diagram of Fig. 5.2b are the analogous of the equivalent magnetic circuit of Fig. 5.1b. The electric circuit comprises the resistance R_i and the internal reactance X_i' of the induction coil, which are not considered in the magnetic circuit.

All impedances of the equivalent circuit are referred to the inductor current, and their values are calculated for a single-turn inductor, i.e. in Ω/turn^2.

The equivalent circuit parameters are calculated as follows:

- Penetration depth in the inductor material, m

$$\delta_i = 503\sqrt{\frac{\rho_i}{f}},$$

with: ρ_i—resistivity of the inductor's material, Ohm m; f—frequency, Hz;
- Resistance and internal reactance of the inductor,

$$R_i = X_i' = \rho_i \cdot \frac{\pi(d_i + \delta_i)}{\ell_i \cdot \delta_i} \cdot \frac{1}{k_{cu}}, \qquad (5.1)$$

with: $d_i = 2r_i$—internal diameter of the inductor coil, m; ℓ_i—axial inductor length, m; $k_{cu} = h_i/p_{it}$—filling factor of the inductor turns; h_i—axial length of one inductor turn, m; p_{it}—axial pitch of inductor turns, m.
- Reactance of no-loaded inductor

$$X_{i0} = 2\pi f \cdot \mu_0 \cdot \frac{A_{i0}}{\ell_i} \cdot k_N, \qquad (5.2)$$

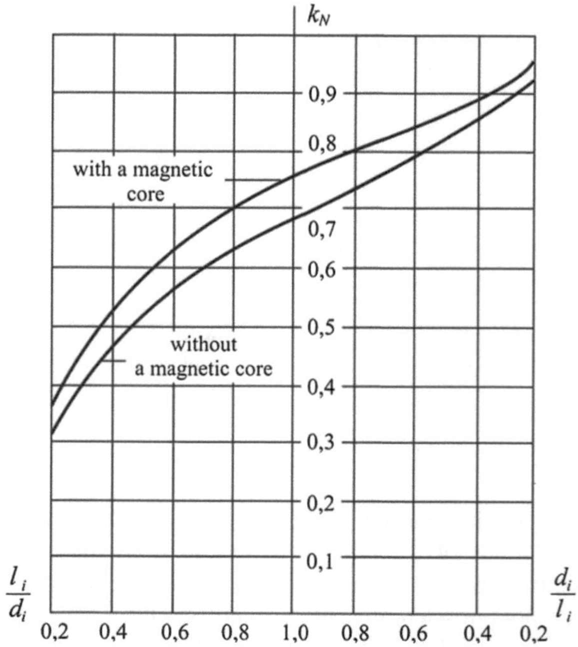

Fig. 5.3 Values of k_N as a function of the ratio d_i/ℓ_i of inductor internal diameter to its axial length (*upper curve*—inductor with external magnetic yoke; *lower curve*—inductor without magnetic yoke)

where: $A_{i0} = \left(\pi \cdot d_i^2\right)/4$—area of the cross-section of the no-loaded inductor, m^2; k_N—coefficient of Nagaoka for the calculation of the inductance of short coils, which takes into account the finite length of the inductor (Fig. 5.3).

- Reactance of the return path of flux

$$X_0 = X_{i0} \cdot \frac{k_N \cdot \ell_i}{\ell_i - k_N \cdot \ell}, \tag{5.2a}$$

with ℓ—load axial length, m.
Note For $\ell > \ell_i$, the calculation is done assuming $\ell = \ell_i$ since the induced power decays very rapidly to zero outside the inductor.

- Leakage reactance

$$X_a = 2\pi f \cdot \mu_0 \cdot \frac{A_{i0} - A_w}{\ell}, \tag{5.3}$$

with: $A_w = \left(\pi \cdot d^2\right)/4$, m^2; $d = 2r_e$—external diameter of the heated charge, m; $(A_{i0} - A_c)$—area of the cross section of the gap between inductor and heated charge, m^2.

- Penetration depth in the load

$$\delta = 503\sqrt{\frac{\rho}{\mu_r f}},$$

with: ρ, μ_r—resistivity and relative magnetic permeability of the load material.

- Dimensionless radius of the charge

$$m = \frac{\sqrt{2}\,r_e}{\delta} = \frac{d}{\sqrt{2}\,\delta},$$

- Resistance and reactance of the charge

$$R'_w + jX'_w = \rho\frac{2\pi\,r_e}{\ell\delta}\sqrt{2}(F + jG) = X_{i0}\left(\frac{r_e}{r_i}\right)^2\frac{\ell_i}{\ell}\mu_r(A + jB) \qquad (5.4)$$

where F, A and G, B are the coefficients of the active and reactive power in the considered load, which are given in Fig. 5.4 for a solid cylindrical charge.

The circuit of Fig. 5.2a can be reduced to the simple series equivalent circuit of Fig. 5.5, referred to the inductor current (i.e. to the inductor's clamps). Its parameters can be calculated with the equations:

$$\left.\begin{aligned} R_e &= R_i + n^2 R'_w \\ X_e &= X'_i - n^2 X'_w \end{aligned}\right\} \qquad (5.5)$$

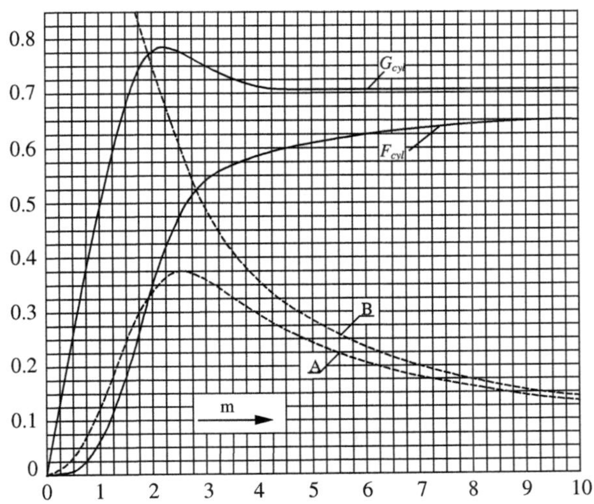

Fig. 5.4 Values of coefficients F_{cyl}, G_{cyl} and A, B for calculation of active and reactive power in a solid cylindrical charge as a function of m [$A = 2F_{cyl}/m$; $B = 2G_{cyl}/m$; for m ≥ 5: $F_{cyl} \approx 1/\sqrt{2} - 1/2m$; $G_{cyl} \approx 1/\sqrt{2}$]

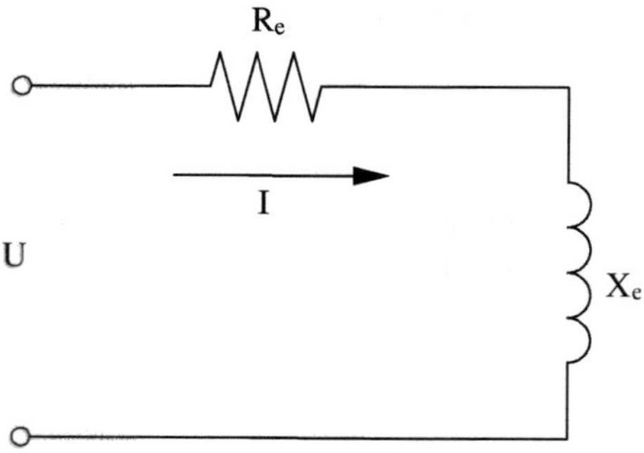

Fig. 5.5 Equivalent series circuit of the inductor-load system

$$Z_e = \sqrt{R_e^2 + X_e^2} \tag{5.6}$$

where:

$$X = -\left[\left(X_a + X'_w\right) + \frac{\left(X_a + X'_w\right)^2 + R'^2_w}{X_0} \right],$$

$$n^2 = \frac{1}{\left(\frac{R'_w}{X_0}\right)^2 + \left(1 + \frac{X_a + X'_w}{X_0}\right)^2}.$$

- The power factor of the inductor-load system therefore is:

$$\cos\varphi = \frac{R_e}{Z_e}. \tag{5.7}$$

- and the electrical efficiency:

$$\eta_e = \frac{n^2 R'}{R_e}. \tag{5.8}$$

Remarks This method is applicable also to limit cases of empty inductor ($\rho \to \infty$, $\ell \to 0$ or $r_e \to 0$) or frequency approaching infinity ($f \to \infty$).

It can be used for magnetic, non-magnetic, homogeneous or layered loads, with cylindrical cross-section or other shapes (oval, tubular, rectangular, etc.) and also in the case of an inductor—without internal magnetic core—located inside a cavity of the body to be heated.

However, this calculation procedure should be applied with particular caution in the following cases.

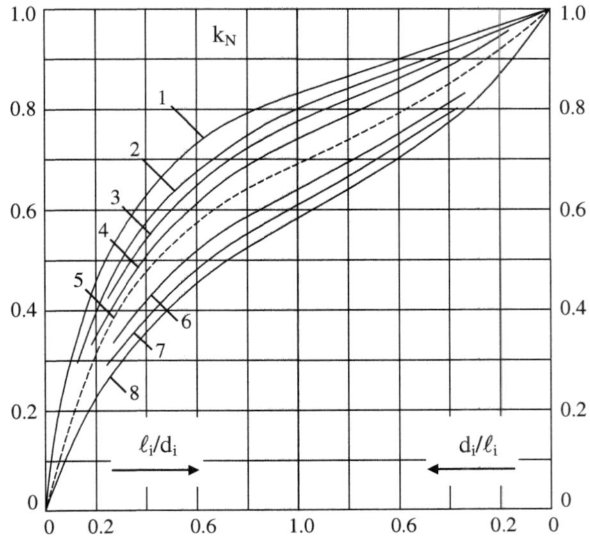

Fig. 5.6 Coefficient k_N for calculation of inductance of cylindrical solenoids with external magnetic yokes (*curves 1–4*); without yoke (*dashed curve 5*); for single-turn coil at high frequency (*curves 6–8*). [*curves 1–4* $d_f/d_i = 1 - 1.1 - 1.2 - 1.5$; *curve 5* as in Fig. 5.3; *curves 6–8* $s/d_i = 0.05 - 0.25 - 0.6$; d_f—yoke internal diameter; s-coil turn radial thickness]

1. *Inductors with small number of turns at high frequency.* The parameters can be strongly influenced by the radial thickness s of the inductor coil conductor. The coefficient k_N can be determined as a function of the ratios d_i/ℓ_i and s/ℓ_i by the curves 6–8 of Fig. 5.6 [2].
2. *Inductors with rectangular cross-section of sides (h) and (w)*—The coefficient k_N can be determined by the diagrams of Fig. 5.7, as a function of the ratios of geometrical dimensions h/ℓ_i and h/w of the coil, whereas for the calculation of the load impedance, reference must be made to coefficients A and B (or F and G), relevant for the case under consideration.
3. *Cylindrical inductors with external magnetic yokes*—The inductors considered here have straight magnetic yokes without pole shoes, such as those used in melting furnaces. The yokes are usually constituted by packets of non-saturated laminations of length ℓ_f longer than the coil ($\ell_f > \ell_i$), uniformly distributed and conveniently spaced outside the coil along a circumference of diameter d_f. The values of k_N can be determined with good approximation by curves 1–4 of Fig. 5.6, which is valid for the case $\ell_f \gg \ell_i$.
4. *Internal inductors without magnetic cores*—In this case, the flux return path is the internal space of the inductor and the reactance X_0 is practically equal to that of the empty inductor. Moreover, the inductor resistance can be notably lower (up to 50 %) than the one given by the Eq. (5.1), since in this case the current is distributed in the conductor cross section both on the external region facing the load, for proximity effect, and the internal one, due to ring effect [1, 2].

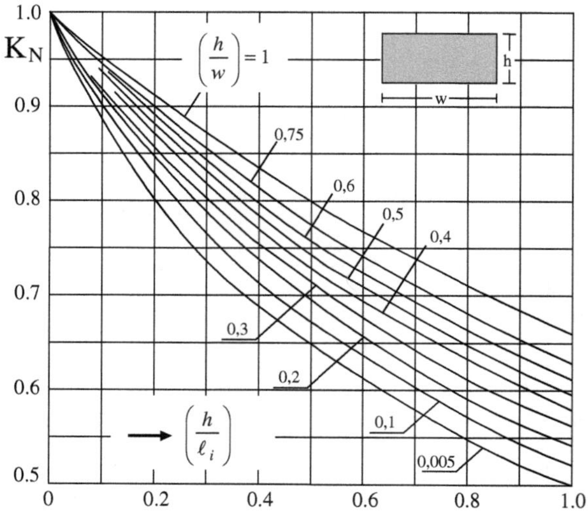

Fig. 5.7 Values of k_N for calculation of inductance of short solenoids with *rectangular* cross-section of sides (h) and (w)

5.2 Calculation of Induction Heating System with Analytical Methods

In some applications the analytical solutions extensively described in Sect. 4.5 can be particularly useful since, through the use of data tabulated in diagrams, they allow a quick evaluation of the system parameters, with an accuracy sufficient for technical calculations.

Among them, the most interesting one for the applications is the case of systems constituted by a "long" cylindrical load (with constant resistivity and magnetic permeability), heated with a single "short" inductor, as sketched in Fig. 5.8. In practice the load can be considered "long" when its length is greater than the sum of the inductor length and two times the radial gap between inductor and load.

With reference to the equations presented in Sect. 4.5, here we focus the attention on the equivalent impedance of the system.

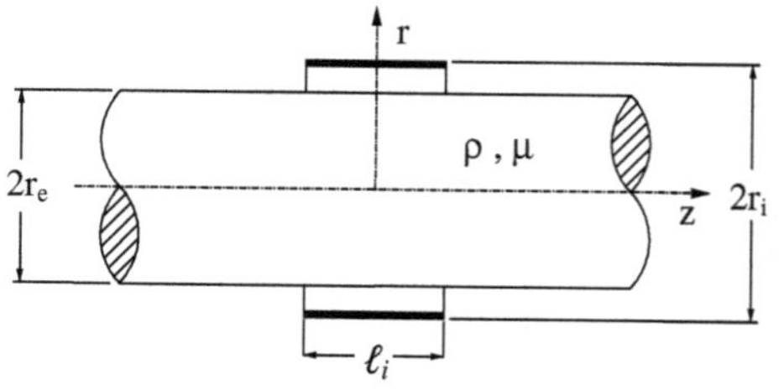

Fig. 5.8 Cylindrical inductor-load system

In Sect. 4.5 the general solution, valid for systems as in Fig. 4.64b, comprising mutually coupled multiple inductors with external magnetic yokes, was written in the form [see Eq. (4.42)]

$$\dot{Z}_e = (R_i + R') + j(X_i - \Delta X) = R_i + jX_i + \frac{X_{i0}}{\alpha^2}(F - jG) \tag{5.9}$$

In the system with multiple inductors, good results for a single coil inductor can be obtained by choosing sufficiently high values of the axial distance between inductor coils and the radius of the external magnetic yoke, so that the mutual coupling between coils and the influence of the yoke are negligible.

Since for a single coil it is $X_i = X_{i0}k_N$, in this case the previous equation can be rewritten in the form:

$$\dot{Z}_e = R_i + \frac{X_{i0}}{\alpha^2}\left[F + j\left(\alpha^2 k_N - G\right)\right] \tag{5.10}$$

with:

R_i, R'	own resistance of induction coil and resistance of the load seen at the coil terminals
X_i, X_{i0}	reactance of one coil of the unloaded system, with and without end effects
ΔX	variation of reactance of one coil of the system from unloaded to loaded conditions
$\alpha = r_i/r_e$	ratio of internal radius of inductor coil and external radius of load
F, G	dimensionless coefficients which account for the influence of load on the impedance $\dot{Z}_e$
k_N	coefficient of Nagaoka (see Fig. 4.6)

The coefficients F and G, for loads consisting of non-magnetic, solid cylinders with $\rho = \cos t.$; $\mu_r = 1$, depend on the dimensionless ratios:

$$m = \frac{\sqrt{2}\, r_e}{\delta}; \quad \alpha = \frac{r_i}{r_e}; \quad \frac{\ell}{2\, r_i}.$$

Their values are given in the diagrams of Figs. 4.65 and 4.66, where—as reference—are also shown the data valid for "long" inductors ($\ell/2\, r_i \; \rightarrow \; \infty$).

For loads consisting of hollow non-magnetic cylinders, the coefficients F and G depend also on a fourth parameter, r_i'/r_e, the ratio between internal and external radii of the load; the corresponding diagrams can be found in Ref. (2, pp. 268–270).

Data for cylindrical ferromagnetic solid workpieces are given in Sect. 3.2; for cylindrical tubes can be found in Ref. (2, pp. 152–155).

5.3 1D Finite Difference Numerical Solution for Induction Heating

The design of induction and direct resistance heating systems is nowadays conveniently carried out by resorting to numerical methods. The possibility to carry out coupled electromagnetic and thermal calculation is the main feature that a commercial code must exhibit to properly compute the performances of induction or direct resistance heaters.

Addressing the reader to the wide literature available on this subject [3], in the following it is presented a simple procedure to solve a coupled EM and Thermal induction heating problem by means of finite difference method in 1D axis-symmetric domain [4].

This procedure is very simple to be implemented in a code that can be very useful for a preliminary design of the heating system, in particular long heating lines for through heating applications. It is able to solve non-linear problems, where all material properties are temperature dependent and the magnetic permeability depends also on the magnetic field intensity.

As usual, the electromagnetic problem is solved to evaluate the induced power density distribution while the thermal problem is solved for the determination of the temperature field at the end of the considered time interval. Both problems are solved alternatively and separately for any elementary time step Δt in which the thermal transient is subdivided [5].

The calculation model makes reference to a cylindrical workpiece of "infinite axial length", which is subdivided into N concentric elements of equal radial thickness Δr, as schematized in Fig. 5.9.

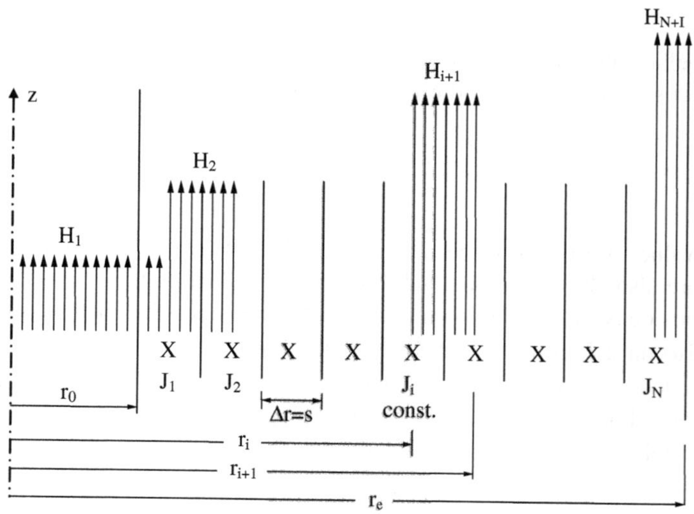

Fig. 5.9 Schematic of subdivision of a cylindrical load [$r_0 \neq 0$ in case of a tube]

5.3.1 Electromagnetic Problem Solution

As shown in the figure, it is assumed that the current density is constant in any element and that the corresponding total current of each element is concentrated in an infinitely thin current layer located in the middle of the element. In this way, the magnetic field intensity varies stepwise and results to be constant between the centerline of one element and the next one. Therefore it is:

$$H_{i+1} = \cos t. \quad \text{for} \quad r_i \leq r \leq r_i + \Delta r$$

$$J_i = \cos t. \quad \text{for} \quad r_i - \frac{\Delta r}{2} \leq r \leq r_i + \frac{\Delta r}{2}$$

The 1-D assumptions means that the magnetic field intensity H has only one spatial component, directed along the z axis in Fig. 5.9, while the induced currents are azimuthally directed. Under these assumptions, assigning an arbitrary value to H_1, the values of the field can be calculated subsequently for any layer, starting from the inner one and moving towards the surface, according to the following scheme:

Calculation of Field Values in Layer 1

- Magnetic permeability in the first layer, dependent on the value arbitrarily assigned to H_1:

$$\mu_1 = f(H_1) \tag{5.11}$$

- Flux linked with the strand-like winding of radius r_1

$$\dot{\Phi}_1 = \pi \dot{H}_1 \left[\mu_0 r_0^2 + \mu_1 \Delta r \left(r_0 + \frac{\Delta r}{4}\right)\right] \tag{5.12}$$

- Induced electric field in the winding of radius r_1

$$\dot{E}_1 = -j\omega \dot{\Phi}_1 \tag{5.13}$$

- Resistance of the layer (evaluated as D.C. resistance, due to the assumption of uniform current density in the layer)

$$R_1 = \rho_1 \frac{2\pi r_1}{\Delta r_1 \ell} \tag{5.14}$$

- Induced current in the layer (supposed to be concentrated at radius r_1)

$$\dot{I}_1 = \frac{-j\,\omega\,\dot{\Phi}_1}{R_1} \tag{5.15}$$

- Current density in layer 1

$$\dot{J}_1 = \frac{-j\omega\,\dot{\Phi}_1}{\rho_1\,2\,\pi r_1} \tag{5.16}$$

Field Values in the Second Layer

Applying to the derivate $\frac{\partial H}{\partial r} = -J$ the forward difference formula, it follows:

$$\left.\begin{aligned} \dot{H}_2 &= \dot{H}_1 - \Delta r \cdot J_1 \\ \mu_2 &= f(H_2) \\ r_2 &= r_1 + \Delta r \\ A_2 &= \pi\,(2\,r_2 - \Delta r) \\ \dot{\Phi}_2 &= \mu_2\,\dot{H}_2\,A_2 \\ \dot{J}_2 &= \frac{-j\,\omega\,(\dot{\Phi}_1 + \dot{\Phi}_2)}{\rho_2 2\pi\,r_2} \end{aligned}\right\} \tag{5.17}$$

For the Generic 'i-th' Layer

$$\left.\begin{aligned} \dot{H}_{i+1} &= \dot{H}_i - \Delta r \dot{J}_i \\ \mu_{i+1} &= f(H_{i+1}) \\ r_{i+1} &= r_i + \Delta r \\ A_{i+1} &= \pi \Delta r_i (2 r_{i+1} - \Delta r) \\ \dot{\Phi}_{i+1} &= \mu_{i+1} \dot{H}_{i+1} A_{i+1} \\ \dot{J}_{i+1} &= \frac{-j\omega(\dot{\Phi}_1 + \dot{\Phi}_2 + \cdots + \dot{\Phi}_{i+1})}{\rho_{i+1} 2\pi r_{i+1}} \end{aligned}\right\} \tag{5.18}$$

The calculation proceeds until the surface layer has been reached.

On the external surface, the magnetic field H_{N+1} is assumed to be constant in the last layer of thickness $\Delta r/2$.

Since the value of H_1 has been chosen arbitrarily, also the value H_{N+1} will not correspond to the surface value H_e, that is supposed to be a given data of the problem.

In case of a workpiece of non-magnetic material, due to linearity, it is only necessary to multiply all field values by the ratio H_e/H_{N+1}.

On the contrary, in the case of ferromagnetic materials, an iterative procedure must be used by modifying the initial value of H_1 at any iteration until the convergence of H_{N+1} to the value H_e has been reached.

This procedure allows the take into account different values of ρ_i (as a function of the temperature), and μ_i (as a function of temperature and local magnetic field intensity) in each layer.

When the current density as well as the values of resistivity of any layer are known, the volumetric specific power $w_i = \rho_i J_i^2$ is computed and applied as the source quantity for the solution of the thermal problem.

5.3.2 Solution of Thermal Problem

The solution allow the determination of the temperature of any layer at the end of the elementary time interval Δt, starting from the temperature at the beginning of the interval.

Therefore, the method allows to update the values of resistivity and permeability in each layer as a function of the temperature and the local magnetic field strength before initiating a new electromagnetic calculation for the successive interval Δt.

With the notations:

- ϑ_{i-1}, ϑ_i, ϑ_{i+1}—temperature of layers $(i-1)$, i, $(i+1)$ at time t (the beginning of time step Δt)
- ϑ'_{i-1}, ϑ'_i, ϑ'_{i+1}—temperature of the same layers at time $t + \Delta t$ (the end of the time interval Δt)
- s—radial thickness of subdivision layers for the thermal problem ($s = \Delta r$, if the same subdivision is assumed in the electromagnetic and thermal problem)
- $k_i = \lambda_i / c_i \gamma$—thermal diffusivity of the material at temperature of the i-th layer

The thermal balance of the generic 'i-th' layer at the generic time step Δt, for unit length of the system, is computed with the following terms:

- Heat flux entering from the surface $2\pi(r_i + s/2)$:

$$Q_{ei} = \lambda_i\, 2\pi(r_i + \frac{s}{2}) \frac{\vartheta_{i+1} - \vartheta_i}{s} \Delta t$$

- Heat flux leaving from the surface $2\pi(r_i - s/2)$:

$$Q_{ui} = -\lambda_i 2\pi(r_i - \frac{s}{2}) \frac{\vartheta_i - \vartheta_{i-1}}{s} \Delta t$$

- Heat produced inside the layer:

$$Q_{wi} = 2\pi r_i \cdot s \cdot w_i \Delta t$$

- Heat quantity increasing the temperature from ϑ_i *to* ϑ_i':

$$Q_i = 2\pi r_i \cdot s \cdot c_i \gamma (\vartheta_i' - \vartheta_i)$$

- Thermal balance of the layer:

$$Q_{ei} + Q_{ui} + Q_{wi} = Q_i \tag{5.19}$$

By substituting the quantities previously computed and the joule losses computed in the field values step, for $i = 2, 3, \ldots, N-1$ we obtain:

$$\begin{aligned} \vartheta_i' = \vartheta_i + \frac{k_i \Delta t}{s^2} \Big\{ (1 + \frac{s}{2r_i}) \cdot \vartheta_{i+1} - [(1 + \frac{s}{2r_i}) + (1 - \frac{s}{2r_i})] \cdot \vartheta_i \\ + (1 - \frac{s}{2\,r_i}) \cdot \vartheta_{i-1} \Big\} + \frac{w_i \Delta t}{c_i \gamma} \end{aligned} \tag{5.20}$$

This equation gives in explicit form the temperature of the '*i*-th' layer at time $(t + \Delta t)$, as a function of the temperature in the same and the adjacent layers at time t.

The thermal balance has to be computed separately for the first and the last (*N-th*) layers, since it depends upon the boundary conditions.

For the first internal layer, the condition $Q_{u1} = 0$, that no thermal flux exists on the axial direction, gives:

$$\vartheta_1' = \vartheta_1 + \frac{k_1 \Delta t}{s^2} (1 + \frac{s}{2r_1}) \cdot (\vartheta_2 - \vartheta_1) + \frac{w_1 \Delta t}{c_1 \gamma} \tag{5.21a}$$

For the external *N-th* layer, the thermal balance has to consider the losses to the ambient due to convection and radiation. These losses can be described by the following equation:

$$Q_{eN} = -\alpha \cdot 2(r_N + \frac{s_N}{2}) \cdot (\vartheta_e - \vartheta_a) \Delta t \tag{5.21b}$$

where:

$$\alpha = \alpha_c + \alpha_i$$

α_c	heat convection coefficient
$\alpha_i = \varepsilon\sigma \frac{T_e^4 - T_a^4}{\vartheta_e - \vartheta_a}$	"linearized" radiation coefficient
$\vartheta_e = \vartheta_N + \frac{\vartheta_N - \vartheta_{N-1}}{2}$	temperature of the external surface,
ϑ_e, ϑ_a	external and ambient temperature
ε	emissivity of the workpiece surface
$\sigma = 5.67 \times 10^{-8}$ [$W/m^2 \cdot K^4$]	Stefan–Boltzmann's constant

Summarizing, in the case of an uniform subdivision with layers of constant thickness s and assuming $\vartheta_e \approx \vartheta_N$ and $\vartheta_a = 0$, we obtain:

$$\left.\begin{aligned} \vartheta_1' &= \vartheta_1 + \frac{2\,\mathrm{k}_1\Delta\mathrm{t}}{\mathrm{s}^2}(\vartheta_2 - \vartheta_1) + \frac{\mathrm{w}_1\Delta\mathrm{t}}{\mathrm{c}_1\gamma} \\ \vartheta_\mathrm{i}' &= \vartheta_\mathrm{i} + \frac{\mathrm{k_i}\Delta\mathrm{t}}{\mathrm{s}^2}\left(1+\frac{\mathrm{s}}{2\mathrm{r_i}}\right)\vartheta_{\mathrm{i}+1} - \left(1-\frac{\mathrm{s}}{2\mathrm{r_i}}\right)\vartheta_{\mathrm{i}-1} + \frac{\mathrm{w_i}\Delta\mathrm{t}}{\mathrm{c_i}\gamma} \\ \vartheta_\mathrm{N}' &= \vartheta_\mathrm{N} + \frac{\mathrm{k_N}\Delta\mathrm{t}}{\mathrm{s}^2}\left\{\left[\left(1-\frac{\mathrm{s}}{2\mathrm{r_N}}\right) - \frac{\alpha\mathrm{s}}{\lambda_\mathrm{N}}\left(1+\frac{\mathrm{s}}{2\mathrm{r_N}}\right)\right]\vartheta_\mathrm{N}\right. \\ &\quad \left. + \left(1-\frac{\mathrm{s}}{2\mathrm{r_N}}\right)\vartheta_{\mathrm{N}-1}\right\} + \frac{\mathrm{w_N}\Delta\mathrm{t}}{\mathrm{c_N}\gamma} \end{aligned}\right\} \quad (5.21c)$$

Since the solution of the problem is obtained in explicit form, its computation is really simple. However, in order to assure stability of the solution, it must be satisfied the condition:

$$\frac{k_i \Delta t}{s^2} \leq \frac{1}{2} \quad (5.22)$$

Therefore, when the thickness s is given, the time step has to satisfy the condition:

$$\Delta t \leq \frac{s^2}{2\,k_i}$$

5.4 Commercial 1D Code *ELTA* for Induction Heating

The available commercial codes are usually ‘general purposes’, not specifically developed for electro-technologies. The 1D code “*ELTA*” which is described in the following is an exception because it is intended only for induction heating.

Obviously, it can be conveniently used when the object of simulation has a simple geometry that can be reasonably assumed to be one-dimensional. This is the main limitation of the software but at the same time it makes *ELTA* very easy and fast in the usage. The software requires just few numerical input data and allows to easily post process the main quantities of interest.

The code provides a very fast and accurate simulation of the electromagnetic and thermal fields using a combination of one-dimensional Finite Difference approach (similar to the one described in Sect. 5.3) for solving the coupled EM and Thermal problems and an analytical one, based on “Total Flux Method” [6], which takes into account the finite lengths of workpiece and induction coil.

It can simulate the majority of basic induction systems: cylindrical or plane, magnetic or non-magnetic, solid or hollow, even with multi-layered load with arbitrary properties and initial temperatures of the layers' materials. *ELTA* has the possibility to take into account specific features of the induction heating installations, like the capacitor tanks for resonance, the connection circuit between power supply and inductor, the cooling stage, either natural or forced, and many others.

The simulation of a multi inductor system can be very easily implemented defining the characteristics of each heating or cooling stage. Finally, it is remarkable that this code allows to supply the inductor not only by an imposed current or voltage, but also to assign a constant active power to the inductor-workpiece system, useful to simulate a process where the resonant generator is controlled by the output power.

5.4.1 Case Study 1: Through Heating of Non-magnetic Steel Billets

In this section the mass heating of a steel billet is presented as a case study developed with *ELTA*.

The workpiece is a cylindrical billet, 700 mm length and 40 mm diameter, of non-magnetic stainless steel. This example makes reference to a long cylindrical load-inductor system, where analytical formulas give accurate results.

The inductor is made of 32 turns, made of copper tube, 18 × 25 × 2.5 mm rectangular cross section, internal radius 35 mm, total inductor length 700 mm.

The inductor is fed with constant current $I = 1\,kA$ at 2 kHz. The electrical resistivity of the billet material is assumed to be of 100 μΩ cm, independent of temperature.

The analytical formulas for a cylindrical system allow to calculate the main system quantities from the knowledge of the external magnetic field, the skin depth and the dimensionless parameter m:

$$H_e = \frac{NI}{\ell} = 45.700,\ \mathrm{A/m}; \quad \delta = 503\sqrt{\frac{\rho}{\mu \mathrm{f}}} = 1.125 \times 10^{-2}, m;$$
$$m = \sqrt{2}r_e/\delta = 2.5.$$

From Eq. (3.11), the active and reactive power in the load can be computed with the values of parameters F_{cyl} and G_{cyl} evaluated for $m = 2.5$. From Fig. 3.5 we have:

$$\text{for}\, m = 2.5: \quad F_{cyl} \approx 0.67 \quad G_{cyl} \approx 1.1.$$

and

$$p = H_e^2 \frac{\rho}{\delta} F_{cyl} = 2 \times 10^{-3} \cdot (N_1 I)^2 \cdot \sqrt{\rho \mu_r f} \cdot F_{cyl} \approx 124\, kW/m^2$$
$$q = H_e^2 \frac{\rho}{\delta} G_{cyl} = 2 \times 10^{-3} \cdot (N_1 I)^2 \cdot \sqrt{\rho \mu_r f} \cdot G_{cyl} \approx 204\, kVAr/m^2.$$

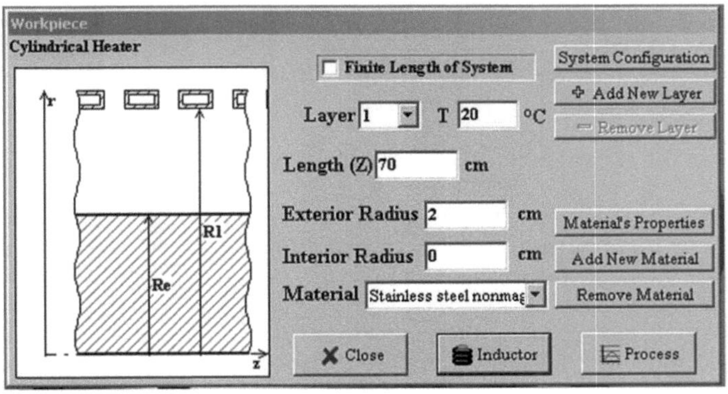

Fig. 5.10 Elta data input interface to entry the load properties

Multiplying by the billet external surface we obtain the total active and reactive power in the billet:

$$P = 2\pi R_e \cdot l \cdot p \approx 10.9\,kW; \quad Q = 2\pi R_e \cdot l \cdot q \approx 18\,kVAr.$$

The above quantities divided by the square of the current intensity in the inductor, give the equivalent resistance and reactance of the load.

From Eq. (4.1), the equivalent resistance of the inductor is

$$R_i = X_i' = \rho_i \cdot \frac{\pi(d_i + \delta_i)}{\ell_i \cdot \delta_i} \cdot \frac{1}{k_{cu}} = 4.5\,m\Omega$$

with: $\rho_i = 2 \times 10^{-8}$ Ωm—resistivity of copper; $\delta_i = 1.59$ mm—skin depth in the inductor; k_{cu}—filling factor of the inductor turns.

The total active power at inductor clamps is the sum of the induced power P, and the power losses in the inductor copper $P_i = 4.5 \times 10^{-3} \times 10^6 = 4.5\,kW$.

The same problem has been solved using *ELTA*.

The pre-processor interface of *ELTA* is shown in Fig. 5.10. In this window, the user has to entry the main geometric quantities, the load external radius, the internal one for tubes and the axial length of the system.

In addition, the user has to select the load material properties from the material's data base, which contains the relevant material characteristics for a coupled electromagnetic and thermal analysis: electrical resistivity, thermal conductivity, specific heat and mass density, all temperature dependent, and magnetic permeability as a function of magnetic field H and temperature.

The material properties of non-magnetic stainless steel are almost linearly temperature dependent: in the *ELTA* database electrical resistivity varies from 70×10^{-8} Ωm at 20 °C up to 125×10^{-8} Ωm at 1,000 °C.

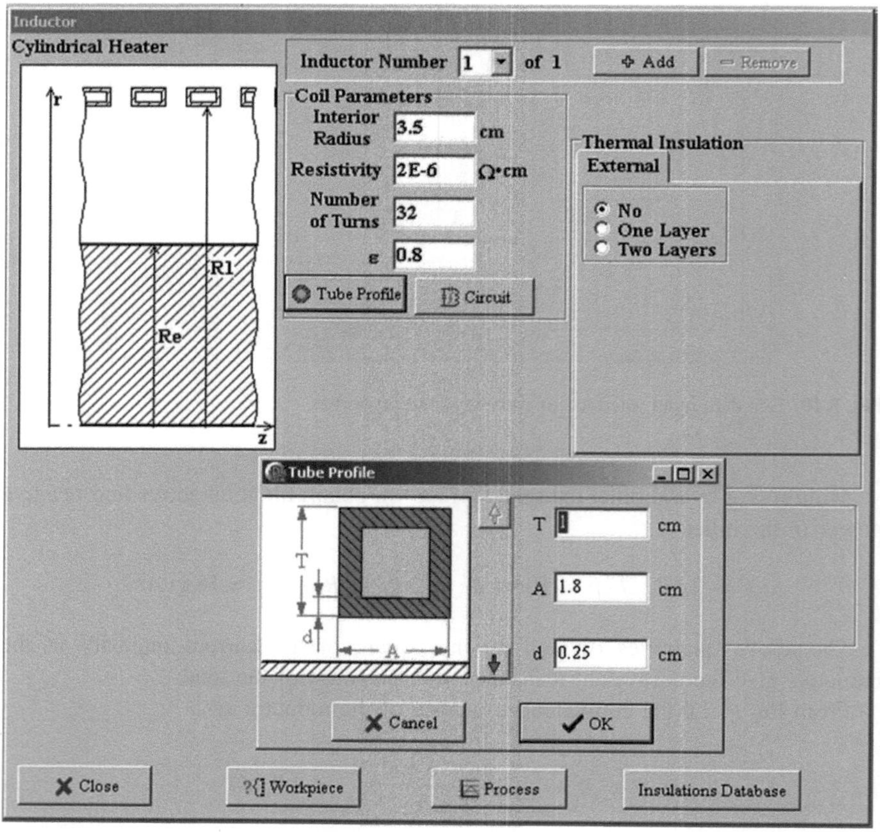

Fig. 5.11 Data input interface to entry inductor characteristics

Figure 5.11 shows the window to entry the characteristics of the inductor. The code allows to take into account the presence of magnetic concentrators and the inductor thermal insulation and to describe the circuit connecting the power generator to the inductor, with series or parallel connection of resonance capacitors.

The entry of the process parameters is carried out through the window interface shown in Fig. 5.12.

This window gives the possibility to describe different heating regimes: e.g. with constant current or voltage applied at the inductor clamps or the generator output, or with constant active power in the workpiece.

The program allows also to define the cooling conditions of the load by radiation or natural or forced convection.

The results of the calculation made with *ELTA* are summarized in Table 5.1 and Fig. 5.13.

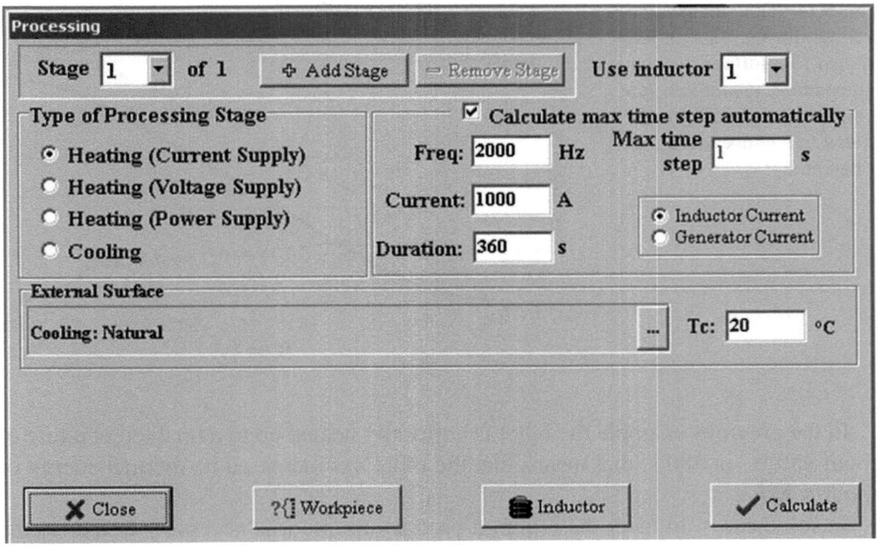

Fig. 5.12 Elta data input interface to entry the process data

Table 5.1 ELTA average integral parameters of the inductor-load system

Field strength	$H_e = 447\,A/cm$	Coil power factor	$\cos\varphi = 0.1855$
Coil current	$I = 1.000\,A$	Workpiece power	$P = 10.374\,W$
Coil voltage	$V = 83.51\,V$	Electrical efficiency	$\eta_e = 0.6736$
Coil power	$P_i = 15.401\,W$	Thermal efficiency	$\eta_{th} = 0.6938$
Coil impedance	$Z_e = 0.08305\,\Omega$	Total efficiency	$\eta = 0.4674$

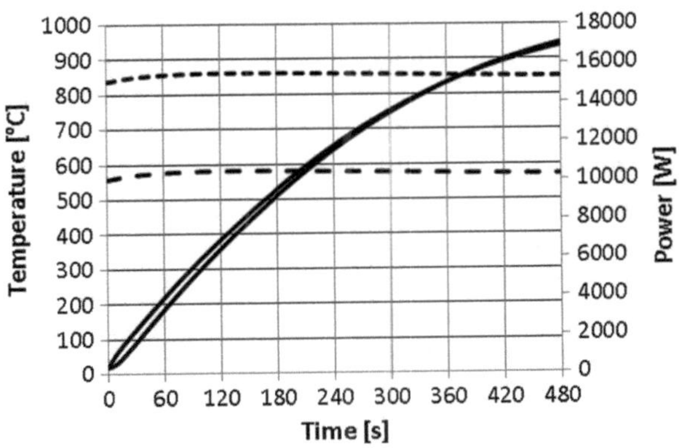

Fig. 5.13 ELTA transient solution [*dotted lines* active power in the workpiece and at inductor clamps; *continuous line* temperatures of axis and surface during heating]

Table 5.1 gives the average integral parameters of the inductor-load system, while Fig. 5.13 shows the transient temperature distributions on the axis and the surface as well the power in the load and the total power at the inductor clamps.

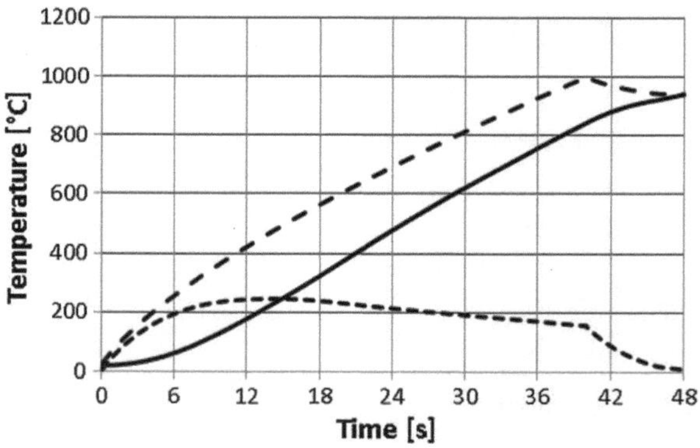

Fig. 5.14 Time-temperature diagram for the 'fast' process. [*dotted line* surface temperature; *continuous line* temperature of axis; *lower dotted line* temperature differential]

In the previous example the billet is uniformly heated up to a final temperature of about 950 °C in 480 s, that means that the billet has increased its thermal energy of about 4 MJ.

In some cases, to meet the required production rate it is necessary to reduce the heating time increasing the induced power. Moreover, the increase of the specific induced power reduces the influence of radiation thermal losses at high temperature on the overall process efficiency.

In order to illustrate this possibility the previous case has been solved with an inductor current $I = 3{,}000$ A.

Since the induced power depends on the square of the surface magnetic field intensity (or the square of the inductor current) when the material is non-magnetic, increasing the inductor current at 3,000 A (rms), all power values become nine times higher than in the previous case.

The computed heating transient of Fig. 5.15 shows that a total process time of 48 s allows to supply almost the same amount of thermal energy of 4 MJ given to the billet as in the previous case.

In order to obtain in the billet at the end of the process the temperature uniformity required by the subsequent hot working, the heating stage is followed by a soaking period, with no induced power.

During this stage the surface temperature decreases while the core temperature increases, providing the required final temperature differential (Fig. 5.14).

In this way the temperature difference of about 150 °C at the end of the heating stage is equalized in about 8 s. The real heating period is in this case of only 40 s, that is 12 times lower than the process time of the previous example, and the overall efficiency of the process increases from 47 to 64 %.

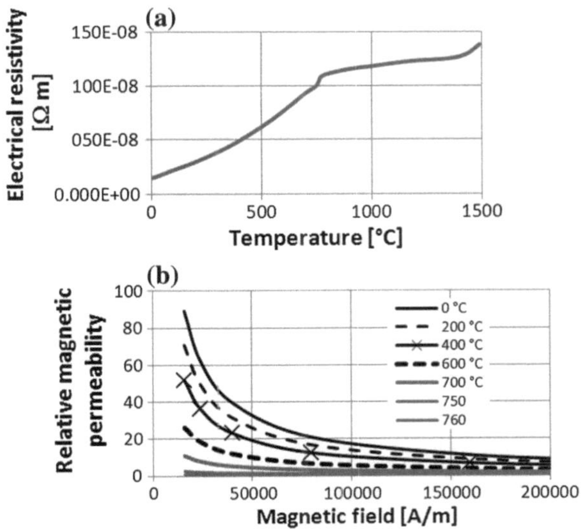

Fig. 5.15 Properties of C40 steel. **a** Electrical resistivity **b** Magnetic permeability, dependent on magnetic field intensity and temperature

5.4.2 Case Study 2: Through Heating of Magnetic Steel Billets

The same inductor of the case study 1 is used to heat a C40 steel billet with same geometry (i.e. cylindrical billet, 700 mm length, 40 mm diameter). The material properties of C40 steel exhibit a strong non linear behavior.

Electrical resistivity (Fig. 5.16a), varies as a linear function of temperature with a higher slope before Curie point. After Curie and before melting temperature, electrical resistivity reaches values around 110–130 × 10^{-8} Ωm, i.e. about the same of stainless steel. At the Curie point (about 760 °C), the material changes abruptly its magnetic characteristic, as shown in Fig. 5.16b which gives the relative permeability as a function of magnetic field and temperature.

Also the thermal properties like thermal conductivity and specific heat (see Fig. 5.17) show a dependence from temperature that strongly affects the induction heating process, in particular at the transition of the Curie point. The peak of the heat capacity for temperature close to Curie is also responsible of the lower heating rate in these heating phase.

Like in previous cases, we consider a heating process with final average workpiece temperature of about 950 °C, almost uniform in the cross section.

As shown in Fig. 5.18, with inductor current I = 1,000 A, the temperature in the magnetic steel increases rapidly until Curie point, while after Curie point, it rises at almost the same, slow, rate of the previous case study for stainless steel. This stage represents again a 'slow rate' heating process, where the temperature distribution in the cross section becomes homogenous after Curie point.

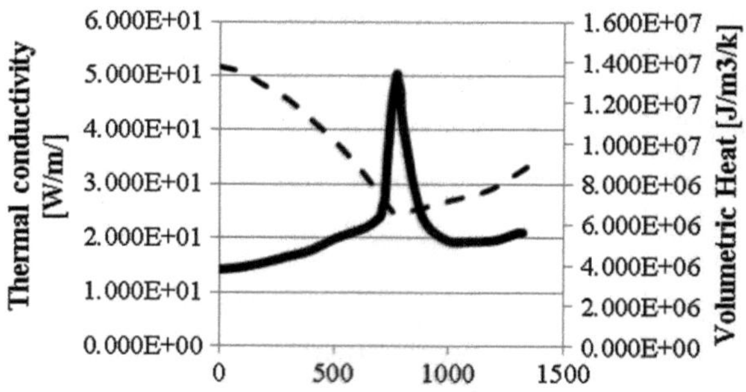

Fig. 5.16 Thermal properties for a C40 steel (*dotted line*—thermal conductivity; *continuous line* volumetric heat)

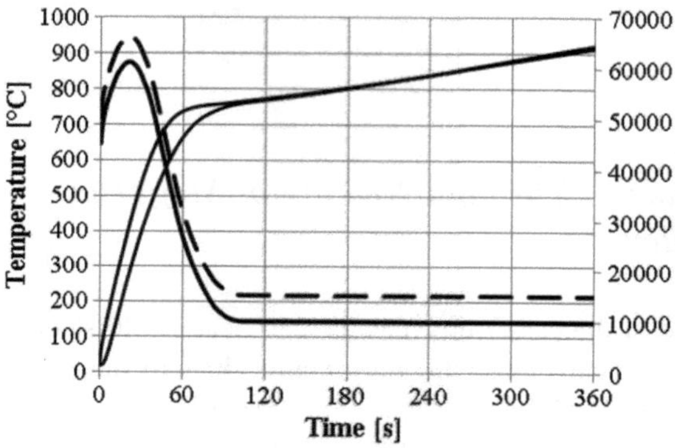

Fig. 5.17 ELTA transient solution with inductor current I = 1 kA rms (*dotted line* active power at inductor clamps; *continuous line* active power in the workpiece: *thin continuous lines* axis and surface temperature)

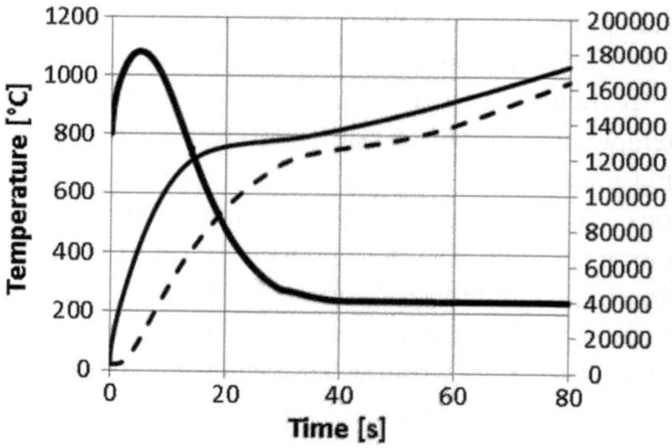

Fig. 5.18 ELTA transient solution with I = 2 kA rms [*thick continuous line* active power in the workpiece; *thin continuous line* surface temperature; *dotted line* axis temperature]

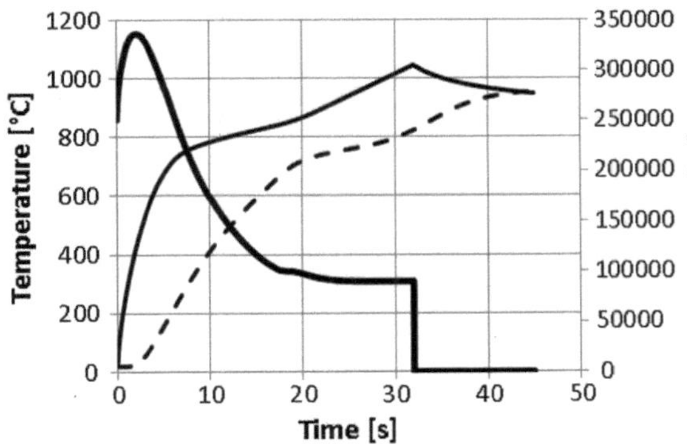

Fig. 5.19 ELTA transient solution with I = 3 kA rms [*thick continuous line* active power in the workpiece; *thin continuous line* surface temperature; *dotted line* axis temperature]

Fig. 5.20 Schematic for the calculation of inductor-load system with the method of the mutual coupled circuits

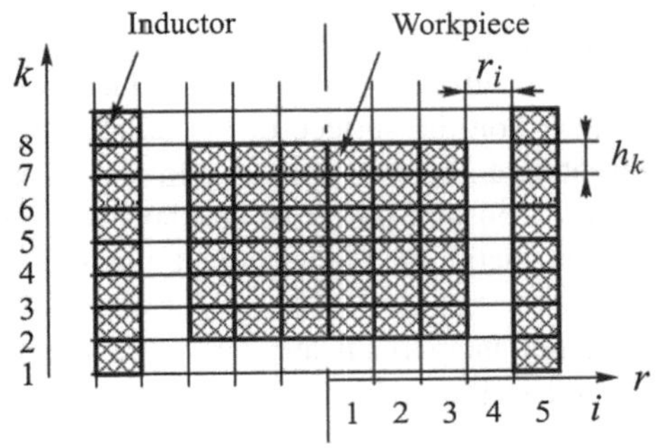

With $I = 2{,}000$ A (Fig. 5.19), the induced power in the workpiece is much lower than 4 times the one of Fig. 5.18 because the variation of the ferromagnetic properties with a higher magnetic field strength, lead to a relatively low increase of the induced power.

With I = 3,000 A (Fig. 5.20), higher saturation occurs and the increase of induced power is further comparatively small. In this case, the heating time is about 32 s, but a homogeneous temperature distribution in the cross section requires a stage of soaking of 13 s.

In Table 5.2 are given the values of the equivalent series resistance of the workpiece R_w calculated in three different heating stages.

The equivalent resistance of the first stage describes the linear heating that occurs until the surface reaches a temperature close to Curie point, when magnetic properties start to vanish. Here for convenience it is assumed that the first stage ends when the induced power is equal to the power induced at the start of process, when the material is cold.

Table 5.2 Equivalent load resistance for ferromagnetic material

	μ_r	R_w (mΩ) (≈0–700 °C)	R_w (mΩ) (≈700–760 °C)	R_w (mΩ) (≈760–1,000 °C)
I = 500 A (He ≈ 22 kA/m)	65	74	28	10.2
I = 1000 A (He ≈ 45 kA/m)	35	50	19.5	10.2
I = 2000 A (He ≈ 90 kA/m)	18	40	18.8	10.2
I = 3000 A (He ≈ 135 kA/m)	12	34	18.3	10.2

The second intermediate stage of heating lasts until the induced power reaches an almost constant value.

In the last stage, the electrical resistivity has small variation that leads to an almost constant induced power. This occurs when the electromagnetic wave decays completely in the non-magnetic surface layer (Table 5.2).

The solutions with constant inductor current, or by imposing the magnetic field intensity on the surface of the billet, are proposed here only as a case study representing a simplified process. In real cases, the power supply is controlled in order to feed constant power or constant voltage to the inductor clamps. In these cases, the magnetic field intensity at the surface of the workpiece varies during the process.

Often, industrial heaters are of the progressive type with inductors constituted by several in-line induction coils, which can even supplied with different frequency and power. In this case the workpieces are moved through the various coils and are progressively heated according to an optimized process, that maximizes the overall efficiency.

5.5 VIM: Volume Integral Method of the Mutually Coupled Circuits

In this section, the Volume Integral Method (VIM) is briefly introduced. VIM is an integral methods that have given good results for the calculation of induction heating systems [7–9].

The main advantage of integral methods is that the field calculation is carried out only in the electrical conductive regions i.e. the 'active' parts of an induction device. Other numerical methods based on differential equations, require to calculate electromagnetic fields also in 'not active' regions, like the air surrounding inductors and load. As a consequence, the study domain of integral methods is significantly reduced and also the number of unknowns is usually smaller in comparison with differential methods. The main drawback of integral methods is that they require the solution of a system of linear equations which lead to fully populated matrix. The numerical treatment of fully populated matrix requires high computational time and memory, that almost growth with the square of the number of unknowns, and this can limit the application of integral method to very large scale problems.

The essence of this method is illustrated in Fig. 5.21. The inductor and the load of a cylindrical system are replaced with a number of elementary annular coils with radial and axial dimensions sufficiently small to consider current density uniformly distributed in their cross section. The inductor-load system can be described by an electrical circuit where each i-th elementary coil is replaced by its resistance, self-inductance, and mutual inductances M_{ik} between the i-th coil and all the other k coils.

These annular elements realize a system of coupled circuits that can be solved applying Kirchhoff's laws. The solution of the resulting system of linear equations allows to calculate the current and power density distributions in the cross section of inductor and load

Assuming a sinusoidal variation in time of currents and voltages, the system of equations with complex coefficients can be written as follows in matrix form:

$$|\dot{Z}| \cdot |\dot{I}| = |\dot{U}| \tag{5.23}$$

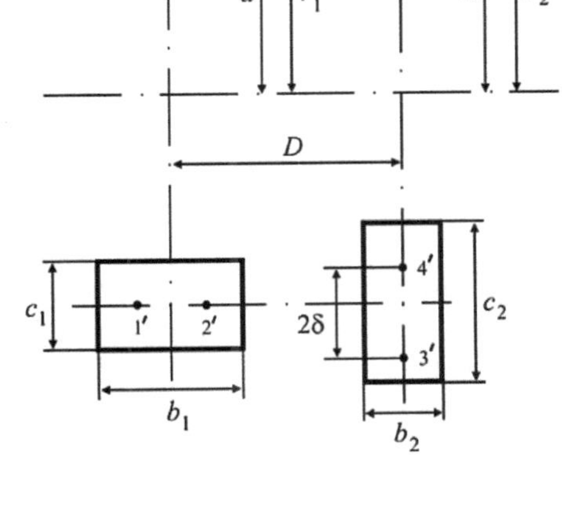

Fig. 5.21 Position of equivalent circular filaments in Lyle's method

where:

$\lvert \dot{Z} \rvert$	$n \times n$ matrix of self impedances $\dot{Z}_{ii} = R_{ii} + j\omega L_{ii}$ and mutual impedances $\dot{Z}_{ik} = j\omega M_{ik}$;
$\lvert \dot{I} \rvert$	vector of the currents $\dot{I}_i$ in the elements;
$\lvert \dot{U} \rvert$	vector of the voltages $\dot{U}_i$ ($\dot{U}_i = 0$ for the load elements);
R_i, L_i, M_{ik}	resistance and inductance of the i-th element and mutual inductance between the i-th and k-th elements;
ω	angular frequency
$i = 1, 2, \ldots, n$	$k = 1, 2, \ldots, n;$
n	total number of subdivision elements.

The subdivision of the inductor turns into annular elements in parallel imposes additional conditions that provide the additional equations necessary for the solution of the system (5.23).

These conditions are:

- the sum of the currents in the subdivision elements of one turn must be equal to the inductor current $\dot{I}$

$$\dot{I} = \sum_{i=1}^{P} \dot{I}_i = \ldots = \sum_{i=(N-1)\cdot P+1}^{N\cdot P} \dot{I}_i \tag{5.24}$$

with: P—number of subdivision elements of each turn; N—number of inductor turns

- the voltage drops $\dot{U}'_k$ in all parallel subdivision elements of one turn must be the same, i.e.

$$\dot{U}'_k = \dot{U}_i = \dot{U}_{i+1} = \ldots = \dot{U}_{i+P-1} \tag{5.25}$$

where: $k = 1, 2, \ldots, N$; $i = 1, P + 1, \ldots, (N-1)\cdot P - 1$; and the indexes k and i are incremented simultaneously

- the total voltage $\dot{U}$ applied to the inductor terminals must correspond to the sum of the voltage drops in the inductor turns:

$$\dot{U} = \sum_{k=1}^{N} \dot{U}'_k \tag{5.26}$$

The resulting system, with $n + N$ complex unknowns, i.e. the currents in the elementary elements of the load and voltages and currents in the elements of the inductor, is solved by separating real and imaginary parts and applying Gaussian elimination with partial pivoting to the real system thus obtained.

The precision of the results and calculation time are strongly influenced by the number of elementary subdivisions, which dimension must be in relationship with the penetration depth, and the requested accuracy in the evaluation of self and mutual inductances.

These criteria are discussed in several publications [9, 10].

In particular, the various implementations of this procedure differ one from another mainly in the method of calculation of inductances.

As regards self inductances, the formula of Weinstein is generally used [11]. Different equations must be adopted only for specific ratios between the sides of the rectangular cross-section of the circular element [10].

The calculation of mutual inductances can present some difficulties, since there are no exact analytical formulas for cylindrical volume elements. In fact, the basic formulae to compute mutual inductances refer to the ideal cases of circular filaments or cylindrical current sheets of negligible radial thickness, for which exact solutions are available in term of elliptical integrals [12–16].

In the case of elements of appreciable cross section, approximate solutions can be obtained only by integration of the basic equations over the cross section [14, 17, 18].

It is interesting to note that the mutually coupled circuit method is also particularly convenient for the evaluation of electromagnetic forces acting on the system. In fact, the same procedures for calculation of mutual inductances can be used for evaluating forces, with consequent advantages for program implementation and computer time requirements [19].

Among these solutions, the most commonly used is the Lyle's method of equivalent filaments, where each element of rectangular cross-section is replaced by two circular filaments conveniently placed in the cross-section, as illustrated in Fig. 5.22 [12], with:

$$r_1 = a \cdot \left(1 + \frac{c_1^2}{24a^2}\right); \quad r_2 = A \cdot \left(1 + \frac{b_2^2}{24A^2}\right); \quad \beta = \sqrt{\frac{b_1^2 - c_1^2}{12}}; \quad \delta = \sqrt{\frac{c_2^2 - b_2^2}{12}}$$

In this way the mutual inductance between the toroidal elements is determined by the arithmetic mean value of the four coefficients of mutual inductance calculated between couples of circular filaments.

$$M = \mu_0 \cdot N_1 \cdot N_2 \cdot \sqrt{a \cdot b} \cdot \left[\left(\frac{2}{k} - k\right) \cdot K(k) - \frac{2}{k} \cdot E(k)\right] \tag{5.27}$$

where (see Fig. 5.22):

$k = \sqrt{\frac{4 \cdot a \cdot b}{D^2 + (a+b)^2}}$; a, b—radii of filaments; D—distance between centres; $K(k)$—complete elliptical integral of 1st kind; $E(k)$—complete elliptical integral of 2nd kind.

Another solution giving a better accuracy for conductors of rectangular cross-section with any ratio of sides dimensions, is the one recently proposed by Akyel and Babic which has proved to be efficient, fast and easy to be implemented [17].

Fig. 5.22 Relative position of two coaxial circular filaments

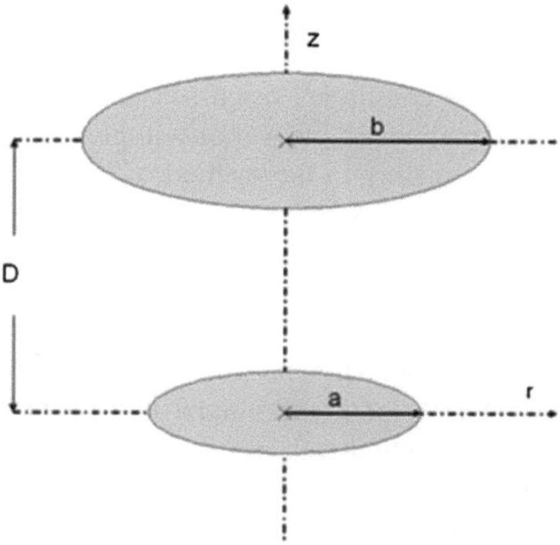

In this solution, the two toroidal elements are further subdivided into a number of elementary rings (cells), and each cell, in turn, is substituted by a central circular filament. Also in this case, the arithmetical mean value of all the mutual inductances of pairs of filaments gives the overall coefficient of mutual induction between the two elements.

Finally, the solution of the system of Eqs. (5.23)–(5.26) gives the currents $\dot{I}_i$ in all the elementary rings, from which we can calculate the inductor current *I,* the total active and reactive power and all other electrical parameters of the inductor-load system:

$$\text{power in elementary rings } P_i = R_i \cdot I_i^2 \tag{5.28}$$

$$\text{inductor current from eqn.} \tag{5.24}$$

$$\text{apparent power } P_a = U \cdot I \tag{5.29}$$

$$\text{active power in the load } P_{load} = \sum_{i \in load} P_i \tag{5.30}$$

$$\text{inductor losses } P_{loss} = \sum_{i \in inductor} P_i \tag{5.31}$$

$$\text{total active power } P = P_{load} + P_{loss} \tag{5.32}$$

$$\text{reactive power } Q = \sqrt{P_a^2 - P^2} \tag{5.33}$$

$$\text{power factor } \cos \varphi = P/P_a \tag{5.34}$$

$$\text{electrical efficiency } \eta_{el} = P_{load}/P \tag{5.35}$$

The main limitation of this method is that it is applicable to linear inductor-load systems where the superposition principle can be used.

The method has been extended to systems containing ferromagnetic regions [9], but today in these cases can be more conveniently used 2D and 3D FEM packages.

5.6 Calculation of Parameters of Direct Resistance Heating Systems

The most common type of direct resistance heating installation in the industry is the batch one, which includes a step down transformer, a high current circuit connecting the transformer to the heating device, the contact system and the heated workpiece (Fig. 4.1).

The basic method of calculating the electrical and energetic parameters is based on the use of its equivalent circuit. The complete equivalent circuit is analogous to the T-shaped circuit of a transformer (Fig. 5.23a).

Such circuit comprises: resistance R_{tr1} and reactance X_{tr1} of the primary windings, R_{tr2}, X_{tr2}—of the secondary windings of supply transformer, R_0, X_0—its magnetizing resistance and reactance, resistance R_{hc} and reactance X_{hc} of the high current circuit (supply bus-bars and/or cables), resistance R_c of the contact system, resistance R_w and the internal reactance X_w of the workpiece.

Since the no-load losses and the magnetization current of transformers used in this type of installations are quite low in comparison with their rated power and current, the magnetization branch can be neglected. Then the equivalent circuit can

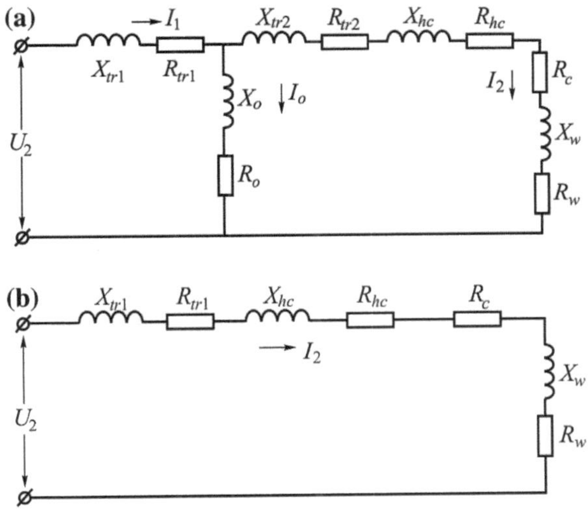

Fig. 5.23 Equivalent circuits of a direct resistance heating installation: **a** with no-load losses of supply transformer, **b** neglecting no-load losses of transformer

be simplified as in Fig. 5.23b, where the impedances of windings of the supply transformer are included in the resistance R_{tr} and the reactance X_{tr}.

The impedance of the transformer can be calculated from its plate data by the equation

$$Z_{tr} = \frac{U_{2n}}{I_{2n}} \cdot u_{\%}, \tag{5.36}$$

where $u_{\%}$—percent short-circuit voltage; U_{2n}—nominal voltage on the low voltage side of transformer, V; I_{2n}—nominal current on the transformer low voltage side, A.

The resistance R_{tr} is given by

$$R_{tr} = P_{sc}/I_{2n}^2, \tag{5.37}$$

with P_{sc}—short-circuit power of supply transformer, W.

The transformer reactance is

$$X_{tr} = \sqrt{Z_{tr}^2 - R_{tr}^2}. \tag{5.38}$$

The short-circuit resistance and reactance of the high current supply circuit can evaluated on the basis of the constructive design of the installation.

As known, the high current circuit consists of bus-bars (or cables) made of copper or aluminum, and a contact system connecting the circuit to the workpiece.

The calculation of resistances of the elements of this circuit must take into account skin and proximity effects [20, 21].

Skin effect can be considered by using the coefficients k_r, given in Chap. 2, while proximity effects can be taken into account by the methods described in [18 ÷ 21].

The reactance of the high current circuit is calculated as the sum of the internal and external reactances X_{hci} and X_{hce}. The internal reactance X_{hci} can be calculated, in a first approximation, using the coefficients k_x, given in Chap. 2. However, the values calculated in this way do not take into account the mutual influence between different elements of the high current circuit, due to proximity effect. More accurate methods to compute the internal reactance taking also into account proximity effect, are given in [22, 23]. The determination of the external reactance X_{hce}, can be made in a first approximation by the diagrams of Figs. 5.24 and 5.25 [24–26].

Figure 5.24 gives the values of X_{hce} as a function of the area of the circuit comprising transformer, high-current circuit and workpiece.

The diagram of Fig. 5.25 gives the reactance per unit length of two parallel bifilar nonmagnetic conductors of rectangular cross-section (20 × 120 mm) as a function of the distance between them.

The contacts which supply current to the workpiece are made usually of copper or bronze, or in special cases (e.g. when heating in vacuum high-temperature materials, like tungsten, molybdenum, etc.) of graphite, high-temperature alloys or other mechanically strong materials [24, 27–29].

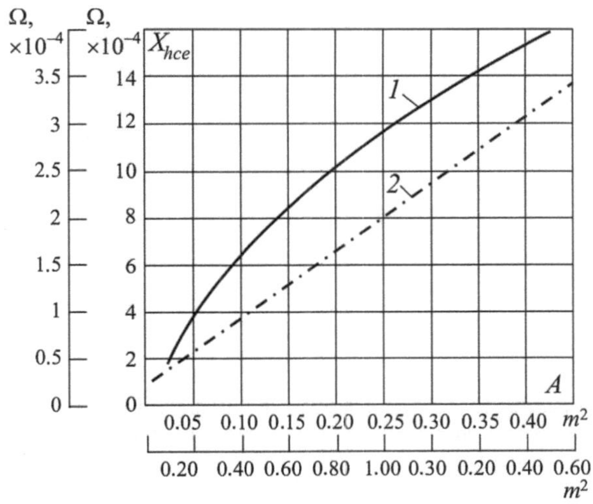

Fig. 5.24 Values of reactance of a circuit as a function of its area A [*1* A < 0.4 m^2, *2* A = 0.1 ÷ 1.6 m^2]

The contact resistance R_c of a copper contact system for steel products is in the range 0.4 ÷ 0.8 × 10^{-4} Ω [24]. For other contact materials, shapes and materials of the workpiece, data can be found in [24, 27] as a function of the force applied between the contact element and the workpiece.

The resistance R_w and the internal reactance X_w of heated non-magnetic workpieces or $R_{w\phi}$ and $X_{w\phi}$ of ferromagnetic workpieces can be evaluated with the relationships

$$R_w = k_r \cdot R_{dc};\, R_{w\phi} = k_{r\phi} \cdot R_{dc};X_w = k_x \cdot R_{dc};X_{w\phi} = k_{x\phi} \cdot R_{dc}$$

where R_{dc}—DC resistance of the workpiece, ohm; $k_r, k_{r\phi}, k_x, k_{x\phi}$—resistance and reactance skin effect coefficients of non-magnetic and magnetic workpieces, respectively.

For non-magnetic materials, the coefficients k_r, k_x can be determined by the following relationships and diagrams:

- solid cylindrical workpieces: Eq. (3.29) and diagram of Fig. 3.10;
- cylindrical tubes: Fig. 3.22;
- thin slabs: Eqs. (2.87) and (2.88), and Fig. 2.24;
- workpieces of rectangular cross-section: Figs. 2.32 and 2.33.

For ferromagnetic materials, the coefficients $k_{r\phi}$, $k_{x\phi}$ can be evaluated as follows:

- solid cylindrical workpieces: Eqs. (3.39)–(3.42) and diagrams of Fig. 3.11;
- cylindical tubes: Fig. 3.25;
- for workpieces of ferromagnetic materials of rectangular or equi-angles cross-sections, $k_{r\phi}$, $k_{x\phi}$ from the diagrams of Fig. 3.11, in Ref. [30]. In this case, χ must be calculated as $\chi = 2A/(\Pi\delta_e)$, where A, Π—area and perimeter of the cross section of the workpiece; δ_e—penetration depth of the electromagnetic wave,

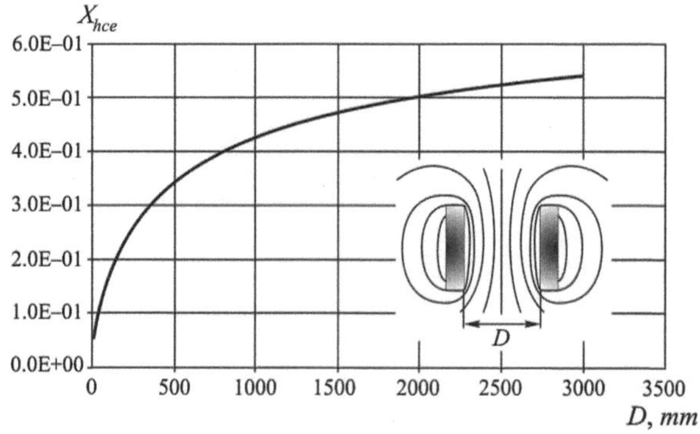

Fig. 5.25 External reactance X_{hce} of 1 m length of a system of two parallel rectangular non-magnetic conductors, cross-section (20 × 120 mm), as a function of the distance between them [25, 26]

calculated with the value of magnetic permeability corresponding to the field intensity at the surface of the workpiece. The penetration depth is calculated as $\delta_e = 503\sqrt{\rho(\vartheta)/\mu_e f}$, where f—frequency; $\rho(\vartheta)$—resistivity of the workpiece material; μ_e—relative magnetic permeability, that is dependent on surface magnetic field.

The dependence of μ_e on temperature and magnetic field intensity can be taken into account by the following relationship [31]

$$\mu_e = 1 + (\mu_{e20} - 1) \cdot \varphi(\vartheta),$$

where μ_{e20}—value of μ_e at 20 °C [32]; for ferromagnetic steels μ_{e20} can be evaluated with the formula $\mu_{e20} = 5 \times 10^5 H_{me}^{-0.896}$ [2]; φ (ϑ)—coefficient given the diagram of Fig. 5.26, which describes the ferromagnetic transition from magnetic to non-magnetic due to temperature. In the diagram, Curie point is at 760 °C.

- Total impedance of secondary circuit

$$Z = \sqrt{(R_{tr} + R_{hc} + R_c + R_w)^2 + (X_{tr} + X_{hc} + X_w)^2}. \tag{5.39}$$

Special transformers are used in DRH installations, which have to meet the production requirements. The following formula give the main data of the installation as a function of secondary voltage U_2.

- Current in the workpiece

$$I_2 = \frac{U_2}{Z}. \tag{5.40}$$

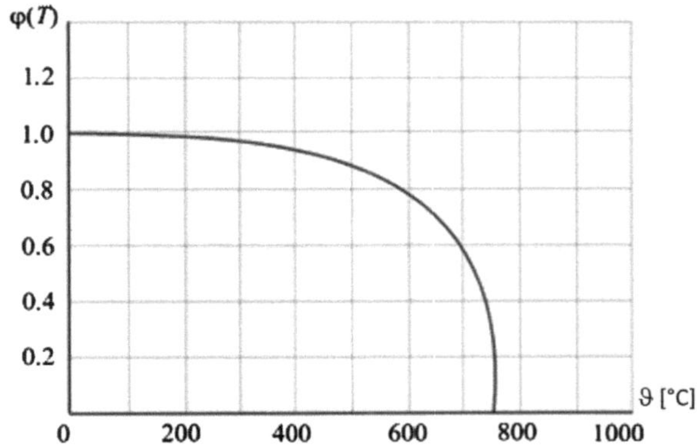

Fig. 5.26 Diagram of the coefficient φ (t) as a function of temperature

- Active power in the workpiece

$$P_w = R_w I_2^2. \tag{5.41}$$

- Total active power of the installation

$$P_{tot} = (R_{tr} + R_{hc} + R_c + R_w) I_2^2. \tag{5.42a}$$

- Total active power taking into account no-load losses of the transformer

$$P'_{tot} = (R_{tr} + R_{hc} + R_c + R_w) I_2^2 + P_0, \tag{5.42b}$$

with P_0—transformer no-load losses.

- Apparent power of the installation

$$P_a = U_2 I_2. \tag{5.43}$$

- Power factor at the high voltage side of the transformer

$$\cos\varphi = \frac{P'_{tot}}{P_a} = \frac{(R_{tr} + R_{hc} + R_c + R_w) I_2^2 + P_0}{U_2 I_2}. \tag{5.44}$$

A correct calculation of the installation power consumption must include all the electrical and thermal losses listed in the following.

- Electrical power losses

$$P_{loss}^{el} = (R_{tr} + R_{hc} + R_c) I_2^2. \tag{5.45}$$

Thermal losses comprise radiation losses P_{loss}^{rad} and convection losses P_{loss}^{conv} from the surface of the workpiece and conduction losses P_{loss}^{cond} from the workpiece to the water cooled contact system.

- Radiation losses can be calculated with the relationship

$$P_{loss}^{rad} = 5.67 \cdot \varepsilon_r \cdot F \cdot \left(\frac{\vartheta_s + 273}{100}\right)^4, \text{ W}, \tag{5.46}$$

with ε_r—emissivity of the workpiece surface (see Table 5.3); F—area of the lateral surface of the workpiece between contacts, m^2; ϑ_s—temperature of the surface of the workpiece, °C.

Table 5.3 Integral emissivity factor ε_r of different metals [27, 33]

Material	Temperature °C	ε_r
C steel		
Oxidized	20–600	0.8
Non oxidized (in vacuum or protective atmosphere)		0.4–0.5
Cuprum		
Oxidized	500	0.6–0.7
Nickel		
Technically clean, polished (in vacuum or protective atmosphere) oxidized	730 930 1130 200–600	0.056 0.063 0.069 0.37–0.48
Platinum	100–1500	0.14–0.18
Silver		
Clean Polished	200–600 200	0.02–0.03 0.05
Tungsten (in vacuum or protective atmosphere)	1330 1730 2130 2530	0.207 0.26 00.296 0.323
Tantalum (in vacuum or protective atmosphere)	1330 1730 2130 2530	0.194 0.232 0.269 0.304
Molybdenum (in vacuum or protective atmosphere)	1330 1730 2130 2530	0.168 0.210 0.248 0.281
Aluminium		
Oxidized at 600 °C	200–600	0.11–0.19
Brass		
Oxidized at 600 °C	200–600	0.61–0.59
Titanium	710–870	0.459

- Convection losses can be calculated with the formula

$$P_{loss}^{conv} = \alpha_c \cdot (\vartheta_s - \vartheta_a), \text{ W}, \quad (5.47)$$

where, $\alpha_c = 2.56 \cdot (\vartheta_s - \vartheta_a)^{1/4}$ W/(m²K)—average value of convection heat transfer coefficient at the workpiece surface. ϑ_a—temperature of the ambient surrounding the workpiece, °C.

Note The above expression of α_c can be used for the natural cooling in still air of horizontal cylindrical workpieces. For workpieces of different shape and when different convection conditions occur, it is necessary to calculate α_c using the method described in Ref. [33].

The specific total power $p_{loss}^{rad+conv}$ lost for radiation and natural convection in air can be evaluated also by the diagrams of Fig. 5.27.

- The losses due to thermal conduction from the workpiece to a single contact point, can be determined by the flow and temperature differential of the water in the cooling circuit with the relationship

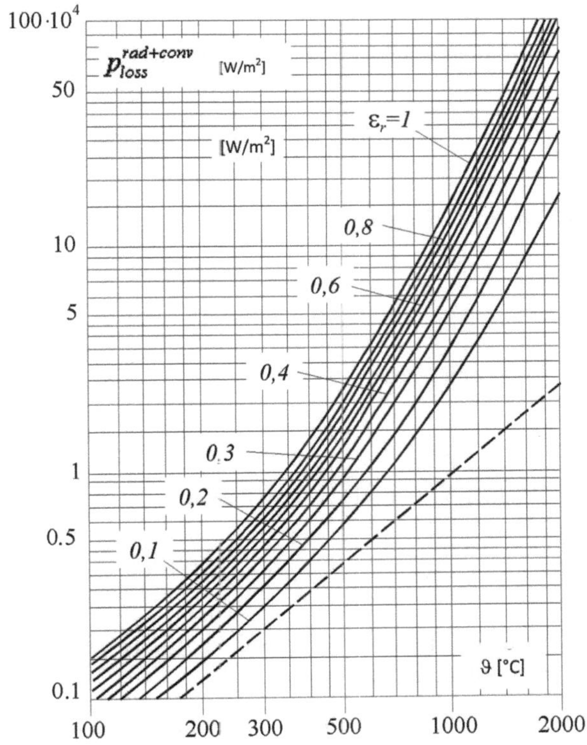

Fig. 5.27 Specific power losses for radiation and convection in air as a function of temperature ϑ and emissivity factor ε_r

$$P_{loss}^{cond} = 1.16 \times 10^3 \cdot G \cdot \Delta\vartheta_c, \text{ W}, \tag{5.48}$$

where G = (25–50) × 10^{-3} m^3/h—water flow in a contact; $\Delta\vartheta_c = (30-40)\,°C$—temperature drop in the cooling water between output and input of a contact point system [33].

- Total thermal losses

$$P_{loss}^{th} = P_{loss}^{rad} + P_{loss}^{conv} + P_{loss}^{cond} = p_{loss}^{rad+conv} \cdot A + P_{loss}^{cond} \tag{5.49}$$

  with A—area of the lateral surface of the heated object.
- Useful power

$$P_u = P'_{tot} - P_{loss}^{el} - P_{loss}^{th} \tag{5.50}$$

- Electrical efficiency of the installation comprising transformer no-load losses

$$\eta_{el} = \frac{P_w}{P'_{tot}} = \frac{R_w I_2^2}{(R_{tr} + R_{hc} + R_c + R_w) I_2^2 + P_0}. \tag{5.51a}$$

- Electrical efficiency neglecting the transformer no-load losses

$$\eta_{el} = \frac{R_w}{(R_{tr} + R_{hc} + R_c + R_w)} = \frac{1}{1 + \frac{(R_{tr}+R_{hc}+R_c)}{R_w}}. \tag{5.51b}$$

- Thermal efficiency

$$\eta_{th} = \frac{P_w - P_{loss}^{th}}{P_w}. \tag{5.52}$$

- Total efficiency

$$\eta = \eta_{el} \cdot \eta_{th}. \tag{5.53}$$

- Production rate of the installation

$$G_m = \frac{m_w}{t_h + t_d}, \tag{5.54}$$

  with m_w—mass of workpiece, kg; t_h—heating time, s; t_d—dead time for loading and unloading workpieces, s.

The dead time t_d for loading and unloading depends on the specific devices and handling system.

Usually, to solve a time dependant problem, the heating time t_h, i.e. the time interval from the initial temperature to the final one, is subdivided in n intervals, or elementary time steps Δt such that in each of them can be assumed constant values of ϑ_{ni}, I_2, R_w and α_Σ. The last assumption is valid when the temperature variations in a single time step do not exceed a certain value, in [9] a temperature interval of 25 °C is proposed.

For a given temperature interval, the corresponding heating time is calculated with the relationship [5]:

$$\Delta t_i = \frac{m_w c_i}{\alpha_{\Sigma i} F} \ln \frac{R_{wi} I_{2i}^2 - \alpha_{\Sigma i} F(\vartheta_{si} - \vartheta_a)}{R_{wi} I_{2i}^2 - \alpha_{\Sigma i} F(\vartheta_{sf} - \vartheta_a)}, \text{ s} \tag{5.55}$$

where: i—indicates the considered calculation interval, c—specific heat of the workpiece material, J/(kg·K); ϑ_{si} and ϑ_{sf}—temperatures of the workpiece surface at the beginning and the end of the i-th interval; α_Σ—total heat transfer coefficient for radiation and convection,

$$\alpha_\Sigma = \alpha_{rad} + \alpha_{conv}$$
$$= 4.54 \frac{\left[\left(\frac{\vartheta_s + 273}{100}\right)^4 - \left(\frac{\vartheta_a + 273}{100}\right)^4\right]}{(\vartheta_s - \vartheta_a)} + 2.56\,(\vartheta_s - \vartheta_a)^{1/4}, \text{ W}/(\text{m}^2\text{K})$$

The total heating time is

$$t_h = \Delta t_1 + \Delta t_2 + \cdots + \Delta t_i + \cdots + \Delta t_n. \tag{5.56}$$

The specific energy consumption during each calculation interval is [5]

$$W_E = \frac{c_i\left[(R_{hc} + R_c + R_w)\, I_{2i}^2 + P_0\right]}{\alpha_{\Sigma i} F} \ln \frac{R_{wi} I_{2i}^2 - \alpha_{\Sigma i} F(\vartheta_{si} - \vartheta_a)}{R_{wi} I_{2i}^2 - \alpha_{\Sigma i} F(\vartheta_{sf} - \vartheta_a)}. \tag{5.57}$$

The total specific energy consumption is the sum of the consumptions in each time step:

$$W_E = \sum_{i=1}^{n} W_{Ei}. \tag{5.58}$$

Temperature intervals are usually assumed in the range 25–50 °C.

One of the parameters of practical interest is the overall efficiency η, which is the product of electrical and thermal efficiency.

Electrical efficiency of high-speed resistance heating processes has always relatively high values, in the range 0.92–0.96 [24, 27]. Therefore, thermal efficiency has a very strong influence on the overall efficiency.

Thermal efficiency is the ratio between the energy required to raise the temperature of the workpiece from the initial to the final temperature, and the total Joule losses in the workpiece:

$$\eta_{th} = \frac{\int\limits_0^{t_h} \left[P_w(t) - P_{loss}^{th}(t)\right] \cdot dt}{\int\limits_0^{t_h} P_w(t) \cdot dt}, \tag{5.59}$$

where: $\left[P_w(t) - P_{loss}^{th}(t)\right]$—useful power for heating the workpiece, W; $P_w(t)$—total power transformed into heat inside the workpiece by the current flowing in it, W; t_h—heating time.

For constant physical characteristics, in a cylindrical workpiece the useful power can be written as

$$P_u(t) = \frac{c\gamma\pi r_e^2 \ell \Delta\vartheta_{av}}{t_h},$$

with c—average specific heat of the workpiece material in the heating interval, J/(kg·K); γ—material density, kg/m^3; r_e—workpiece external radius, m; ℓ—workpiece length, m; $\Delta\vartheta_{av}$—average increase of workpiece temperature in the heating time t_h,°C.

Since the total power delivered to the workpiece is

$$P_w = R_w \cdot I_2^2 = k_r \rho \frac{\ell}{\pi r_e^2} I_2^2$$

the thermal efficiency can be calculated with the relationship:

$$\eta_{th} = \frac{c\,\gamma\,\pi r_e\,\ell\,\Delta\vartheta_{av}}{P_w\,t_h} \tag{5.60a}$$

which can be re-written in the form

$$\eta_{th} = 1 - \frac{P_{loss}^{th}}{P_w} = 1 - \frac{2\,\pi\,r_e\,\ell\,p_{loss}^{th}}{p_w}, \tag{5.60b}$$

with P_{loss}^{th}—power of the thermal losses for radiation and convection from the surface of the workpiece (excluding conduction losses at the contacts), W; $p_{loss}^{th} = P_{loss}^{th}/(2\,\pi\,r_e\,\ell)$—specific power losses from the surface of the workpiece, W/m^2; $p_w = P_w/(2\,\pi\,r_e\,\ell)$—specific surface power delivered to the workpiece.

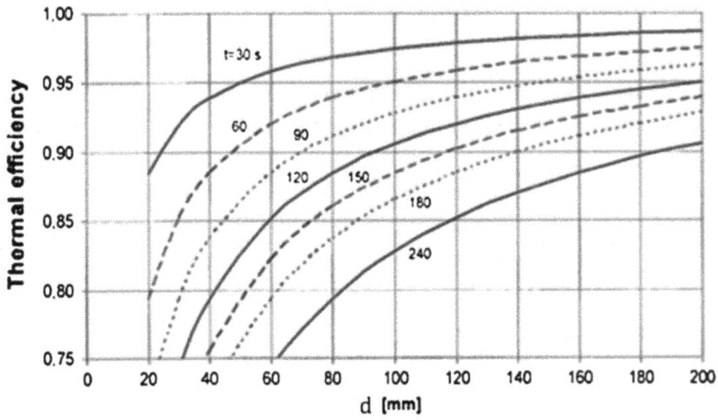

Fig. 5.28 Thermal efficiency as a function of bar diameter, for heating times 30–60–90–120–150–180–240 s

In a more convenient form the thermal efficiency can be rewritten as follows

$$\eta_{th} = \frac{1}{1 + \frac{2 p_{loss}^{th} \, t_h}{c \gamma \, r_e \, \Delta\vartheta_{av}}}. \quad (5.60c)$$

The diagrams of Fig. 5.28 give the values of η_{th} calculated with the above equation, as a function of the diameter, in the heating of steel bars up to 1250 °C and heating times in the range 30–240 s.

They show that rapid heating cycles allow for high thermal efficiency. They also show that the efficiency is generally high and, all other things being equal, it increases when the bar diameter increases.

5.7 1D Finite Difference Numerical Model for DRH

The procedure considered in the previous section is very useful for a preliminary calculation of the main parameters of DRH installation, but it lacks to provide an accurate evaluation of the transient heating process, particularly in the case of steel magnetic workpieces.

Numerical methods can provide an accurate solution of the coupled thermal and electromagnetic process also in the case of direct resistance heating.

In this section a simple 1-D finite difference procedure is presented for calculating an axis-symmetric "long" DRH system.

Taking into account that the procedure is analogous to the one presented in Sect. 5.3 for induction heating [4], here we present only the solution of the electromagnetic problem, addressing the reader to the above mentioned section for the solution of the thermal one.

5.7.1 Electromagnetic Problem

In an axisymmetric geometry, the workpiece (e.g. long bar, billet or tube) is conveniently subdivided into N concentric elements of uniform radial thickness $s_i = \Delta r$, as in Fig. 5.29.

As shown in the figure, assuming that the function representing the magnetic field intensity H is constant in each element, while the one representing the current density J varies stepwise at the centre of each layer, it is:

$$\dot{H}_i = const., \text{ for } r_i - \frac{\Delta r}{2} \leq r \leq r_i + \frac{\Delta r}{2}$$
$$\dot{J}_i = const., \text{ for } r_{i-1} \leq r \leq r_i$$

The field quantities in the various elements can be calculated in sequence one at a time, starting from the value of the current density in the innermost layer where an

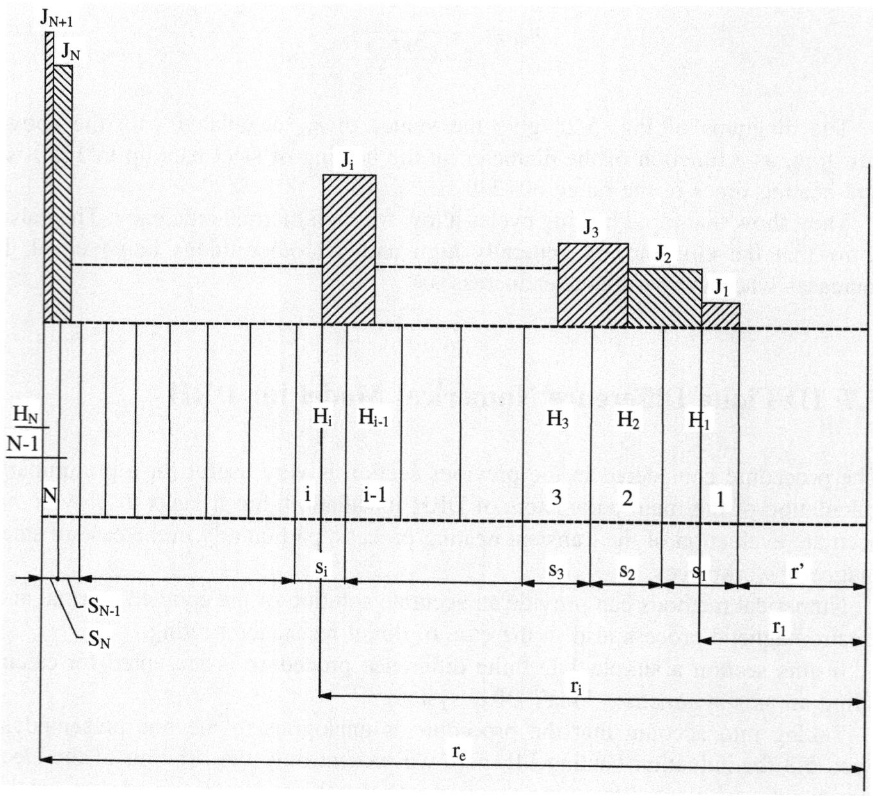

Fig. 5.29 Subdivision of a cylindrical bar or a tube. [r′ = 0 for cylindrical bar]

arbitrary value of J_1 is fixed, and proceeding toward the surface according to the following scheme:

- Current $\dot{I}_1$ in the layer between the radii r' and r_1 ($r' = 0$ for solid cylinder):

$$\dot{I}_1 = \pi\left(r_1^2 - r'^2\right)\dot{J}_1. \tag{5.61}$$

- Magnetic field intensity $\dot{H}_1$ at radius r_1:

$$\dot{H}_1 = \frac{\dot{\mathrm{I}}_1}{2\,\pi\,\mathrm{r}_1} \tag{5.62}$$

- Relative permeability μ_{r1} of layer 1, evaluated as a constant scalar quantity and neglecting hysteresis losses:

$$\mu_{r1} = f(H_1) \tag{5.63}$$

- Current density $\dot{J}_2$, obtained from the equation $d\dot{J}/dr = j(\omega\mu_r\mu/\rho)\dot{H}$ by substituting the data of previous layer (see Eq. (3.15)):

$$\dot{J}_2 = \dot{J}_1 + j\frac{\omega\,\mu_{r1}\,\mu_0}{\rho_1}\Delta r \cdot H_1 \tag{5.64}$$

- Iteration of the previous steps, with i = 2, 3,…,N applying the relationships:

$$\left.\begin{array}{l} \dot{I}_i = \pi\left(r_i^2 - r_{i-1}^2\right)\cdot \dot{J}_i \\ \dot{H}_i = \frac{1}{2\pi\,r_i}\sum\limits_{k=1}^{i}\dot{I}_k \\ \mu_{ri} = f(H_i) \\ \dot{J}_{i+1} = \dot{J} + j\frac{\omega\mu_{ri}\mu_0}{\rho_i}\Delta r\cdot\dot{H}_i \end{array}\right\} \tag{5.65}$$

with:
I_i current in the i-th layer between radii r_{i-1} and r_i
H_i magnetic field intensity at radius r_i
ρ_i, μ_{ri} resistivity and relative permeability of the i-th layer
$\dot{J}_{i+1}$ current density between the layers r_i and r_{i+1}

- Current I_{N+1} in the layer between the radii r_N and r_e:

$$\dot{I}_{N+1} = \pi\left(r_e^2 - r_N^2\right)\cdot\dot{J}_{N+1} \tag{5.66}$$

- Total current I in the workpiece:

$$\dot{I} = \sum_{i=1}^{N} \dot{I}_i \tag{5.67}$$

- Magnetic field intensity H_0 at the bar surface:

$$\dot{H}_0 = \frac{\dot{I}}{2\pi r_e} \tag{5.68}$$

Since the current density J_1 in the innermost layer has been arbitrarily fixed as initial tentative value, the calculated total current I does not correspond, in general, to the prescribed value I_0.

In the case of non-magnetic bars, the solution is obtained by multiplying all the calculated field quantities by the ratio between I_0 and the total calculated current.

On the contrary, in the case of magnetic steels, the solution must be determined by an iterative procedure, repeating the calculation several times with initial values J_1 appropriately updated at each iteration, until the assigned value I_0 (or H_0) is reached.

As shown by the equations, this procedure allows to consider values of ρ_i and μ_{ri} different from layer to layer, which can be updated according to the temperature of each subdivision element during the calculation of the heating transient.

5.7.2 Examples of Coupled Numerical Solutions

The proposed numerical method allows to evaluate with great precision the temperature transient in the workpiece (in particular in steel ferromagnetic bars) and to analyse the influence of the installation design parameters.

Among the others, these parameters are of particular importance: the external impedance ($X_e = X_{tr} + X_{hc}$) of the supply circuit and the secondary voltage U_2 of the power transformer.

The voltage U_2 is imposed in order to produce in the secondary circuit a current sufficient to obtain the desired production rate.

As already stated, the "internal" impedance of the workpiece depends on frequency, geometrical dimensions, current flowing through it and the material characteristics which, in most cases, undergo significant changes during heating.

Consequently, at constant supply voltage, a large variability of the absorbed current and power can occur, which is more or less pronounced depending on the bar dimensions and the value of the "external" impedance of the supply circuit, which constitutes a "ballast" impedance of the circuit.

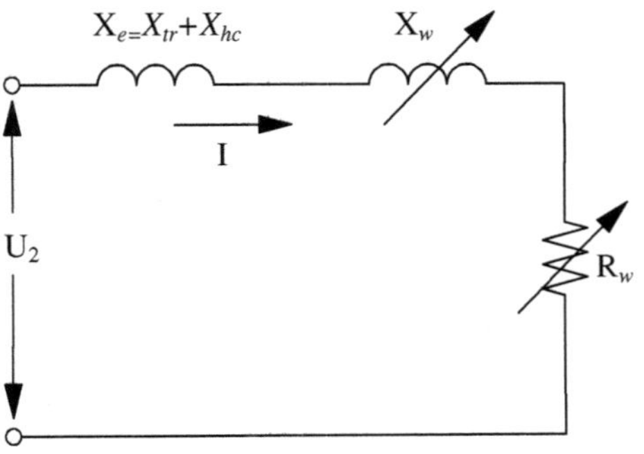

Fig. 5.30 Simplified equivalent circuit of DRH installation

With reference to the simplified equivalent circuit of Fig. 5.30, this complex behaviour is illustrated by some examples given in the following diagrams with reference to the heating of C45 steel bars, 6 m length, from 20 to 1250 °C [23, 34].

In particular, the diagrams show:

- The relative distributions along the radius, at the beginning of heating, of the specific power per unit volume *w* for different bar diameters and magnetic field intensities *He* of 100 and 300 kA/m, respectively (Fig. 5.31a, b).

It can be noted that these distributions are significantly different from the analogous distributions calculated with the classical solutions with constant permeability in the workpiece cross section (see Sect. 3.1). Moreover, the comparison of the curves in (a) and (b) for d = 37 and d = 40 mm shows that the power distribution is strongly modified when the current flowing in the workpiece varies.

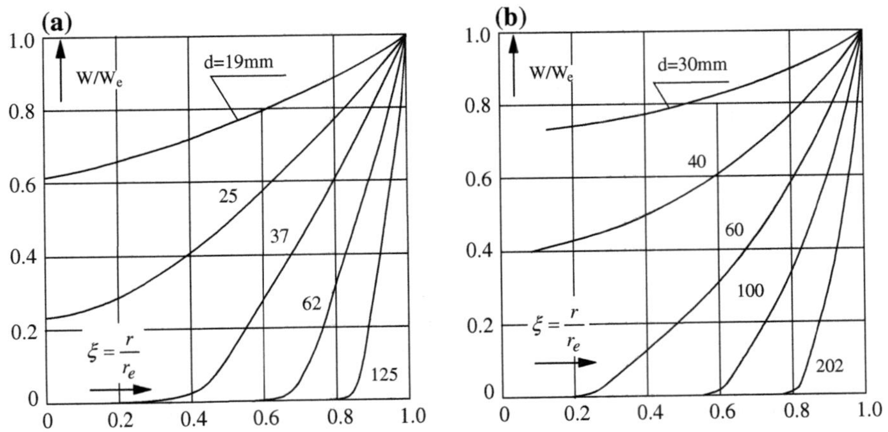

Fig. 5.31 Radial distributions of specific power per unit volume at beginning of heating [Steel UNI-C45; **a** H_e = 100 kA/m, **b** H_e = 300 kA/m]

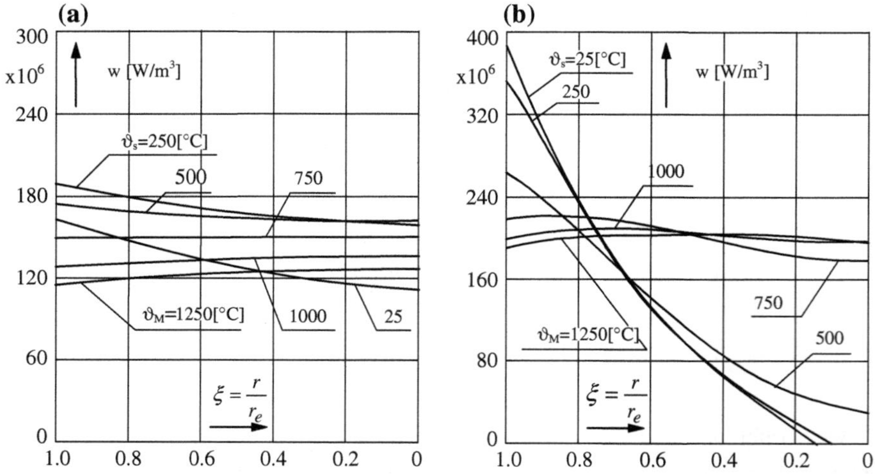

Fig. 5.32 Radial distributions of specific power per unit volume w at different instants of heating transient, when the surface temperature is ϑ_s [Steel UNI-C45; $\ell = 6$ m; $X_e/R_{dc} = 2$; **a** d = 20 mm, $U_2 = 80$ V, **b** d = 60 mm, $U_2 = 100$ V; R_{dc}–DC resistance at 20 °C]

- Distributions along the radius of the specific power per unit volume at various instants of the heating transient (Fig. 5.32a, b).

The diagrams show that, in workpieces of small diameter, the distributions of specific power are nearly uniform at the beginning of heating, while, with increasing diameter or with lower current intensity, this occurs only in the temperature range above Curie point, where the distributions are influenced mainly by the variations of resistivity with temperature. The influence of these changes is particularly evident near the workpiece surface, where the temperature values are lower due to the heat transfer to the surrounding environment for radiation.

- The variations during the heating transient of the "internal" impedance of the bar and the total impedance of the secondary circuit (Fig. 5.33).

Therefore the distributions of the specific power in the cross-section and in time may be very different from case to case. As a consequence also the corresponding temperature transients can have different shapes, which depend on the load dimensions and also on the ratio between the external reactance of the supply system and the internal impedance of the bar.

This is illustrated in the following example, which refers to the heating at a constant voltage of a bar of 40 mm diameter and length 6 m, with different values of the external reactance of supply circuit, i.e. with ratios between external reactance X_e and DC resistance R_{dc} of the "cold" workpiece equal to 1, 2 and 3, respectively.

The corresponding heating transients have the shapes shown in Fig. 5.34, and are characterized by the increase of the heating time when the reactance X_e increases and final temperature differences $\Delta\vartheta = \vartheta_s - \vartheta_a$ practically unchanged in the three cases.

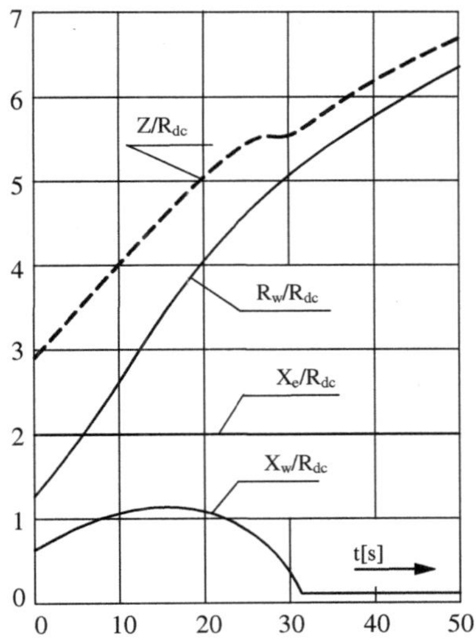

Fig. 5.33 Variations during heating transient of bar resistance, "internal" reactance and total circuit impedance [Steel UNI-C45; $\ell = 6\,\text{m}$; d = 40 mm; $X_e/R_{dc} = 2$; $U_2 = 80$ V; R_{dc}—DC resistance at 20 °C]

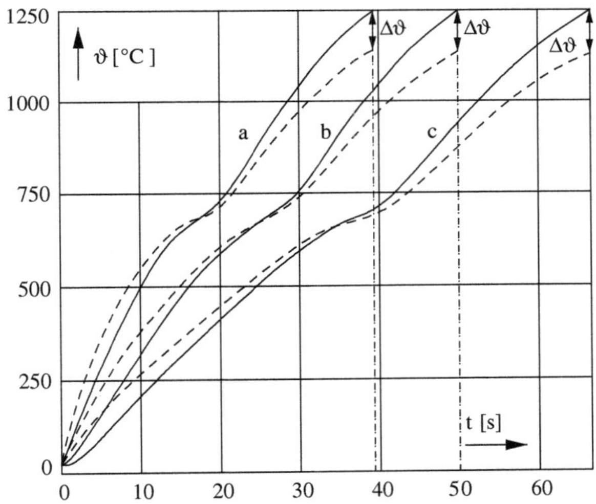

Fig. 5.34 Transient distributions of surface *dashed line* and axis *solid line* temperature. [Steel UNI-C45; $\ell = 6\,\text{m}$; d = 40 mm; $U_2 = 80$ V; **a** $X_e/R_{dc} = 1$, $\Delta\vartheta = 115$ °C, **b** $X_e/R_{dc} = 2$, $\Delta\vartheta = 117$ °C; **c** $X_e/R_{dc} = 3$, $\Delta\vartheta = 119$ °C]

The relative variations of apparent power and power factor during the above heating transients are given in the Fig. 5.35a.

In particular, it can be observed that in case of steel workpieces, the power factor has very low values (0.4–0.6) at the beginning of heating, while it grows up to (0.85–0.95) above Curie point.

On the contrary, in case of non-magnetic materials the power factor remains at rather high values (0.8–0.9) throughout the heating transient. Figure 5.35 also shows that when the reactance X_e increases, the values of power factor are lower and the variations of the current during the heating cycle are less pronounced.

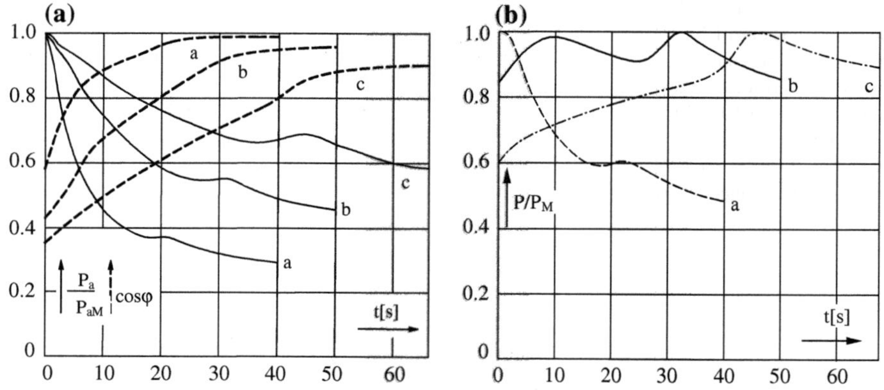

Fig. 5.35 **a** Relative variations of apparent power and power factor (on the *left*), **b** of active power during heating (on the *right*). [circuit data as in Fig. 5.36; **a** P_{aM} = 3472 kVA, $H_e' \approx$ 345 kA/m, $\cos\varphi_m$ = 0.911; P_M = 2073 kW, **b** P_{aM} = 2190 kVA, $H_e' \approx$ 218 kA/m, $\cos\varphi_m$ = 0.801; P_M = 1,123 kW, **c** P_{aM} = 1628 kVA, $H_e' \approx$ 162 A/m, $\cos\varphi_m$ = 0.700; P_M = 960 kW]

The active power varies according to the curves of Fig. 5.35b; they show that a convenient choice at the design stage of the ratio X_e/R_{dc} allows to reduce the variations of the power transformed into heat in the bar.

Finally, once defined the value of the reactance X_e it is possible to obtain the required heating time and production rate by adjusting the value of the supply voltage U_2.

To this end, the supply power transformer is generally equipped with additional taps for a stepwise regulation of the secondary voltage, which also can be used for adapting the heater to workpieces of different diameters and lengths.

This is possible by the use of diagrams similar to those given in the Fig. 5.36, which refer to UNI-C45 steel bars, 6 m length, heated to a surface temperature of 1,250 °C.

The diagrams refer to different value of the ratio X_e/R_{dc} and different supply voltages; they provide the following quantities:

- average values of the power transformed into heat during the heating, referred to the unit surface of the bar (Fig. 5.36a);
- heating times (Fig. 5.36b);
- temperature differential $\Delta\vartheta = \vartheta_s - \vartheta_a$ between surface and axis at the end of the heating (Fig. 5.36c);
- process thermal efficiency (Fig. 5.36d).

Finally, Fig. 5.37 gives the total efficiency in the heating of cylindrical steel C45 bars, 6 m length, to 1,250 °C, as a function of the diameter and different heating times; the values refer to a ratio $X_e = (X_{tr} + X_{hc})/R_{dc} = 2$.

For comparison, in the same figure solid lines show the values of efficiency for resistance heating installations, the dashed lines for induction heating installations.

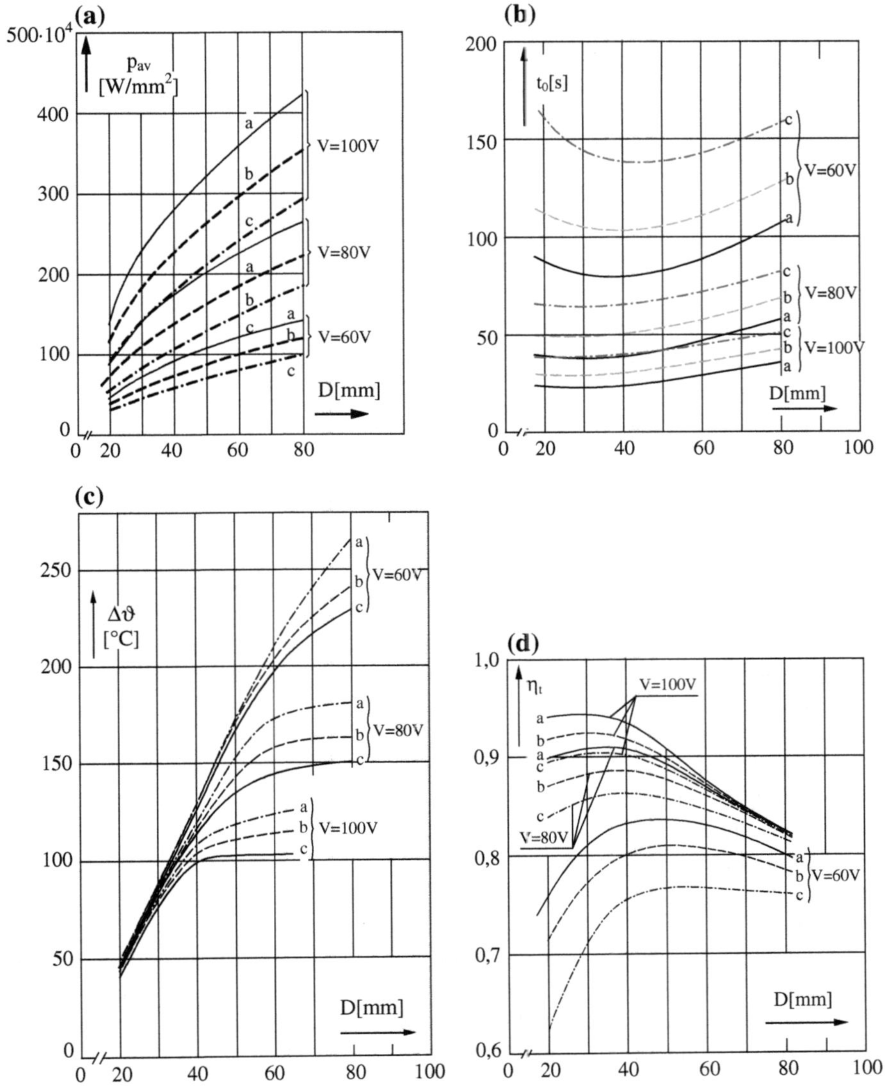

Fig. 5.36 Process parameters as a function diameter in DRH of steel bars of Steel UNI-C45; $\ell = 6$ m; a $X_e/R_{dc} = 1$, b $X_e/R_{dc} = 2$, c $X_e/R_{dc} = 3$ [**a** Average specific power during heating cycle, **b** heating time, **c** final temperature differential, **d** thermal efficiency]

5.8 FEM: Finite Element Method

Finite Element Model (FEM) is the most widespread numerical method applied in engineering problems. In electrotechnogy, it is applied mostly to solve the electromagnetic and thermal field distributions, thanks to the possibility of direct or

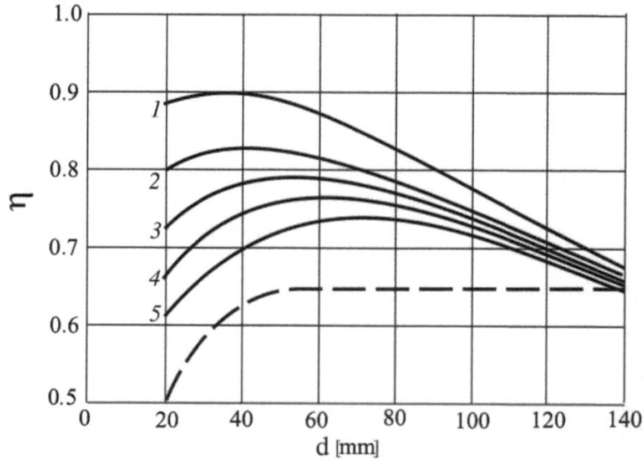

Fig. 5.37 Total efficiency in the heating of steel C45 bars to 1,250 °C, as a function of diameter and heating time [l = 6 m; $(X_{tr} + X_{hc})/R_{dc} = 2$; *solid line* Resistance heating; *dashed line* Induction heating; heating time: 1–30; 2–60; 3–90; 4–120; 5–150 s]

indirect coupling of the problems. FEM allows also to calculate other field quantities that are of interest in induction and direct resistance heating, like mechanical deformations in solid workpiece or fluid velocity in liquid materials (even if fluid dynamic problems are mostly solved by resorting to finite integral volume techniques).

There are several commercial codes based on finite element technique able to solve the EM (at low frequency, i.e. neglecting the displacement currents) and thermal coupled problems. Among them the most diffused for electro-thermal applications are ANSYS© Multiphysics, CEDRAT Flux©, Comsol© Multiphysics, JMag©, QuickField©. There are also several packages that are 'open source', i.e. free available, mostly developed for research and teaching purposes, like Agros2D and Elmer.

Finite element modeling is a powerful tool to design electromagnetic devices. The use of numerical modeling can speed-up the realization of new devices or the optimization of existing ones, thanks to the possibility of creating virtual prototypes that can be easily and quickly modified.

In fact, one of the most interesting feature of numerical analysis consists in the parametric solution of a problem, that means that the designer can modify almost every quantity of the model, for instance, a geometrical dimension, frequency or amplitude of supplied voltage or currents, type of materials. Parametrical modifications of the model allow the designer to find the relative effect due to a certain variation of the system: this is a very effective way to realize the expected performances of an electromagnetic device.

Usually commercial codes have also embedded specific tools of numerical optimization that automatically launch several simulations where design variables are modified to find the 'mathematical' optimal solution. On the other hand, numerical modelling can lead to severe mistakes in design, that usually can occur when the results of a calculation are taken for granted without a careful verification. This issue is due to the inherent complexity of realizing a FE model, that, for instance, requires several input data. Some recommendations for modeling in a proper way are suggested in the following, nevertheless the theoretical approach to the solution of field problems presented in this book is a valuable support to discriminate the correctness of a numerical solution. In other words, the user of a simulation tool should have in mind the expected result of a FEM thanks to a preliminary study based on the classical theoretical or/and analytical tools.

Due to the wide possibilities of modeling, a simulation strategy must be defined. The main purposes of a simulation must be always clear to avoid excessive complications in developing it. For instance, a 3D EM-thermal coupled simulation gives probably the most exhaustive information about an induction or resistance heating process. It gives the variation during the process of the electrical quantities (current, voltage, active and reactive power) due to the modification of material properties with temperature or a precise temperature distribution, that takes into account the end/edge effect due to the finite lengths of inductors and workpieces. This is the last kind of simulation that should be carried out, because it is real time consuming and no or just few modifications of the model are reasonably possible. Probably, a 3D coupled analysis should be used only as a final assessment of the design. If the goal of the simulation is simply the sizing of resonance circuit, probably a simplified 2D only electromagnetic model, solved with 'cold' and 'hot' material properties of the workpiece give all the necessary information for the specific aim. But also when the full 3D coupled analysis is required, the solution of simplified 2D models must be considered, starting from the solution of the mere EM eddy current problem. In this way it is possible to analyze in a parametric way some basic quantities and also to have a preliminary estimation of the results expected from the 3D solution.

Usually a FEM code consists of several modules, each of them requires the user to complete some tasks. The '*preprocessing*' module is where the user has to draw the geometry, create the finite element subdivision, or mesh, define materials, field sources and boundary conditions. The '*solving*' module is where the solution is carried out while '*post-processing*' module is used to visualize field quantities and compute integral parameters. In the following, some information about FEM are given with reference to the different simulation steps, without claiming to provide an exhaustive treatise on this numerical method.

5.8.1 Preprocessor

FEM commercial codes are usually capable of treating 2D or 3D geometries.

2D geometry can be used when one component of the field, e.g. the magnetic field, can be neglected. Figure 5.38 shows a simple case where the system length along z axis is predominant with respect of the cross section dimensions. In this case the currents can flow only along the z direction while the magnetic and induction fields lye on the [x,y] 2D domain. In this case 2D planar models give results with a good accuracy, also in terms of integral values (total power, induced power, reactive power) that depend upon the actual length along the z axis. Of course, this model cannot take into account edge or end effects.

The axisymmetric model describes the cross section of the geometry, usually only the part in the semi plane of positive radius, where currents have only the azimuthal direction while magnetic fields lie only on the 2D domain. An example of this kind of models is presented in Fig. 5.39.

3D models are usually much more complex to draw, in particular when they refer to the solution of induction hardening process. In fact, the geometries of parts that are typically induction hardened are often complex. Complex geometries can be imported from a draw realized with a CAD software, but it is suggested to avoid to use detailed drawings of the parts that are not directly affected by the process. All the existing symmetries should be also taken into account in order to simplify the geometrical complexity.

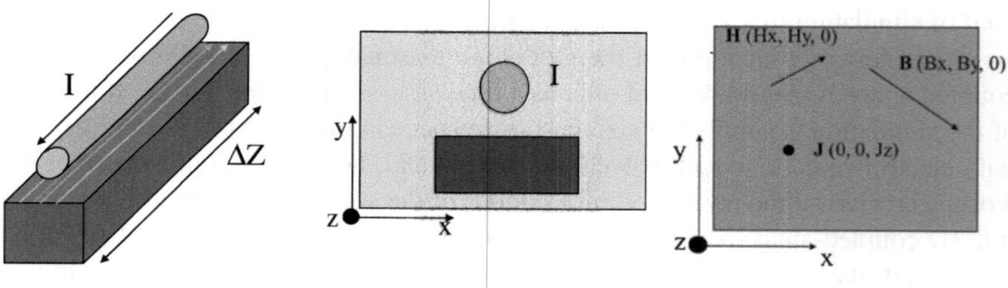

Fig. 5.38 A typical case of a 2D model [35]

Fig. 5.39 Example of axisymmetric geometry

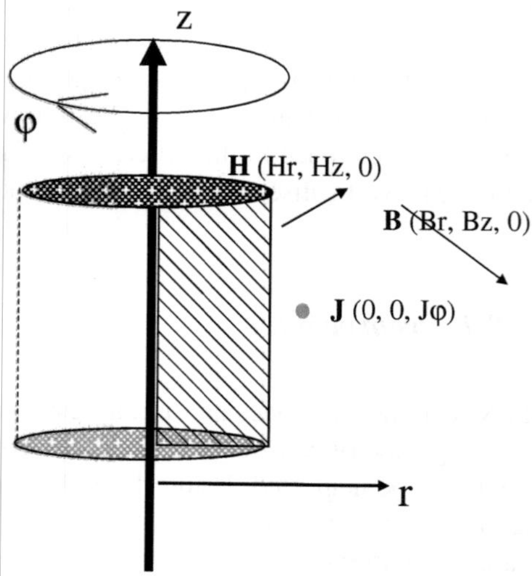

The geometry of a FEM model contains points, lines, faces and eventually volumes. The solution is obtained in a discretized space, the *mesh*. 2D elements of the mesh are usually triangles, for free meshes, or rectangles, for realizing a mapped discretization. The vertex of the elements are called nodes, while the sides are named edges.

The values of the unknowns, normally the values of a vector or scalar potential, are calculated on the nodes or the edges of the mesh. 3D elements are tetrahedral (all the faces are triangular), hexahedra (all the faces are quadrilateral, sometimes called 'bricks'), prisms (3D elements built by an extrusion: the base face can be triangular, quadrilateral, or, for more complex mesh, a general polygonal while the sides are rectangular) and pyramids (one quadrilateral face and four triangular faces).

The choice of the element dimensions is crucial for obtaining good results. For induction and resistance heating or, more generally, for eddy current problems, the mesh dimension has to be related to the penetration depth of electromagnetic field inside a conducting body. The golden rule to apply is to have at least 2 s order elements in 2–3 external layers of a thickness equal to the skin depth. Mapped mesh is preferred to the automatic tetrahedral mesh because all the elements have the same shape and dimensions, so the mesh dimension rule is easily satisfied everywhere.

In the pre-processor the user has to assign material properties to the different geometric entities (points, lines, faces, volumes), defining some physical regions. A physical region is characterized by the same material and can be realized by several geometric entities of the same type.

The relevant material properties for electromagnetic problems are magnetic permeability and electrical conductivity (or resistivity), for thermal problems, thermal conductivity and, for transient solution, heat capacity.

The field sources must be defined: currents flowing in the inductor are the typical source of magnetic fields while the power density distribution, produced by Joule effect of direct impressed or magnetically induced current densities, represents the source of thermal transient problem (and also the coupling term of a magneto-thermal simulation). FEM can consider two types of current sources: stranded conductors describe regions where an uniform current density is imposed, massive or solid conductors are regions where the effective current distribution is calculated. When inductors with massive copper tube are described as stranded conductors, the solution can be affected by significant errors and the calculation of the AC resistance and electrical efficiency is not possible.

Boundary Conditions (BCs) are used to define the symmetries on the domain edges and to reduce the problem dimensions. The solution of electromagnetic problems requires to impose at least one Dirchlet BC, but also Infinite Boundary Conditions IBC and Neumann BC are used. IBC impose zero value of magnetic field at infinite distance from the source, Dirichlet and Neumann boundary conditions are used to set the value or normal component of the unknown variable, typically a scalar or vector potential, on the boundary. Usually the user of commercial codes has to set normal or tangent field conditions without caring about the real condition on the unknowns.

5.8.2 Solver

FEM is a numerical method which requires to solve a linear system of n equations, where n is the number of unknowns. The linear system can be with real coefficient when the unknown variables are real or with complex coefficients for complex unknowns. When the model contains materials with nonlinear properties, the nonlinear solution is carried out usually by Newthon-Raphson method that requires the iterative solution of a linear system until a certain precision on the evaluation of nonlinear properties has been reached.

A linear system of n equations and n unknowns can be written under the following matrix form:

$$[\mathbf{A}][\mathrm{x}] = [\mathrm{b}] \tag{5.69}$$

where:

[A] is a square matrix of $n \times n$ size (with n lines and n columns)
[x] is a column vector of size n, representing the unknowns (DoFs)
[b] is a column vector of size n, with known components, usually called Right Hand Side

The matrix [A] is always a sparse matrix, i.e. the zero coefficient in [A] are much more than the non-zero ones. Depending on the adopted formulation and the possible coupling with an external circuit, [A] can exhibit also symmetry or some fully populated lines. The matrix characteristics are important in order to choose the type of frontal solver. Usually both iterative method, like ICCG, Incomplete Cholesky Conjugate Gradient or others, and direct method, like Sparspak or SuperLU, are available for diagonalizing the matrix, while multi-frontal solvers, like the ParDiSo, Parallel Direct Solver, or MUMPS Multi-frontal Massively Parallel sparse direct Solver, allow to parallelize the solution job. The choice of the solver is also driven by dimension of the matrix and hardware resources, notably RAM amount is usually the main limiting factor on maximum matrix size, that is the maximum number of unknowns.

The solver usually allows to run parametric analysis.

5.8.3 Post Processor

The analysis of results is the last step of the numerical modelling. Post-processor module allows the user to visualize the field distribution, rendering them by means of iso-values plots or colour shade maps. An example of the induction hardening of a spindle gear is given in Fig. 5.40. The colour map of the induced current distribution is presented in Fig. 5.41 while the temperature field for the same problem is shown in Fig. 5.42.

Post processor allows also to compute integral values, like magnetic energy or joule losses, evaluate a certain quantity along different geometrical supports, like paths or grids of points, export the results, as images or numerical values, for further analysis and preparation of reports.

The computation of global quantities, like the total active power or the magnetic energy in the domain, allows to evaluate the equivalent lumped electrical parameter of the workpiece—inductor system.

Integral values are obtained by integration of the basic quantities, e.g. specific active power or magnetic energy on specific support.

The total Power P_{all} is calculated as the integral of the specific Joule losses all over the domain Ω that can be the entire volumetric space of the model for 3D analysis or the entire surface for 2D models.

$$P_{all} = \int_{\Omega} \rho \dot{J} \cdot J^{*} \, d\Omega$$

The magnetic energy E_{mag}, must be also computed on the whole domain.

$$E_{mag} = \frac{1}{2} \int_{\Omega} \dot{B} \cdot H^{*} d\Omega$$

The equivalent resistance and inductance are:

- for series connected equivalent R_S and L_S:

$$R_S = \frac{P_{all}}{I^2} \quad L_S = \frac{2 \cdot E_{mag}}{I^2}$$

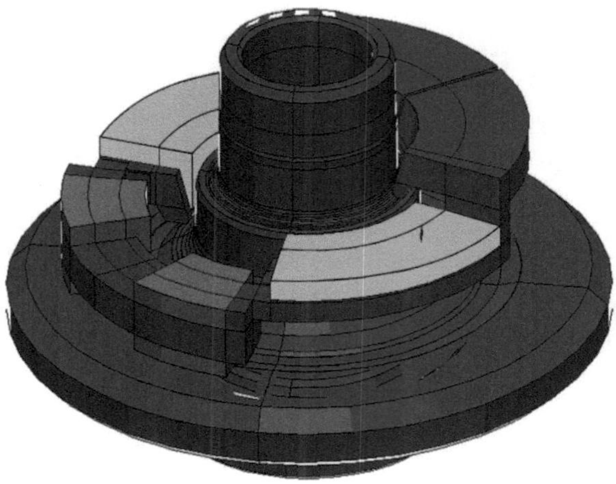

Fig. 5.40 3D FE model of induction hardening of a spindle gear. [*Blue*—inductor (*copper*), *Yellow* and *Magenta* —Magnetic concentrator, *Red*—workpiece]

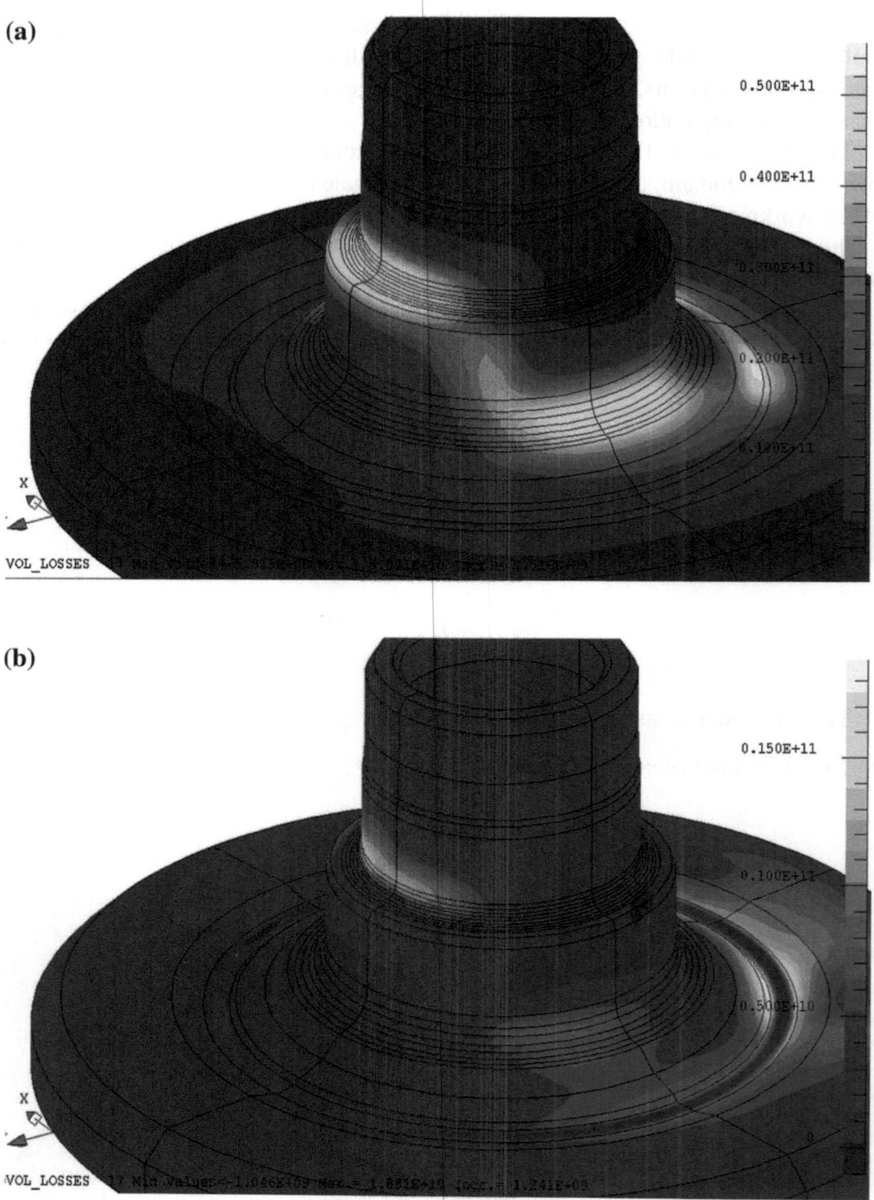

Fig. 5.41 Power density distribution in the spindle. **a** below Curie point, **b** above Curie point (lighter regions correspond to more intensive power density)

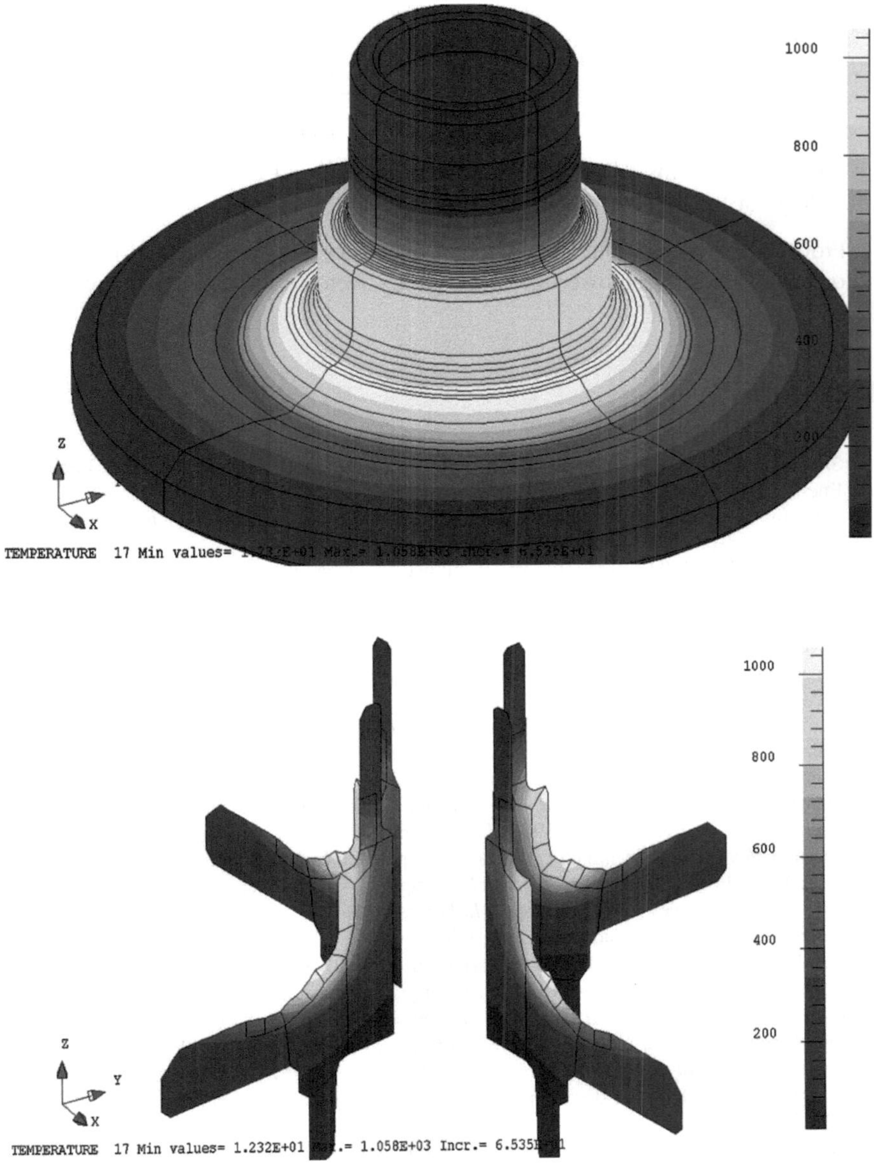

Fig. 5.42 Temperature distribution at the end of heating (after 6 s) on the external surface and in the cross-sections of the workpiece (lighter regions correspond to higher temperatures)

- for parallel connected equivalent R_P and L_P:

$$R_P = \frac{U^2}{P_{all}} \quad L_P = \frac{U^2}{2 \cdot E_{mag}}$$

where I is the rms values of the inductor current and U is the rms value of the voltage at the inductor clamps.

From these values, other fundamental quantities for the design of electrotechnological systems, like power factor or quality factor, can be easily computed.

References

1. Sluhockiĭ A. E., & Ryskin S. E. (1974). *Inductors for induction heating* (p. 264). Leningrad: Energhia (in Russian).
2. Nemkov V. S., & Demidovich V. B. (1988). *Theory and calculation of induction heating installations* (p. 280). Leningrad: Energoatomizdat (in Russian).
3. Lavers, J. D. (2007). State of art of numerical modeling for induction processes. In: *Proceedings of HES-07—heating by electromagnetic sources* (pp. 13–24). Padua (Italy), June 19–22, n.A-1. ISBN 88-89884-07-X.
4. Lavers, J. D., & Biringer, P. P. (1974). An improved method of calculating the power generated in an inductively heated load. *IEEE Transactions on Industry Applications, 10*(2), 273–278.
5. Lupi, S. (2005). *Elettrotermia* (453 pp.). Padua: Libreria Progetto (Italy) (in Italian).
6. Bukanin, V., Kuchmasov, D., Nemkov, V., & Zenkov, A. (1997). Computer simulation—an effective creative tool for development of the induction heat treatment a process. In *1st international induction heat treating symposium*. Indianapolis, Indiana, 785–792.
7. Kolbe, E., & Reiss W. (1963). Eine methode zur numerisce bestimmung der stromdichteverteilung in induktiv erwärmten körpern unterschliedlicher geometrischer Form. In Z. Wiss (Ed.), *Hochschule Elektrotechnik* (pp. 311–317). Ilmenau, J.9, H.3.
8. Nemkov, V., & Polevodov, B. (1980). *Mathematical modeling on computer of devices for RF heating* (48 pp.). Leningrad: Mashinstroenie.
9. Nemkov, V., Polevodov, B., & Gurevich, S. (1991). *Mathematical modelling of high-frequency heating installations* (60 pp.). Leningrad: Politecnika (in Russian).
10. Ceolini, F., & Lupi, S. (1978). The mutually-coupled circuits method for calculations regarding inductors for induction heating with special configurations. In *Proceedings of IEEE-IAS annual meeting conference Rec* (Vol. 41B, pp. 1151–1157). Toronto, Canada.
11. Weinstein, B. (1884). Zur Berechnung des Potentials von Rollen. *Annalen der Physik, 257*(2), 329–360.
12. Grover, F. W. (1962). *Inductance calculations working formulas and tables*. New York: Dover Publ. Inc.
13. Garret, M.W. (1963). Calculations of fields, forces and mutual inductances of current systems by elliptic integrals. *Journal of Applied Physics, 34*(9), 2567–2573.
14. Thompson, M. (1999). Inductance calculation techniques—Part I: classical methods; Part II: approximations and handbook methods. *Power Control and Intelligent Motion, 25*(12), 40–50.
15. Hurley, W. G., & Duffy, M. C. (1995). Calculation of self and mutual impedances in planar magnetic structures. *IEEE Transactions on Magnetics, 31*(4), 2416–2422.
16. Hurley, W. G., & Duffy, M. C. (1997). Calculation of self and mutual impedances in planar sandwich inductors. *IEEE Transactions on Magnetics, 33*(3), 2282–2290.

17. Akyel, C., & Babic, S. I. (2006). New analytic-numerical solutions for the mutual inductance of two coaxial circular coils with rectangular cross section in air. *IEEE Transactions on Magnetics, 42*(6), 1661–1669.
18. Forzan, M., Lupi S., & Toffano, E. (2011). Compensation of induction heating load edge effect by space control. *COMPEL, 30*(5), 1558–1569.
19. Lupi, S. (1979). The numerical calculation of forces in induction heating systems. In IEEE-IAS *annuals meeting conference proceedings* (pp. 1226–1231). Cleveland (USA).
20. Strunskiĭ, B. M. (1962). *Short circuit network of electrical furnaces* (p. 335). Moscow: Metallurgizdat (in Russian).
21. Mukoseev, Y. P. (1959). *AC current distribution in conductors*. Moscow: Gosenergoizdat (in Russian).
22. Aliferov, A., & Lupi, S. (2007). *Direct resistance heating of metals* (p. 223). Novosibirsk: NGTU. ISBN 5-7782-0475-2 (in Russian).
23. Lupi, S., & Nunes, M. F. (1990). *Riscaldamento dei metalli mediante conduzione diretta di corrente* (p. 86). Padova: CLEUP (Italy). ISBN 88-7178-338-7.
24. Romanov, D. I. (1981). *Direct resistance heating of metals* (p. 280). Moskow: Mashinostrenie (in Russian).
25. Lupi, S., Forzan, M., & Aliferov, A. (2008). Characteristics of installations for direct resistance heating of ferromagnatic bars of square cross-section. In *Proceedings of the international scientific colloquium MEP— « modelling for electromagnetic processing* (pp. 43–49). Hannover (Germany). ISBN 978-3-00-026003-2.
26. Lupi, S., Forzan, M., & Aliferov, A. I. (2009). Integral parameters of installations for direct resistance heating of ferromagnetic steel bars of square cross-section. In *Proceedings of international IEEE conference for the 150-anniversary of A.S. Popov* (pp. 1581–1587), May 18–23, 2009. Saint Petersburg (Russia); IEEE Copyrights, IEEE Service Center, Piscataway (USA), Hoes Lane.
27. Kovrev, G. S. (1975). *Electro-contact heating for processing non-ferrous metals* (p. 312). Moskow: Metallurghia (in Russian).
28. Neimark, B. E. (1967). *Physical properties of steels and alloys used in the energy field* (240 pp.). Moscow: Energia (in Russian).
29. Kazanciev, E. A. (1975). *Industrial furnaces: A Reference Manual for analysis and design* (312 pp.). Moscow: Metallurgia (in Russian).
30. Neiman, L. R. (1949). *Skin effect in ferromagnetic bodies* (190 pp.). Leningrad: Gostehnizdat (in Russian).
31. Valeev, A. X., & Beljaeev, A. M. (1968). *Calculation and selection of operating modes of electro-contact heating installations of steel products* (pp. 2–13). Moskow: VNIIEM, Informelektro (in Russian).
32. Hegewaldt, F. (1961). *Induktives Oberflächenhärten* (pp. 434–456). Induktionsöfen: BBC-Nachrichten, July/August.
33. Svenčanskiĭ, A. D. (1975). *Electrical industrial furnaces*. Moscow: Energia Chapt.1 (in Russian).
34. Crepaz, G., & Lupi, S. (1982). Influence of the installations electrical parameters on the transient temperature distribution in the direct resistance heating of ferromagnetic billets. In: *Proceedings of the Conference Electroheat for Metals "UIE-BNCE"*. Cambridge, England, 21–23 September, Paper 3.4.
35. Rudnev, V., & Totten, G. E. (Eds.). (2014). ASM Handbook—Vol. 4c, induction heating and heat treatment. In *ASM international materials park* (820 pp.). Ohio, USA. ISBN-13 978-1-62708-012-5.

Index

S. Lupi et al., *Induction and Direct Resistance Heating*,
DOI 10.1007/978-3-319-03479-9

B

C

Q

R

S

T

MIX
Papier aus verantwortungsvollen Quellen
Paper from responsible sources
FSC® C105338

If you have any concerns about our products,
you can contact us on
ProductSafety@springernature.com

In case Publisher is established outside the EU,
the EU authorized representative is:
Springer Nature Customer Service Center GmbH
Europaplatz 3, 69115 Heidelberg, Germany

Printed by Libri Plureos GmbH
in Hamburg, Germany